Water on Sand

Water on Sand

Environmental Histories of the Middle East and North Africa

EDITED BY ALAN MIKHAIL

OXFORD
UNIVERSITY PRESS

OXFORD
UNIVERSITY PRESS

Oxford University Press is a department of the University of Oxford.
It furthers the University's objective of excellence in research, scholarship,
and education by publishing worldwide.

Oxford   New York
Auckland   Cape Town   Dar es Salaam   Hong Kong   Karachi
Kuala Lumpur   Madrid   Melbourne   Mexico City   Nairobi
New Delhi   Shanghai   Taipei   Toronto

With offices in
Argentina   Austria   Brazil   Chile   Czech Republic   France   Greece
Guatemala   Hungary   Italy   Japan   Poland   Portugal   Singapore
South Korea   Switzerland   Thailand   Turkey   Ukraine   Vietnam

Oxford is a registered trademark of Oxford University Press in
the UK and certain other countries.

Published with the assistance of the Frederick W. Hilles
Publication Fund of Yale University.

Published in the United States of America by
Oxford University Press
198 Madison Avenue, New York, NY 10016

© Oxford University Press 2013

Library of Congress Cataloging-in-Publication Data
Water on sand : environmental histories of the middle east and north africa /
edited by Alan Mikhail.
       p. cm.
Includes bibliographical references and index.
ISBN 978–0–19–976867–7 (hardcover : alk. paper) —
ISBN 978–0–19–976866–0 (pbk. : alk. paper)
1. Human ecology—Middle East—History.
2. Human ecology—Africa, North—History.
3. Middle East—Environmental conditions.
4. Africa, North—Environmental conditions.
I. Mikhail, Alan
   GF670.W48 2013
   304.20956—dc23
                        2012022534

ISBN 978–0–19–976867–7
ISBN 978–0–19–976866–0

وَجَعَلْنَا مِنَ الْمَاءِ كُلَّ شَيْءٍ حَيٍّ

We made from water every living thing.

Quran 21:30

Contents

Acknowledgments

This book is first and foremost a collective effort, and I therefore very happily and profusely thank all of the contributors to this project. Their professionalism and patience and, much more importantly, their ideas and hard work make this book possible. For their early advice and support of this book, I thank Edmund Burke III, Beshara Doumani, Timothy Mitchell, and Amy Singer. My colleagues at Yale University make being a historian a true pleasure every day. I thank Abbas Amanat, Ned Blackhawk, Laura Engelstein, Valerie Hansen, Naomi Lamoreaux, Mary Lui, Joe Manning, Joanne Meyerowitz, Peter C. Perdue, Paul Sabin, Francesca Trivellato, and Charles Walton. At Yale, I must also thank Seven Ağır, Frank Griffel, Kishwar Rizvi, James C. Scott, and Harvey Weiss for their collegiality and generosity of spirit. I am very fortunate to also benefit from the support and friendship of Christine Philliou and Edith Sheffer.

Richard P. Tucker and two anonymous reviewers each read earlier versions of this book and helped to shape it in significant ways. In Istanbul, I thank Arzu Öztürkmen of Boğaziçi University and Esra Müyesseroğlu of the Topkapı Palace Museum Library for their work on behalf of this project. Publication of this book was generously supported by the Frederick W. Hilles Publication Fund of Yale University. Thank you. Stacey D. Maples of the Yale University Map Department skillfully and patiently made the wonderful maps for this book. Maureen Cirnitski's keen eye and hard work helped make this book a reality. *Water on Sand* only exists because of the expert shepherding of Susan Ferber at Oxford University Press. From beginning to end, she has improved this book immeasurably, and it has been a true pleasure working with her.

Alan Mikhail
Istanbul
August 2012

Contributors

Jessica Barnes is a postdoctoral associate at the Yale Climate and
Energy Institute and Yale School of Forestry and Environmental
Studies. She holds a Ph.D. in sustainable development from
Columbia University and a M.A. in environmental management
from the Yale School of Forestry and Environmental Studies. Her
research focuses on the intersection of water management, agricultural policy, and climate change impacts in the Middle East. She
has conducted over twenty months of ethnographic research in
the region, funded by the Wenner-Gren Foundation and the Yale
Climate and Energy Institute. She is currently completing a book
manuscript entitled *Cultivating the Nile: The Everyday Politics of Water
in Egypt*. Her work has been published in *Social Studies of Science* and
Geopolitics and will be published in a forthcoming issue of *Geoforum*.

Richard W. Bulliet is professor of Middle Eastern History at
Columbia University where he also directed the Middle East Institute
of the School of International and Public Affairs for twelve years.
He has been a Guggenheim and a Carnegie Corporation Fellow.
His most recent scholarly work is *Cotton, Climate, and Camels in
Early Islamic Iran: A Moment in World History* (New York: Columbia
University Press, 2009). His earlier books include *Hunters, Herders,
and Hamburgers: The Past and Future of Human-Animal Relationships*
(New York: Columbia University Press, 2005); *The Case for
Islamo-Christian Civilization* (New York: Columbia University

Press, 2004); *Islam: The View from the Edge* (New York: Columbia University Press, 1994); *Conversion to Islam in the Medieval Period: An Essay in Quantitative History* (Cambridge, MA: Harvard University Press, 1979); *The Camel and the Wheel* (Cambridge, MA: Harvard University Press, 1975), which won the Dexter Prize of the Society for the History of Technology; and *The Patricians of Nishapur: A Study in Medieval Islamic Social History* (Cambridge, MA: Harvard University Press, 1972). He has also written five novels, beginning with *Kicked to Death by a Camel* (New York: Harper and Row, 1973) and ending with *The One-Donkey Solution: A Satire* (Bloomington: iUniverse, 2011), and is coauthor of the world history textbook *The Earth and Its Peoples: A Global History*, 5th edn. (Boston: Wadsworth, 2010).

Diana K. Davis, a geographer and veterinarian, is associate professor of history at the University of California, Davis. Her first book, *Resurrecting the Granary of Rome: Environmental History and French Colonial Expansion in North Africa* (Athens: Ohio University Press, 2007), was awarded the George Perkins Marsh Prize by the American Society for Environmental History as well as the Meridian Prize and the Blaut Award by the Association of American Geographers. She recently published *Environmental Imaginaries of the Middle East and North Africa* (Athens: Ohio University Press, 2011), edited with Edmund Burke III. She has also published numerous articles and book chapters and is the recipient of a Guggenheim Fellowship and a Ryskamp Fellowship from the American Council of Learned Societies for her new book project *Imperialism and Environmental History in the Middle East* (Cambridge University Press).

Suraiya Faroqhi is professor of history at Istanbul Bilgi University. She taught previously at Middle East Technical University in Ankara and the Ludwig Maximilians Universität in Munich. She is the author of *Towns and Townsmen of Ottoman Anatolia: Trade, Crafts, and Food Production in an Urban Setting, 1520–1650* (New York: Cambridge University Press, 1984); *Kultur und Alltag im Osmanischen Reich: Vom Mittelalter bis zum Anfang des 20. Jahrhunderts* (Munich: Verlag C. H. Beck, 1995); *Approaching Ottoman History: An Introduction to the Sources* (Cambridge: Cambridge University Press, 1999); *The Ottoman Empire and the World Around It* (London: I. B. Tauris, 2004); *Artisans of Empire: Crafts and Craftspeople under the Ottomans* (London: I. B. Tauris, 2009). She is also the editor, most recently, of volume 3 of *The Cambridge History of Turkey*, entitled *The Later Ottoman Empire, 1603–1839* (Cambridge: Cambridge University Press, 2006), and *Animals and People in the Ottoman Empire* (Istanbul: Eren, 2010).

Toby C. Jones is a historian of the modern Middle East who teaches at Rutgers University, New Brunswick. His scholarship focuses primarily on the political history of Saudi Arabia and the Persian Gulf. Before joining the Department of History at Rutgers University, he taught at Swarthmore College; was a Fellow at Princeton University's Project on Oil, Energy, and the Middle East; and worked as an analyst of the Persian Gulf for the International Crisis Group. He is the author of *Desert Kingdom: How Oil and Water Forged Modern Saudi Arabia* (Cambridge, MA: Harvard University Press, 2010), and is currently working on a new book project entitled *America's Oil Wars*, also to be published by Harvard University Press. His articles have appeared in the *International Journal of Middle East Studies*, *Middle East Report*, the *New York Times*, *The Atlantic*, *Foreign Affairs*, *Foreign Policy*, the *Arab Reform Bulletin*, *Strategic Insights*, and the *CTC Sentinel*.

Arash Khazeni teaches Middle Eastern and North African history at Pomona College in Los Angeles, California. He received his Ph.D. from the Department of History at Yale University. His research focuses on the environmental history of Islamic Eurasia and the Persianate world between the sixteenth and nineteenth centuries. He is the author of *Tribes and Empire on the Margins of Nineteenth-Century Iran* (Seattle: University of Washington Press, 2010), recipient of the Middle East Studies Association Houshang Pourshariati Book Award, and "Across the Black Sands and the Red: Travel Writing, Nature, and the Reclamation of the Eurasian Steppe, circa 1850," published in the *International Journal of Middle East Studies*. He is currently working on a global environmental history of the turquoise trade, under contract with the University of California Press.

Karim Makdisi teaches international politics and international environmental policy in the Department of Political Studies and Public Administration at the American University of Beirut (AUB). He is the associate director of AUB's Issam Fares Institute for Public Policy and International Affairs, where he oversees its climate change and environment research project. He also coordinates the environmental policy component of AUB's Interfaculty Graduate Environmental Sciences Program. Prior to joining AUB, he worked at the United Nations Economic and Social Commission for Western Asia on projects related to the implications and outcomes for the Arab World of the World Summit on Sustainable Development. His current research projects include reconceptualizing environmentalism and international relations in Lebanon and the Arab World and investigating the shifting role of the United Nations since the end of the Cold War.

J. R. McNeill is university professor at Georgetown University, where he has taught since 1985 in the Department of History and School of Foreign Service. He has held two Fulbright Awards, a Guggenheim Fellowship, a MacArthur Grant, a Fellowship at the Woodrow Wilson Center, and a visiting appointment at L'École des hautes études en sciences sociales. His books are *Atlantic Empires of France and Spain: Louisbourg and Havana, 1700–1763* (Chapel Hill: University of North Carolina Press, 1985); *The Mountains of the Mediterranean World: An Environmental History* (New York: Cambridge University Press, 1992); *Something New under the Sun: An Environmental History of the Twentieth-Century World* (New York: W. W. Norton, 2000), which has been translated into nine languages, was listed by the *London Times* as among the best science books ever written, and was awarded the World History Association Book Prize and the Forest History Society Book Prize; *The Human Web: A Bird's-Eye View of World History* (New York: W. W. Norton, 2003), coauthored with his father, William H. McNeill, and translated into seven languages; and, most recently, *Mosquito Empires: Ecology and War in the Greater Caribbean, 1620–1914* (New York: Cambridge University Press, 2010), which won the Beveridge Prize from the American Historical Association and a PROSE Award from the Association of American Publishers, and was listed by the *Wall Street Journal* as among the best books in early American history. He has edited or coedited eight more books. In 2010 he was awarded the Toynbee Prize for "academic and public contributions to humanity." In 2012–2015 he served as a vice-president of the American Historical Association and in 2011–2013 as president of the American Society for Environmental History.

Alan Mikhail is assistant professor of history at Yale University. His first book, *Nature and Empire in Ottoman Egypt: An Environmental History* (New York: Cambridge University Press, 2011), won the Roger Owen Book Award from the Middle East Studies Association and the Samuel and Ronnie Heyman Prize for Outstanding Scholarly Publication from Yale University. His article in the *International Journal of Middle East Studies* won the Ömer Lütfi Barkan Article Prize from the Turkish Studies Association. His articles have also appeared in the *Journal of the Economic and Social History of the Orient, Comparative Studies in Society and History, History Compass, Bulletin of the History of Medicine, Akhbār al-Adab, Wijhāt Naẓar,* and elsewhere. He is currently writing a book about the changing relationships between humans and animals in Ottoman Egypt.

Nancy Y. Reynolds is associate professor of history at Washington University in St. Louis. Her research concentrates on the cultural and social history of twentieth-century Egypt. She is the author of *A City Consumed: Urban Commerce,* *the Cairo Fire, and the Politics of Decolonization in Egypt* (Stanford: Stanford University Press, 2012). Her work on Egyptian department stores has appeared in the *International Journal of Middle East Studies, Journal of Women's History, European Review of History,* and *Arab Studies Journal.* In 2011, she received a fellowship from the American Council of Learned Societies for her project "The Politics of Environment, Culture, and National Development in the Building of the Aswan High Dam in Egypt, 1956–1971."

Sam White is assistant professor of environmental history at Oberlin College. He is the author of *The Climate of Rebellion in the Early Modern Ottoman Empire* (New York: Cambridge University Press, 2011). His chapters and articles have appeared in *Environmental History* and the *International Journal of Middle East Studies,* among other publications. He is the recent recipient of the American Society for Environmental History Leopold Hildy Prize, the Agricultural History Society Wayne D. Rasmussen Award, and the British-Kuwaiti Friendship Society prize for the best book in Middle East Studies. In 2011–2012 he was a fellow at the John Carter Brown and the Huntington libraries, where he began a new project on the Little Ice Age and early Atlantic history.

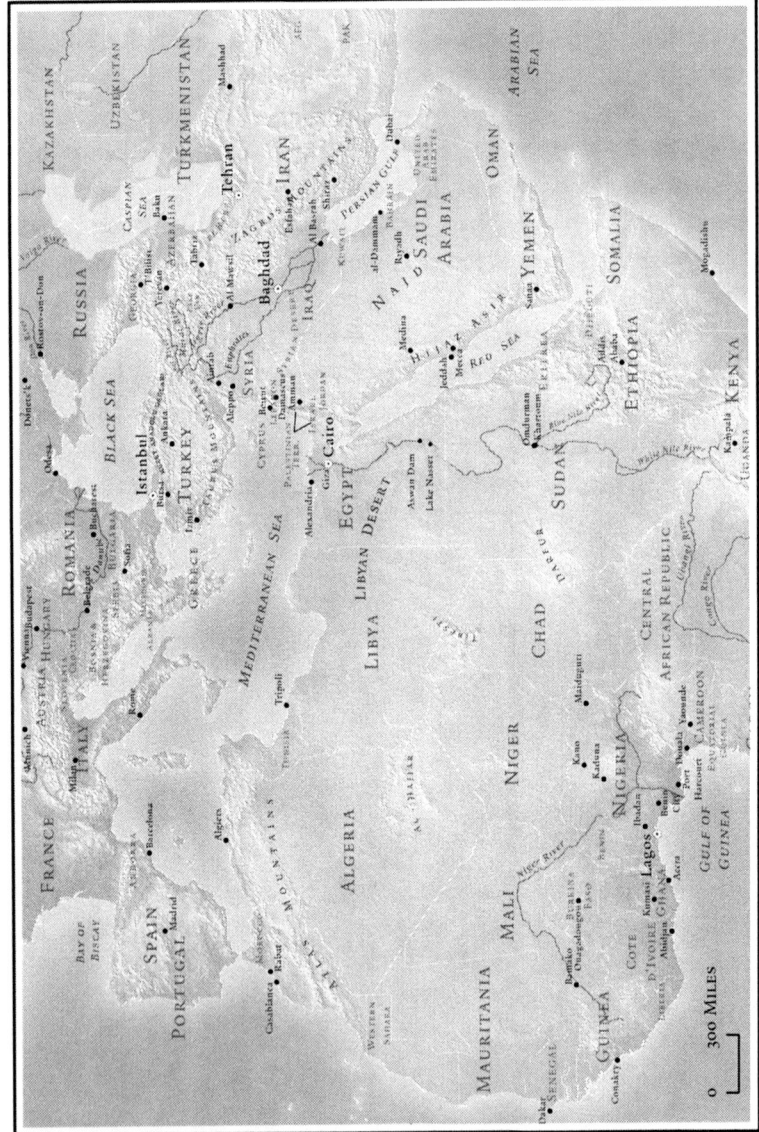

MAP 1. Modern Middle East and North Africa

xvi

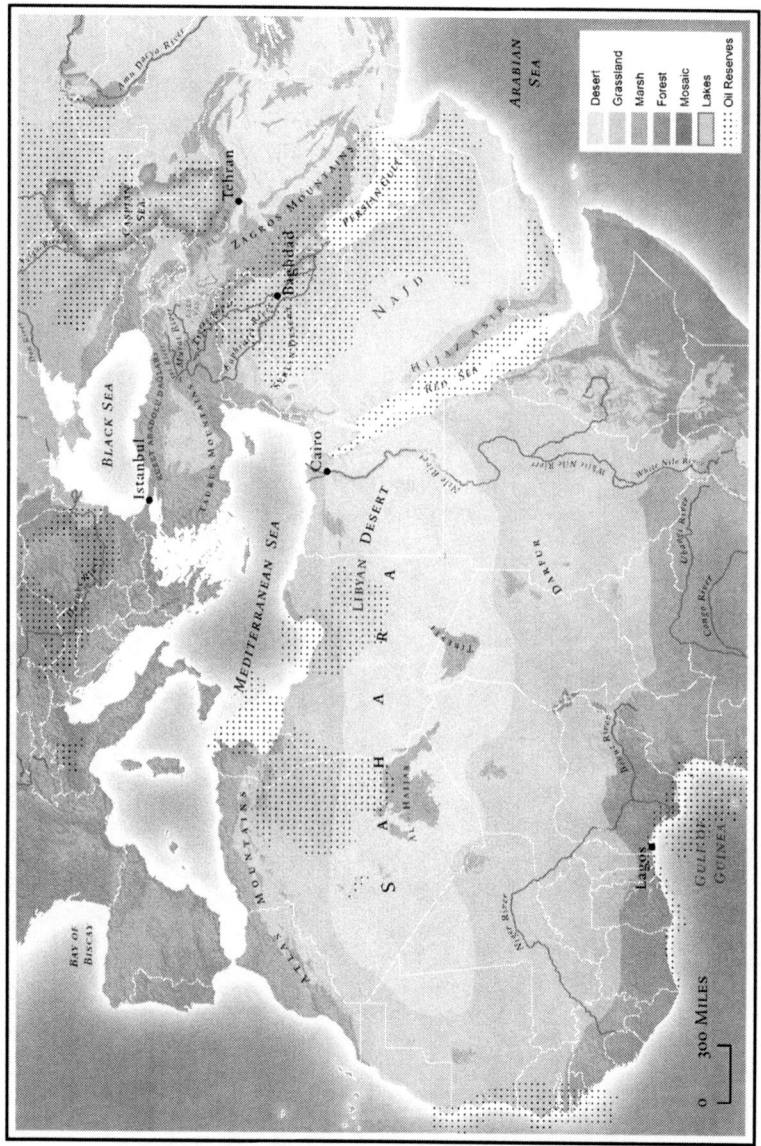

MAP 2. Natural Resources in the Middle East and North Africa

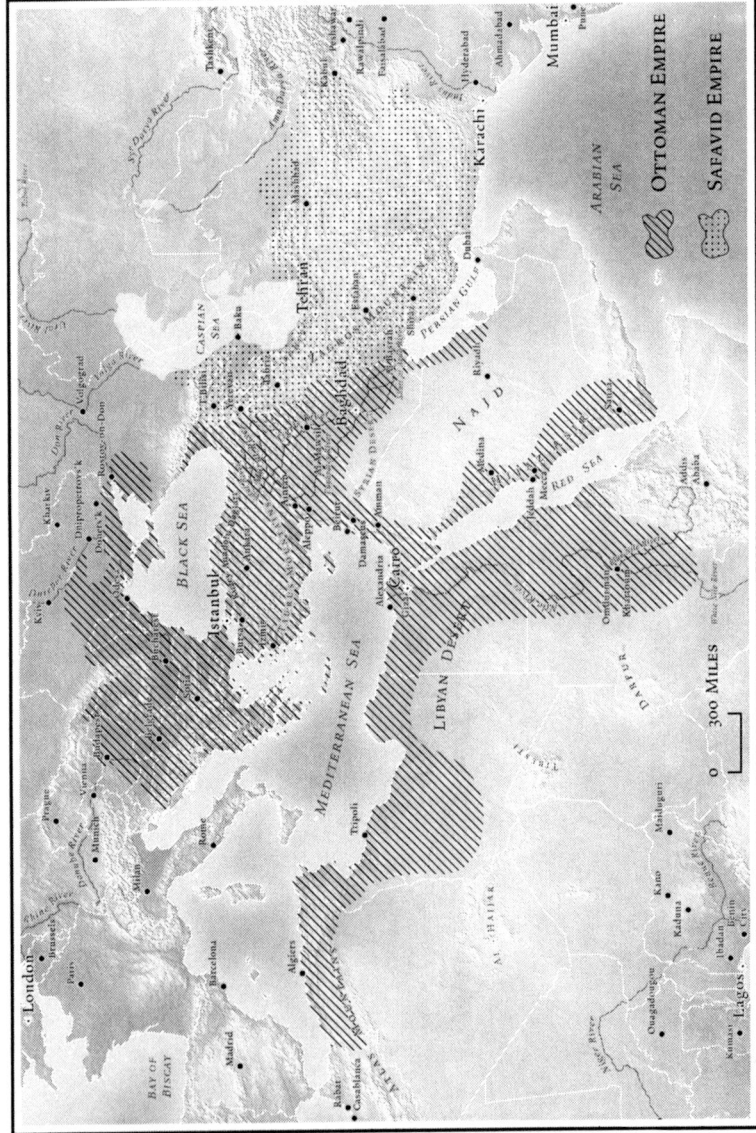

MAP 3. Early Modern Middle East and North Africa, c. 1600

Water on Sand

Introduction

Middle East Environmental History: The Fallow between Two Fields

Alan Mikhail

Our understanding of the global history of the environment necessarily advances through the study of local situations—how local populations deal with large-scale environmental change; how market integration affects the ecologies of specific locales; and how the movement of peoples, commodities, ideas, and microbes impacts areas far from their original homes. One of the gaping holes in the global story of the environment thus far has been the history of the Middle East and North Africa (MENA).[1] Likewise, one of the gaping holes in the study of the MENA has been the history of its environment. Yet from antiquity—if not earlier—the region has been the crucial zone of connection between Europe, the Mediterranean, and Africa on the one hand, and East, Central, South, and Southeast Asia on the other.[2] It was an arena of contact where European and South Asian merchants traveled and conducted commerce, and where religious pilgrims from Mali to Malaysia came each year not only to visit Mecca and Medina but also to buy goods in the many bazaars in cities throughout the MENA. Pastoral nomads moved from Northern India and Central Asia through Iran and into Anatolia.[3] Ships carrying goods, peoples, vermin, and ideas sailed from India, China, and Southeast Asia to ports on the Red Sea and Persian Gulf. Beginning in the late medieval and early modern periods, the world from the western Mediterranean to China was characterized by intense circulation, interconnection, and movement.[4] All the threads of these connections ran through the MENA, with important environmental implications.[5]

This book aims to tell some of the environmental history of the MENA and, in so doing, position the MENA on the agenda of environmental history and show the utility of an environmental perspective for the study of the region. Water in the MENA meets much more than just sand. Although much of the region is indeed arid or semiarid, it is home to a wide variety of ecologies: plains in Turkey, Iran, and Syria; mountain ranges in Morocco and Lebanon; fertile river deltas in Iraq and Egypt; thick forests in Turkey, Lebanon, and Iran. Each of these broad categories of landscape is obviously unique and vastly complex, and few generalizations can be made about forests, river valleys, or mountain habitats across the whole of the MENA. Each of the following chapters thus advances the goals of this book through a fine-grained empirical study of Middle East environmental history in a specific ecology and time. Arranged in roughly chronological order, they combine to show how the history of Middle Eastern environments crucially affects understandings of various historical and global environmental processes and how environmental history offers some essential tools for understanding the MENA's many histories.

What emerges from these studies is a picture of how populations in the MENA—like populations everywhere—have lived within their specific ecological constraints over the past half millennium. The day-to-day concerns of farmers, fishermen, and foresters stood alongside attempts by political powers to make clear their legitimacy through the harnessing, control, and management of environmental resources. Various scholars and other observers have offered grand narratives of civilizational worth or epochal change based on the perceived fragilities, aberrations, or contradictions of the MENA's ecologies. The facades of these predictably ahistorical overarching explanations inevitably fracture when one begins to look for the details holding them together. All of the chapters in this book thus do the needed work of pushing on these grand narratives to open up a space for the flowering of new, more historically grounded interpretations of complex phenomena.

This introduction first examines some of the broad outlines of an older archaeological and geographical scholarly literature that forms the bedrock that makes possible much of the analysis of this book and of Middle East environmental history more generally. It then points to what global environmental historians and those of regions beyond the MENA stand to gain from a consideration of its ecological pasts, as well as how Middle East Studies can benefit from adopting some of the methodologies of environmental history.

Beneath the Fields: Continuity or Collapse?

The most prominent area of research in the existing literature on the history of Middle Eastern environments is in the realm of geoarchaeology and cultural geography and ecology, and this book necessarily builds on this groundwork.[6] Most important for environmental historians of the MENA, this earlier work provides a historic picture of ecological systems and their operations. Because scholars have been writing about the pasts of Middle Eastern environments for some time, the following survey is not intended to be exhaustive but rather to highlight some of the differences between this older literature and the newer environmental history.[7]

Earlier works can roughly be divided into two major categories: first, descriptions of the continuity of human settlement and community in the face of the delicate stability of environments around them and, second, examinations of human groups inevitably doomed to collapse because of inherent contradictions in their ecological systems. Both of these modes of narrating the environmental past—either as continuity or collapse—fall short of showing how humans were integrally connected to environments or how the many relationships between humans and nature changed over time. Moreover, these works sometimes paint a picture of the MENA as a zone of ecological exceptionalism or, to use political theorist Timothy Mitchell's words, as a place of "unnatural nature."[8] In the first story of continuity, the picture is one of dynamic stasis—of human communities constantly working and struggling to survive in the face of enormous environmental challenges. In this story of society constantly running only to stay in place, there is no consideration of human agency's effects on ecological systems. The second narrative of eventual and inevitable collapse also lacks the explanatory power to account for historical specificities and contingencies. If certain human communities are predetermined to decline because of a fundamental flaw or contradiction inherent in their local world, then what, other than documenting this downward spiral, is left for the historian to do? Nevertheless, these works do prove extremely useful because their descriptions of ecological systems serve as a backdrop for thinking about how various human communities dealt with the ecological constraints of their specific environments.

Two instructive examples from this older geographical literature are Karl Butzer's *Early Hydraulic Civilization in Egypt* and Peter Christensen's *The Decline of Iranshahr*.[9] Both books sketch a picture of the ecological limits within which human communities were made to live and survive. In both of these cases, and in others as well, the picture is one of continuity—of systems in overall equilibrium on the timescale of generations despite massive fluctuations from year to year. It is no coincidence that Egypt and the Iranian plains—the respective subjects of these studies—are reliant on the annual floods of massively complex river systems. Floods could be bad or good from one season to the next, but overall, through the development of various technologies of environmental manipulation—chiefly irrigation and agriculture—human communities were able to ride out these vagaries to achieve some semblance of ecological balance. Other examples of this kind of description of ecological systems that prevailed throughout the MENA before 1500 include J. M. Wagstaff's general study *The Evolution of Middle Eastern Landscapes*; Carlos Cordova's *Millennial Landscape Change in Jordan*; Russell Meiggs's and J. V. Thirgood's examinations of timber supplies around the Mediterranean, respectively *Trees and Timber in the Ancient Mediterranean World* and *Man and the Mediterranean Forest*; and Robert M. Adams's study of southern Iraq, *Land Behind Baghdad*.[10]

One of the most important differences between this earlier literature and the kind of environmental history presented in this book is their radically divergent timescales. In the former geographic literature, the temporal unit of analysis is usually the millennium or, less often, the century. In contrast, most environmental histories—the majority of those in this book included—deal with years, decades, and sometimes centuries. This is a difference between geological and historical time. As a historical discipline, environmental history is therefore concerned with *both* perturbations from year to year *and* the overall state of ecological systems rather than only trying to explain how an overall system functions. Hence, through a detailed consideration of a specific place in a specific time, environmental history allows one to see how human dealings with constantly changing environmental circumstances came to affect a particular social and ecological system in fundamental ways.

A related difference between the older geographical literature about the MENA and newer environmental histories of the region is the sources used.[11] The former rely largely on geoarchaeological and biological data—morphology, dendrochronology, pollen cores, and the like. Apart from the occasional anecdote culled from classical sources (Herodotus, Josephus, the Quran, the Bible, etc.), these geographical studies employ very little human-authored source material. This is both a strength and a weakness. Physical data allow for comparisons over time and provide proof of weather events, agricultural settlements, and natural phenomena, but they do not offer much in the way of examining how humans thought about, harnessed, used, or feared the natural worlds around them.[12]

Reflecting the wider field of environmental history, this book's chapters use both physical sources and historical, literary, and administrative accounts. On balance, though, the bulk of the material comes from the mass of human observations, anecdotes, poetry, narratives, bureaucratic records, and chronicles recorded on paper about relationships between humans and other parts of the environment. Almost all of these sources take as their explicit subject something other than a region's ecology, and thus histories of the relationships between peoples and the natural world must be carefully assembled through close readings of many different kinds of sources. When used judiciously—and often in conjunction with physical data—these sources offer insight into the history of human engagements with other parts of nature that empirical data alone miss. They also allow historians a way to test broad theories like continuity or collapse to see if they hold up on the ground in specific instances. In short, these written sources afford the most immediate access to the history of humans in nature.[13] Thus, one of the most important differences between environmental history and older geographical literatures is the use of social, political, cultural, and economic written source materials to uncover a concern for the human causes and effects of environmental processes and phenomena.

To put it in terms more familiar to environmental historians, earlier studies of the MENA's geography and cultural ecology do not, for the most part, recognize the dialectical relationships between humans and environments. This work is very good at setting the ecological stage on which human history takes place, but it does not engage with the ways in which humans move the props and rearrange the set in fundamental ways. By contrast, a dialectical understanding of humans and environments recognizes, in William Cronon's well-known formulation, that "environment may initially shape the range of choices available to a people at a given moment, but then culture reshapes environment in responding to those choices. The reshaped environment presents a new set of possibilities for cultural reproduction, thus setting up a new cycle of mutual determination."[14] Thus, Cronon continues, environmental history seeks "to locate a nature which is within rather than without history, for only by so doing can we find human communities which are inside rather than outside of nature."[15] In building upon but departing from older literatures on Middle Eastern landscapes and environments, this book—and indeed the larger project of Middle East environmental history that it reflects—seeks to locate the kind of nature that is within rather than outside of Middle Eastern history.

Why Environmental History Needs the Middle East

The story of the Middle Eastern environment addresses three main avenues of research in the field of environmental history. First, there is the simple fact that understanding the full histories of global environmental phenomena, from climate change to soil erosion, means comprehending the Middle Eastern component of the histories of these processes in Eurasia and North Africa. Second, because robust state systems have existed in the region for multiple millennia, various areas of the MENA enjoy a documentary record far longer than that of most other parts of the world, the chief exception perhaps being China. This record contains much information about environmental change and the relationships between humans and nature that reveals the effects of longer-term environmental processes and manipulation. Finally, the MENA is home to various religious, ethnic, and linguistic traditions. Foremost among these in the last millennium and a half are the cultures of Islam. For those seeking to understand how the people of one of the world's largest religious traditions thought about, dealt with, and otherwise related to the natural world, the MENA offers much material.

Just as global movements of trade were deeply intertwined with what was happening in the MENA, so too large-scale trends in climate, disease, and crop diffusion impacted and were impacted by the histories of the MENA's many environments. For example, one of the results of the rise and spread of Islam in the seventh and eighth centuries was the diffusion of new kinds of crops and agricultural technologies across Eurasia. In what one scholar has termed the "Islamic green revolution," the unity provided by the new religion brought disparate parts of the world together for the first time into a unified ecological contact zone.[16] In the medieval period, it was most likely through the Middle East that plague moved westward to the Mediterranean basin.[17] In another medieval example of environmental exchange, knowledge of irrigation technologies and waterworks was transferred from the Muslim world to Spain and Italy.[18] In the fifteenth and sixteenth centuries, coffee grown in the soils of Yemen, using techniques borrowed from East Africa, soon filled cups in Istanbul, Vienna, and Isfahan.[19] Trans-Saharan caravan networks ensured a steady circulation of animals, salt, paper, disease, and human slaves between the Middle East, North Africa, and West Africa.[20] It was most likely through the Ottoman Empire that maize and other New World crops moved east.[21] Since the early twentieth century, the extraction, refinement, and distribution of Middle Eastern oil clearly have been of enormous global environmental and geopolitical significance.[22] So too has the emergence of a worldwide environmental movement that was crucial for sparking environmental consciousness and activism in the region.

The histories of various parts of the MENA were also profoundly shaped by different instances of global climate change. Whether the Little Ice Age in the early modern period that greatly reduced agricultural yields in the Ottoman Empire, Icelandic volcano eruptions at the end of the eighteenth century that affected Nile flood levels, El Niño-induced famines at the end of the nineteenth century in Anatolia and Iran, or today's warnings of global warming, the MENA has clearly been profoundly impacted by these instances of climatic alteration, and studying these cases offers a deeper understanding of the global history of climate change.[23] In other words, the MENA was a region that served to diffuse agricultural and natural products, knowledge of environmental manipulation techniques, climatic trends, and disease agents across large swaths of Eurasia and the world. Therefore, to understand the history of these crops, diseases, commodities, weather patterns, and technologies means necessarily understanding the Middle Eastern components of their global histories.[24] This is perhaps patently clear and unsurprising, but much of this history nevertheless continues to be under-researched.

One of the reasons most commonly cited for this lack of research is a paucity of sources. As this book, the older geophysical literature, and other recent works show, however, the MENA offers a wealth of continuous source materials from antiquity to the present.[25] An unbroken documentary record of Egyptians' interactions with the Nile exists from at least as early as 3000 BCE to the present.[26] Similarly, archaeological and written sources from antiquity forward bring to light environmental management techniques in Anatolia, the Iranian Plateau, the Mesopotamian lands of Iraq, and elsewhere. The MENA's available published and unpublished sources thus offer environmental historians nearly unparalleled empirical depth and detail to track environmental change, landscape manipulation, and human–environment interactions over the three timescales famously identified by Fernand Braudel: events (singular moments in history that alone perhaps say little about anything larger), conjunctures (the unique interweaving and coalescing of various phenomena, historical actors, and political and economic forces to bring forward a never-before-experienced historical epoch), and the *longue durée* (broad historical trends that over centuries create discourses and practices that shape the worldview of large groups of people).[27] Having the sources to operate in such great detail at all these levels of historical time is a luxury not enjoyed by historians of most other regions of the globe.

Historians have successfully used this evidence of the long-term effects of human manipulation of the natural world to examine, for example, how and with what techniques extremely ecologically fragile environments were able to survive for millennia and what perturbations to these environments made them no longer viable. Thus, for instance, various enclave communities on the Iranian Plateau were able to manage their environmental vulnerability for centuries through the careful use of irrigation technologies, agriculture, and nomadic pastoralism.[28] Similarly, while the fortunes of any one community in the Mediterranean mountain environments of Turkey and Morocco might rise and fall drastically from year to year, when seen over the course of centuries, these environments were actually quite resilient and able to deal with intense change.[29] Only because of the presence of such a long record can the function of these environments be seen over time. Without this record, one might mistakenly judge a year of acute food shortages to be the norm in the mountains.

The presence of such a long and detailed record of human interaction with the environment in the MENA also affords environmental historians the opportunity to offer new perspectives on some long-debated issues in the field of environmental history. One of the best-known historiographical interventions of the MENA in environmental history revolves around Karl A. Wittfogel's thesis of Oriental despotism, which states that highly complex and coordinated systems of irrigation lead to authoritarian forms of government because only a central power can oversee and coordinate such a diffuse network of actors, interests, and limited resources.[30] Several of his examples of the emergence of such despotic forms of authoritarian government come from the Middle East—Mesopotamia and the Nile Valley. However, a full examination of the documentary record shows that it was the very people Wittfogel claims were exploited by sultans, emperors, and kings who were actually in control of the day-to-day functioning and maintenance of the large-scale irrigation networks that are so crucial to his argument.[31]

The fact that the available sources provide both the potential for *longue durée* environmental analyses and the opportunity to construct an environmental history of much more concentrated periods of time and of particular places allows for overarching narratives like continuity, collapse, or Oriental despotism to be tested over centuries in empirically nuanced and specific ways. In some of the older geographic literature, human agency is limited to a constant fight against the inevitability of environmental circumstances—a kind of environmental determinism.[32] Taking advantage of the breadth and depth of both published and unpublished written sources, more recent histories of Middle Eastern environments, by contrast, focus more on the push and pull of relationships between humans and nature.[33]

Examples include recent works on the history of climate change in the Middle East that challenge various propositions put forth about this global phenomenon.[34] One of these propositions is that the Middle Ages were a period of global warming. Evidence for this worldwide weather trend is based mainly on European and Chinese sources. Persian and Arabic sources from medieval Iran, however, show that in fact the opposite seems to have been occurring in Iran and Central Asia—a "Big Chill," in Richard W. Bulliet's words, that led to enormous reductions in levels of cotton cultivation.[35] More so than the medieval centuries, the early modern period—especially the seventeenth century—has been identified as a time in which climate change played a crucial role in shaping global history, greatly affecting the political and economic fortunes of the Ottoman Empire in the eastern Mediterranean and North Africa.[36]

Environmental historians interested in various other topics will also find much to offer in the history and historiography of the MENA. Some work exists on the history of natural disasters, most notably earthquakes, and their impact on human communities in the MENA.[37] The environmental history of the MENA also allows for the exploration of subjects that have garnered much attention in other geographies, such as the rise of commercial agriculture, indigenous land tenure rights, changes to transportation systems, and the emergence of megacities.[38] The history of public works is similarly usefully addressed. As with elsewhere in the world (India and the American Southeast, for instance), the middle of the twentieth century was a period of intense dam building throughout the Middle East—at Aswan, in Iran, and in Turkey.[39] What was the connection between such major public works projects in the twentieth century and earlier instances of this kind of environmental work—the Suez Canal, the Ottoman Don-Volga canal building project, or Safavid plans to build a dam to divert the course of the Karun river toward their new capital at Isfahan—or current ones—the Red to Dead pipeline project in Jordan or Turkey's dam projects in its southeast?[40] Records of such projects and social, economic, cultural, and political commentaries about them offer a rich source base for thinking about the environmental history of public works in the MENA.

Finally, although some work has been done on the subject of Islam and nature, there is much more left to research about what various Islamic texts from different places and times say about the natural world, environmental manipulation, plants, animals, and the history of how Muslims have put into real world effect their religious ideas about nature.[41] Islam and its various traditions clearly deserve as much attention as has been devoted to Jewish, Buddhist, Christian, Hindu, and Shamanist views of nature.[42]

Because of the MENA's extreme demographic heterogeneity, environmental historians interested in how different cultural groups have thought about the environments they shared also stand to gain much from a consideration of the MENA. Both in the early modern period and today, many linguistic, religious, and ethnic groups shared the same Middle Eastern environments.

In general, were there differences in how various religious, ethnic, and linguist communities thought about and interacted with their common landscapes and ecological constraints? If so, how did this heterogeneity manifest itself? In Lebanon, for instance, Shi'a, Druze, Maronite, Sunni Muslim, Orthodox Christian, Armenian, and other cultural and religious groups all lived on the same mountains and in the same valleys. How did differences between groups in their access to environmental resources or in their environmental manipulation techniques affect their ideas about nature? Similar questions could be asked about a place like southeastern Turkey with its various communities of Turks, Kurds, Arabs, Armenians, and others.[43]

Given the MENA's enormous diversity of cultural traditions; its centrality to the global traffic of economics, politics, and ecology; and the vast quantities of source materials available from various parts of the MENA from all periods, it should be no surprise that the region's history helps to explain and illuminate a range of aspects of global environmental history. Add to these facts the presence of the world's longest river, along with other important watersheds; the global significance of the region's oil; the Suez Canal and other crucial past and present public works; the historic importance of cotton, coffee, and other agricultural commodities from the region; and the presence of a vast diversity of landscapes and ecologies, and it becomes clear that there are countless topics to address in the environmental history of the MENA and that these histories are crucial to any study of the global environment.

Why the Middle East Needs Environmental History

The study of the MENA has regularly borrowed methodologies from other historiographies to advance work on the region. In the 1960s and 1970s, for example, Middle East historians turned to social history and political economy to study peasant agriculture, the history of workers, interactions among different social groups, and state and society relations. In the 1980s and 1990s, they benefited from a consideration of histories of gender and sexuality, nationalism, and colonialism and postcolonialism. In each of these and in other cases, Middle East historians have not simply borrowed wholesale the ideas and theoretical insights of other fields but have advanced them in important and novel ways. Most Middle East historians have so far ignored the perspectives offered by environmental history.[44]

While the MENA can usefully nudge several strands of environmental history onto slightly different paths, so too can environmental history be adopted as a useful new methodology in Middle East Studies. An environmental perspective helps address some of the largest and most intractable problems of Middle East historiography, among them issues of geographic demarcation and periodization. It also allows for the introduction of historical actors who have been crucial to the region's history but remain missing from the historiographical literature, including animals, microbes, wind, and silt. Finally, it offers fresh perspective on some key debates in the field—imperial decline in the early modern period, the transition from colonialism to postcolonial states, and the role of oil in twentieth-century politics.

A familiar tension in the literature of environmental history is that between political and environmental boundaries. Ecosystems or migratory peoples and animals clearly do not recognize the artificial borders of nation–states or earlier imperial provinces, though these political demarcations obviously impact ecosystems, nomadic peoples, migratory animals, and environmental processes. For example, twentieth-century states generally focused their resource exploitation within their borders with little concern for any "downstream" effects on other nations. Likewise, political conflicts often emerge when states threaten to withhold natural resources from other states. Perhaps one of the clearest examples of the tensions inherent in international resource use and management is the Nile watershed, where disputes between Egypt and upstream riparian states have been escalating for decades.[45] In the Jordan River valley, competition over limited water supplies has emerged as a central part of the conflict between Israel and the Palestinians.[46] Similarly, Turkey, Iran, and Syria have been locked for decades in a struggle over the use of the waters of the Tigris and Euphrates.[47] These historical and contemporary struggles support former United Nations secretary general Boutros Boutros-Ghali's well-known admonition that the next wars in the Middle East will be not over oil but water.[48]

Current political exigencies notwithstanding, the tension between ecology and politics is nothing new to students of environmental history. Scholars of the MENA have for the most part accepted the region's artificial geographic borders, those of newly created twentieth-century nation–states and those of provinces of the Ottoman and Safavid Empires as well. However, by taking a river system, a climate pattern, or a migratory group of humans or animals as the unit of analysis, the traditional concentration of historians on political or administrative territorial divisions can be bypassed, or at the very least broadened and balanced. There is no way of getting around the fact that the Nile flows through most of the countries of East Africa or that the history of pastoralism must be studied across the political divides of imperial provinces and nation–states.[49] Indeed, a proper analysis of the ecology of the Nile or of the effects of pastoralism on the Qajar and Ottoman Empires demands going beyond politico-territorial categories.

This tension between political demarcations of space and ecological systems that do not overlay neatly offers other research opportunities as well. Perhaps most obvious is how nineteenth-century imperial powers and twentieth-century nation–states constructed representations of nature within the borders of their possessions to further their political ambitions.[50] How did particular notions of Lebanese, Qajar, or Moroccan nature emerge? What are their specific characteristics? In the same way that various political, cultural, and other discourses constructed a gendered notion of an Algerian or Turkish man or woman, so too is it useful to examine similar constructions of nature, landscape, and resources.[51] Focusing either on ecological processes that traverse geographic borders or on cultural and political processes that accept these divisions as a precondition for constructions of new discourses about nature, an environmental perspective helps avoid reifications of artificially created spatial demarcations.

In thinking about how various societies and polities have come to construe geographic territory, historians of the MENA have also focused a great deal of attention on how to periodize the region's history. Here again environmental history is useful as it allows for the reconceptualization of many of the temporal divisions conventionally employed in the field. Although the bulk of the chapters in this book focus on the period after 1500, the authors fully recognize the artificiality of this starting point for discussions of ecological processes like climate change, disease etiologies, and crop cycles. A long-term ecological perspective thus permeates many of the following discussions of specific historical moments and places. For example, neither 1501—the year marking the rise of the Safavid Empire in Iran—nor 1517—when the Ottoman Empire captured most of the Arab World—represents a distinct breaking point in the history of plague and the many other environmental phenomena associated with it. Likewise, the assumed transition from a colonial regime to a postcolonial national one in North Africa did not necessarily change how national parks land was used to alienate nomadic populations and raise state revenue. These and other examples shed light on how temporal divisions usually associated with political change are often not particularly useful in explaining chronological beginning and ending points.

Indeed, historical periodization could usefully be based on climatic shifts rather than political ones. Since the medieval period, the MENA has experienced numerous climate cycles that have been wholly important to the political, social, economic, and even religious history of the region. The Big Chill in the eleventh and twelfth centuries facilitated Turkic migrations west into Iran and eventually Anatolia and perhaps even contributed to the emergence of Shi'ism as the dominant form of religiosity in Iran.[52] This period of cooling decimated large landholdings and much of the economic and social base of

Iran, which likely made the Iranian Plateau vulnerable to Mongol and other incursions. After this period, the MENA then enjoyed a few centuries of temperate weather patterns that in turn likely contributed to the emergence of the great empires of the early modern period—the Ottomans, Safavids, Mughals, Uzbeks, and Mamluks.

In the sixteenth and seventeenth centuries, however, a well-known period of cool set into the MENA and elsewhere. The Little Ice Age contributed to various economic and political troubles for the Ottoman Empire, forcing a realignment of resources and sovereignty.[53] A key component of the general crisis of the seventeenth century, consideration of this climatic event sheds new light on the historiographically vexed issue of the decline, decentralization, or realignment of the Ottoman Empire. It also allows for fresh analyses of early modern political and social transitions as products not just of governance and economics, but also of weather, soil chemistry, and climate.

Weather fluctuations—albeit less dramatic, but no less crucial—were features of the last two decades of the eighteenth century as well. The 1783 and 1784 eruptions of the Laki fissure in Iceland triggered a climate event of enormous proportions.[54] Monsoon levels decreased over the Indian Ocean; temperatures plunged around the Mediterranean; flood levels of various river systems in the MENA were dramatically reduced; and a twenty-year period of famine, drought, disease, and hardship set in. This climatic flux no doubt contributed to political turmoil at the end of the eighteenth century, resulting in, among other things, various European incursions into the MENA and eventually the period of Ottoman reform known as the Tanzimat.[55] In the latter half of the nineteenth century, another period of climatically induced famine and food shortages contributed to the discontent of provincial notables in the Ottoman Empire and to calls for political reform.[56] In short, there is good reason to conceive of the last millennium of Middle Eastern history along the lines of climate and weather patterns rather than political states. Indeed, it would seem that periods of cooling or climatic fluctuations often had greater impact on rural peoples in the MENA than did the political powers ostensibly ruling over them.

Another utility provided by environmental history for studying the pasts of the MENA is its ability to give voice to multiple vital historical actors. Efforts to write subaltern histories of the MENA, often inspired by models developed in South Asian historiography, have so far concentrated mostly on the historical importance of peasants.[57] This is clearly of the utmost significance since the vast majority of people living in the MENA were—and largely still are—peasants. A generation of historians using Islamic court records, land surveys, and other sources has detailed how peasants throughout the MENA interacted and dealt with various political and economic regimes.[58] Most of these studies have attempted to insert peasants into larger frameworks—imperial systems of governance, global commercial networks, and wider religious or cultural communities. Building on these analyses, much of this book also focuses on rural communities. Yet an environmental perspective demonstrates that peasants were not merely junior partners in larger political and economic processes and forces; rather, they often played a leading role in bringing agricultural goods to market, managing natural resources, and controlling understandings of rural ecologies.[59] Because so much political power in the MENA over the last five hundred years has been invested in and derived from productively using the agricultural products of the land, imperial, colonial, national, and international powers throughout the region have necessarily relied on the knowledge and experience of local populations to manage the natural resources on which their economic and political power depended.

While revealing the power invested in peasant cultivators as workers on and stewards of the land, environmental history also demonstrates how nonhuman elements of the natural world shaped Middle Eastern history. Domesticated animals are clearly central nonhuman actors in any history of the MENA and should be afforded the historiographical significance they deserve.[60] Before fossil fuels, animals were the engines that powered the countryside. They brought goods from fields to markets; pulled waterwheels and plows; provided heat, milk, and sometimes food; transported peoples, information, and goods; and were exchanged as status gifts. For these and other reasons, animals were some of the most coveted and valuable possessions a peasant could own in the countryside and hence were involved in myriad economic transactions.[61] As the primary purveyors of usable energy until the twentieth century, animals' histories must be understood in order to fully capture the MENA's pasts.

Environmental history also illuminates the historical centrality and agency of other nonhuman and non-animal actors. The position, size, weight, and other physical characteristics of granite in the Nile Valley, for example, helped determine mid-twentieth-century national and geological narratives of Egypt's past that were harnessed in the service of building the Aswan High Dam and displacing Nubian populations in Egypt's south. Rocks thus came to affect the history of the dam as much as the geopolitical concerns of the Cold War or 'Abd al-Nasir's domestic agenda. Water's particular physical properties and manifestations also clearly shaped the contours of Middle Eastern history. How and in what amounts the region's rivers flooded determined the well-being and life span of more people over the course of the MENA's history than any imperial or national state or any economic development scheme. Thus only by attending to the historical import of actors like water, microbes, donkeys, and silt do the full complexities of the MENA's pasts emerge.

Finally, an environmental perspective in the study of the MENA allows for revisiting, reinterpreting, and perhaps even resolving some of the most important debates in Middle East Studies. Consider, for example, the political transformation of the Ottoman Empire in the early modern period. The Celali peasant revolts at the end of the sixteenth and the beginning of the seventeenth centuries, often cited as evidence of the empire's waning ability to control rural populations and food supplies, were partly a result of ecological pressures precipitated by the Little Ice Age.[62] Some of the manifestations of this climatic change were violence and banditry that eventually coalesced into an organized rebellion against the Ottoman state. Thus, the middle period of Ottoman history, interpreted as the crucial hinge between the rise and sixteenth-century efflorescence of the empire and its eventual efforts at reform in the nineteenth century, might be usefully thought of as a time when the empire was attempting to manage massive climatic fluctuations and their attendant ecological effects on agricultural production, human and animal populations, and flood levels.[63]

Furthermore, the MENA, with the largest known petroleum supplies in the world, clearly played a central role in the one major energy regime shift in human history—from solar energy to fossil fuels.[64] Understanding how humans cultivated and then exploited their energy supplies must be at the heart of any history of human communities.[65] Since the latter half of the nineteenth century, much of these energy supplies have come from the MENA; thus, the region must be included in any history of energy since this period, just as fossil fuels must be taken as central to the region's political history.[66] There is, however, a far longer story of energy in the region that predates the discovery of oil. Indeed, putting petroleum utilization and consumption in the larger historical context of transitions between energy regimes reveals that oil and the supposed need to seek out alternative sources of energy in the MENA are only the latest iterations in a much longer history of low-cost energy.[67] In the medieval and early modern periods, animals— abundant, effective, and cheap—remained the primary means of transport and power in the MENA and Central Asia, precluding any need to search for alternative sources of energy. So too there has been little incentive for Middle Eastern oil states to seek out other forms of energy as long as petroleum has remained similarly abundant and cheap. Thus, by making energy outputs (whether caloric or carbon organic) the unit of analysis, the politics of oil in the MENA becomes part of a considerably longer and more comprehensive history of energy and society.

Toward Middle East Environmental History

The main goal of this book is to offer the broad outlines of the current state of Middle East environmental history and some ideas about the rich potential of this line of inquiry. Neither this introduction nor the book as a whole is meant to exhaustively represent all the work that has been done, is currently being done, or could potentially be done on environmental topics in the MENA. Important subjects not covered in this book in any great depth include the environmental impacts of war, gender and the environment, pollution, science and nature, the extraction of resources other than oil, literary representations of nature, and the role of environmental symbolism and other kinds of environmental thought in the MENA.[68] The following chapters thus represent a snapshot of much of the current work being done on various topics in Middle East environmental history and offer case studies engaging with the themes outlined here. We hope they start a much larger conversation between the fields of environmental history and Middle East Studies.

Acknowledgment

For their very helpful comments on earlier versions of this chapter, I thank Paul Sabin, Valerie Hansen, Peter C. Perdue, Daniel R. Headrick, Fabian Drixler, Diana K. Davis, Sam White, Ranin Kazemi, Helen Curry, Robin Scheffler, and Oxford University Press's two anonymous reviewers.

Notes

1. I use the terms MENA and the Middle East interchangeably throughout this chapter.

2. For works that discuss the Middle East as a Eurasian contact zone, see Janet L. Abu-Lughod, *Before European Hegemony: The World System A.D. 1250–1350* (New York: Oxford University Press, 1989); Richard W. Bulliet, *Cotton, Climate, and Camels in Early Islamic Iran: A Moment in World History* (New York: Columbia University Press, 2009); Richard M. Eaton, "Islamic History as Global History," in *Islamic and European Expansion: The Forging of a Global Order*, ed. Michael Adas (Philadelphia: Temple University Press, 1993), 1–36; Gagan D. S. Sood, "Pluralism, Hegemony and Custom in Cosmopolitan Islamic Eurasia, ca. 1720–90, with Particular Reference to the Mercantile Arena" (Ph.D. diss., Yale University, 2008).

3. For instructive treatments of pastoralism in the Middle East, see Reşat Kasaba, *A Moveable Empire: Ottoman Nomads, Migrants, and Refugees* (Seattle: University

of Washington Press, 2009); Arash Khazeni, *Tribes and Empire on the Margins of Nineteenth-Century Iran* (Seattle: University of Washington Press, 2010); Andrew Gordon Gould, "Pashas and Brigands: Ottoman Provincial Reform and Its Impact on the Nomadic Tribes of Southern Anatolia, 1840–1885" (Ph.D. diss., University of California, Los Angeles, 1973).

4. For a study of some of these connections in the early modern Muslim World, see Muzaffar Alam and Sanjay Subrahmanyam, *Indo-Persian Travels in the Age of Discoveries, 1400–1800* (Cambridge: Cambridge University Press, 2007).

5. The most ambitious work to tackle this subject is John F. Richards, *The Unending Frontier: An Environmental History of the Early Modern World* (Berkeley: University of California Press, 2003). In a telling indication of the absence of the MENA in the global history of the environment, this magisterial work has very little to say about the region.

6. For a representative sampling of some of this work, see William C. Brice, ed., *The Environmental History of the Near and Middle East Since the Last Ice Age* (London: Academic Press, 1978).

7. A great concentration of work focuses on the ancient Middle Eastern environment and addresses topics such as Persian Gulf seafloor changes, desertification, irrigation, and salinization. For some of this literature, see for example P. Kassler, "The Structural and Geomorphic Evolution of the Persian Gulf," in *The Persian Gulf: Holocene Carbonate Sedimentation and Diagenesis in a Shallow Epicontinental Sea*, ed. B. H. Purser (Berlin: Springer-Verlag, 1973), 11–32; Elazar Uchupi, S. A. Swift, and D. A. Ross, "Gas Venting and Late Quaternary Sedimentation in the Persian (Arabian) Gulf," *Marine Geology* 129 (1996): 237–269; Michael Brookfield, "The Desertification of the Egyptian Sahara during the Holocene (the Last 10,000 Years) and Its Influence on the Rise of Egyptian Civilization," in *Landscapes and Societies: Selected Cases*, ed. I. Peter Martini and Ward Chesworth (Dordrecht: Springer, 2010), 91–108; Arie S. Issar, *Water Shall Flow from the Rock: Hydrogeology and Climate in the Lands of the Bible* (Berlin: Springer-Verlag, 1990); Thorkild Jacobsen and Robert M. Adams, "Salt and Silt in Ancient Mesopotamian Agriculture," *Science* 128 (1958): 1251–1258; Thorkild Jacobsen, *Salinity and Irrigation Agriculture in Antiquity: Diyala Basin Archaeological Report on Essential Results, 1957–58*, Bibliotheca Mesopotamica, vol. 14 (Malibu: Undena Publications, 1982).

8. Timothy Mitchell, "Are Environmental Imaginaries Culturally Constructed?" in *Environmental Imaginaries of the Middle East and North Africa*, ed. Diana K. Davis and Edmund Burke III (Athens: Ohio University Press, 2011), 266.

9. Karl W. Butzer, *Early Hydraulic Civilization in Egypt: A Study in Cultural Ecology* (Chicago: University of Chicago Press, 1976); Peter Christensen, *The Decline of Iranshahr: Irrigation and Environments in the History of the Middle East, 500 B.C. to A.D. 1500* (Copenhagen: Museum Tusculanum Press, 1993).

10. J. M. Wagstaff, *The Evolution of Middle Eastern Landscapes: An Outline to A.D. 1840* (London: Croon Helm, 1985). Wagstaff's book does address the post-1500 period, but the bulk of the work (over two-thirds of it) concerns the earlier period. Carlos E. Cordova, *Millennial Landscape Change in Jordan: Geoarchaeology and Cultural Ecology* (Tucson: University of Arizona Press, 2007); Russell Meiggs, *Trees and Timber in the*

Ancient Mediterranean World (Oxford: Clarendon Press, 1982); J. V. Thirgood, *Man and the Mediterranean Forest: A History of Resource Depletion* (London: Academic Press, 1981); Robert McC. Adams, *Land Behind Baghdad: A History of Settlement on the Diyala Plains* (Chicago: University of Chicago Press, 1965).

11. For a discussion of source materials and Middle East environmental history, see Alan Mikhail, "Global Implications of the Middle Eastern Environment," *History Compass* 9 (2011): 952–970.

12. Moreover, as Diana K. Davis has shown, many of these scientific sources have a long and deep relationship with European colonialism and its notions of environmental decline in North Africa and elsewhere. Diana K. Davis, *Resurrecting the Granary of Rome: Environmental History and French Colonial Expansion in North Africa* (Athens: Ohio University Press, 2007), 131–176; idem., "Potential Forests: Degradation Narratives, Science, and Environmental Policy in Protectorate Morocco, 1912–1956," *Environmental History* 10 (2005): 211–238.

13. For recent examples of the use of this copious primary source material to write environmental histories of the MENA, see Sam White, *The Climate of Rebellion in the Early Modern Ottoman Empire* (New York: Cambridge University Press, 2011); Arash Khazeni, *Tribes and Empire*; Alan Mikhail, *Nature and Empire in Ottoman Egypt: An Environmental History* (New York: Cambridge University Press, 2011).

14. William Cronon, *Changes in the Land: Indians, Colonists, and the Ecology of New England*, 1st rev. edn. (New York: Hill and Wang, 2003), 13.

15. Ibid., 15.

16. Andrew Watson, *Agricultural Innovation in the Early Islamic World: The Diffusion of Crops and Farming Techniques, 700–1100* (Cambridge: Cambridge University Press, 1983). For a critique of certain aspects of the "Islamic green revolution" thesis, see Michael Decker, "Plants and Progress: Rethinking the Islamic Agricultural Revolution," *Journal of World History* 20 (2009): 187–206.

17. Michael W. Dols, *The Black Death in the Middle East* (Princeton, NJ: Princeton University Press, 1977); William H. McNeill, *Plagues and Peoples* (Garden City, NY: Anchor Press/Doubleday, 1976); Alan Mikhail, "The Nature of Plague in Late Eighteenth-Century Egypt," *Bulletin of the History of Medicine* 82 (2008): 249–275; Michael W. Dols, "The Second Plague Pandemic and Its Recurrences in the Middle East: 1347–1894," *Journal of the Economic and Social History of the Orient* 22 (1979): 162–189.

18. Thomas F. Glick, *Irrigation and Hydraulic Technology: Medieval Spain and Its Legacy* (Brookfield, VT: Variorum, 1996); idem., *Irrigation and Society in Medieval Valencia* (Cambridge, MA: Harvard University Press, 1970). For a comparative study of irrigation techniques in Iran, Egypt, and Spain, see Abigail E. Schade, "Hidden Waters: Groundwater Histories of Iran and the Mediterranean" (Ph.D. diss., Columbia University, 2011).

19. Ralph Hattox, *Coffee and Coffeehouses: The Origins of a Social Beverage in the Medieval Near East* (Seattle: University of Washington Press, 1985); Michel Tuchscherer, ed., *Le commerce du café avant l'ère des plantations coloniales: espaces, réseaux, sociétés (XVᵉ-XIXᵉ siècle)* (Cairo: Institut français d'archéologie orientale, 2001); William Gervase Clarence-Smith and Steven Topik, *The Global Coffee Economy*

in Africa, Asia, and Latin America, 1500–1989 (Cambridge: Cambridge University Press, 2003).

20. Ghislaine Lydon, *On Trans-Saharan Trails: Islamic Law, Trade Networks, and Cross-Cultural Exchange in Nineteenth-Century Western Africa* (Cambridge: Cambridge University Press, 2009); idem., "Writing Trans-Saharan History: Methods, Sources and Interpretations Across the African Divide," *Journal of North African Studies* 10 (2005): 293–324.

21. On maize, see J. R. McNeill, *The Mountains of the Mediterranean World: An Environmental History* (Cambridge: Cambridge University Press, 1992), 89–90; Faruk Tabak, *The Waning of the Mediterranean, 1550–1870: A Geohistorical Approach* (Baltimore: Johns Hopkins University Press, 2008), 255–269. On the diffusion of maize from Egypt and North Africa to other parts of Africa, see James C. McCann, *Maize and Grace: Africa's Encounter with a New World Crop, 1500–2000* (Cambridge, MA: Harvard University Press, 2007).

22. In this regard, see, for example, Timothy Mitchell, *Carbon Democracy: Political Power in the Age of Oil* (London: Verso, 2011).

23. On the Little Ice Age and the Ottoman Empire, see Sam White, *Climate of Rebellion*. On the Icelandic eruptions, see Luke Oman, Alan Robock, Georgiy L. Stenchikov, and Thorvaldur Thordarson, "High-Latitude Eruptions Cast Shadow over the African Monsoon and the Flow of the Nile," *Geophysical Research Letters* 33 (2006): L18711. On El Niño famines, see Mike Davis, *Late Victorian Holocausts: El Niño Famines and the Making of the Third World* (London: Verso, 2001). On the famines in Anatolia, see Mehmet Yavuz Erler, *Osmanlı Devleti'nde Kuraklık ve Kıtlık Olayları, 1800–1880* (Istanbul: Libra Kitap, 2010).

24. For an instructive example in this regard, see the following work on the history of malaria in Egypt: Timothy Mitchell, "Can the Mosquito Speak?" in *Rule of Experts: Egypt, Techo-Politics, Modernity* (Berkeley: University of California Press, 2002), 19–53.

25. Such is also the case with China and South Asia. On the environmental histories of China from antiquity to the present, see Mark Elvin, *The Retreat of the Elephants: An Environmental History of China* (New Haven, CT: Yale University Press, 2004); Mark Elvin and Liu Ts'ui-jung, eds., *Sediments of Time: Environment and Society in Chinese History* (Cambridge: Cambridge University Press, 1998). On South Asia, see Mahesh Rangarajan and K. Sivaramakrishnan, eds., *India's Environmental History*, 2 vols. (Ranikhet: Permanent Black, 2012); Arun Agrawal and K. Sivaramakrishnan, eds., *Agrarian Environments: Resources, Representations, and Rule in India* (Durham, NC: Duke University Press, 2000); Mahesh Rangarajan, ed., *Environmental Issues in India: A Reader* (New Delhi: Pearson Education, 2009); Deepak Kumar, Vinita Damodaran, and Rohan D'Souza, eds., *The British Empire and the Natural World: Environmental Encounters in South Asia* (Oxford: Oxford University Press, 2011); Richard H. Grove, Vinita Damodaran, and Satpal Sangwan, eds., *Nature and the Orient: The Environmental History of South and Southeast Asia* (Delhi: Oxford University Press, 1998).

26. For some of this history, see Robert O. Collins, *The Nile* (New Haven, CT: Yale University Press, 2002).

27. These different notions of historical time are delineated in Braudel's epic work: Fernand Braudel, *The Mediterranean and the Mediterranean World in the Age of Philip II*, trans. Siân Reynolds, 2 vols. (Berkeley: University of California Press, 1995). Richard W. Bulliet has recently advanced the notion of a "moment" in world historical time as conceptually more capacious than Braudel's temporal divisions. Bulliet, *Cotton, Climate, and Camels*, 143–144.

28. Christensen, *Decline of Iranshahr*.

29. See the relevant sections of McNeill, *Mountains of the Mediterranean*.

30. Wittfogel's seminal work is *Oriental Despotism: A Comparative Study of Total Power* (New Haven, CT: Yale University Press, 1957). For a more abbreviated form of his main arguments, see Karl A. Wittfogel, "The Hydraulic Civilizations," in *Man's Role in Changing the Face of the Earth*, ed. William L. Thomas, Jr. (Chicago: University of Chicago Press, 1956), 152–164. For an example of the use of Wittfogel's ideas outside of a Middle Eastern context, see Donald Worster, *Rivers of Empire: Water, Aridity, and the Growth of the American West* (Oxford: Oxford University Press, 1992).

31. This argument is advanced more thoroughly in the case of Egypt in Mikhail, *Nature and Empire*.

32. On some of the dangers of environmental determinism in the study of the Middle Eastern environment, see Diana K. Davis, "Power, Knowledge, and Environmental History in the Middle East and North Africa," *International Journal of Middle East Studies* 42 (2010): 657–659.

33. See works such as White, *Climate of Rebellion*; Khazeni, *Tribes and Empire*; Sandra M. Sufian, *Healing the Land and the Nation: Malaria and the Zionist Project in Palestine, 1920–1947* (Chicago: University of Chicago Press, 2007); Wolf-Dieter Hütteroth, "Ecology of the Ottoman Lands," in *The Cambridge History of Turkey, Volume 3: The Later Ottoman Empire, 1603–1839*, ed. Suraiya N. Faroqhi (Cambridge: Cambridge University Press, 2006), 18–43; Selçuk Dursun, "Forest and the State: History of Forestry and Forest in the Ottoman Empire" (Ph.D. diss., Sabancı University, 2007); Mikhail, *Nature and Empire*. See also several of the essays in the following volume: Suraiya Faroqhi, ed., *Animals and People in the Ottoman Empire* (Istanbul: Eren, 2010).

34. Two of the earliest attempts to integrate climate change into a history of the Middle East are Rhoads Murphey, "The Decline of North Africa Since the Roman Occupation: Climatic or Human?" *Annals of the Association of American Geographers* 41 (1951): 116–132 and William Griswold, "Climatic Change: A Possible Factor in the Social Unrest of Seventeenth Century Anatolia," in *Humanist and Scholar: Essays in Honor of Andreas Tietze*, ed. Heath W. Lowry and Donald Quataert (Istanbul: Isis Press, 1993), 37–57. See also Arie S. Issar and Mattanyah Zohar, *Climate Change— Environment and Civilization in the Middle East* (Berlin: Springer, 2004). The best recent studies of the impacts of climate change in the MENA are Bulliet, *Cotton, Climate, and Camels* and White, *Climate of Rebellion*.

35. Bulliet, *Cotton, Climate, and Camels*, 69–95.

36. Early modern climate change is usually understood as part of what is termed the general crisis of the seventeenth century. The literature on this topic began in the

1950s with the following pair of articles: E. J. Hobsbawm, "The General Crisis of the European Economy in the 17th Century," *Past and Present* 5 (1954): 33–53; idem., "The Crisis of the 17th Century—II," *Past and Present* 6 (1954): 44–64. For a recent series of studies on the topic, see the articles that make up the following forum: Jonathan Dewald, Geoffrey Parker, Michael Marmé, and J. B. Shank, "*AHR* Forum: The General Crisis of the Seventeenth Century Revisited," *American Historical Review* 113 (2008): 1029–1099. On the impacts of early modern climate change and the general crisis of the seventeenth century on the Ottoman Empire, see White, *Climate of Rebellion*.

37. On earthquakes, see Mohamed Reda Sbeinati, Ryad Darawcheh, and Mikhail Mouty, "The Historical Earthquakes of Syria: An Analysis of Large and Moderate Earthquakes from 1365 B.C. to 1900 A.D.," *Annals of Geophysics* 48 (2005): 347–435; Elizabeth Zachariadou, ed., *Natural Disasters in the Ottoman Empire* (Rethymnon: Crete University Press, 1999); N. N. Ambraseys and C. F. Finkel, *The Seismicity of Turkey and Adjacent Areas: A Historical Review, 1500–1800* (Istanbul: Eren, 1995).

38. For various works that address these many topics, see Kenneth M. Cuno, *The Pasha's Peasants: Land, Society, and Economy in Lower Egypt, 1740–1858* (Cambridge: Cambridge University Press, 1992); idem., "Commercial Relations between Town and Village in Eighteenth and Early Nineteenth-Century Egypt," *Annales Islamologiques* 24 (1988): 111–135; McNeill, *Mountains of the Mediterranean*; Richard W. Bulliet, *The Camel and the Wheel* (Cambridge, MA: Harvard University Press, 1975); Jim Krane, *City of Gold: Dubai and the Dream of Capitalism* (New York: St. Martin's Press, 2009), 223–249.

39. On Aswan, see Yusuf A. Shibl, *The Aswan High Dam* (Beirut: The Arab Institute for Research and Publishing, 1971); Hussein M. Fahim, *Dams, People and Development: The Aswan High Dam Case* (New York: Pergamon Press, 1981). See also the relevant sections of John Waterbury, *Hydropolitics of the Nile Valley* (Syracuse: Syracuse University Press, 1979); Collins, *The Nile*. On the construction of dams during what came to be known as Iran's First Seven Year Plan from 1948 to 1955, see Peter Beaumont, "Water Resource Development in Iran," *The Geographical Journal* 140 (1974): 418–431; Gordon R. Clapp, "Iran: A TVA for the Khuzestan Region," *Middle East Journal* 11 (1957): 1–11. On dams in Turkey, see J. R. McNeill, *Something New under the Sun: An Environmental History of the Twentieth-Century World* (New York: W. W. Norton, 2000), 123.

40. On the Suez Canal, see D. A. Farnie, *East and West of Suez: The Suez Canal in History, 1854–1956* (Oxford: Clarendon Press, 1969); John Marlowe, *World Ditch: The Making of the Suez Canal* (New York: Macmillan, 1964). On the Don-Volga Canal, see Halil İnalcık, "The Origins of the Ottoman-Russian Rivalry and the Don-Volga Canal 1569," *Les annales de l'Université d'Ankara* 1 (1946–1947): 47–106; A. N. Kurat, "The Turkish Expedition to Astrakhan and the Problem of the Don-Volga Canal," *Slavonic and East European Review* 40 (1961): 7–23. On the Karun river project, see Khazeni, *Tribes and Empire*, 23–25. On the construction of a pipeline between the Red and Dead Seas, see Basel N. Asmar, "The Science and Politics of the Dead Sea: Red Sea Canal or Pipeline," *Journal of Environment and Development* 12 (2003): 325–339. On Turkey's Southeast Anatolia Development Project (GAP), see Ali Çarkoğlu and Mine Eder,

"Development *alla Turca*: The Southeastern Anatolia Development Project (GAP)," in *Environmentalism in Turkey: Between Democracy and Development?* ed. Fikret Adaman and Murat Arsel (Aldershot: Ashgate, 2005), 167–184; idem., "Domestic Concerns and the Water Conflict over the Euphrates-Tigris River Basin," *Middle Eastern Studies* 37 (2001): 41–71; Leila Harris, "Postcolonialism, Postdevelopment, and Ambivalent Spaces of Difference in Southeastern Turkey," *Geoforum* 39 (2008): 1698–1708.

41. Mawil Izzi Dien, *The Environmental Dimensions of Islam* (Cambridge: Lutterworth Press, 2000); idem., "Islam and the Environment: Theory and Practice," *Journal of Beliefs and Values* 18 (1997): 47–57; Yusuf al-Qaradawi, *Ri'ayat al-Bi'ah fi Shari'at al-Islam* (Cairo: Dar al-Shuruq, 2001); Harifyah Abdel Haleem, ed., *Islam and the Environment* (London: Ta-Ha Publishers, 1998); Ziauddin Sardar, ed., *An Early Crescent: The Future of Knowledge and the Environment in Islam* (London: Mansell, 1989); Richard C. Foltz, Frederick M. Denny, and Azizan Baharuddin, eds., *Islam and Ecology: A Bestowed Trust* (Cambridge, MA: Harvard University Press, 2003); Fazlun M. Khalid with Joanne O'Brien, eds., *Islam and Ecology* (New York: Cassell, 1992); Richard Foltz, ed., *Environmentalism in the Muslim World* (New York: Nova Science Publishers, 2005); idem., "Is There an Islamic Environmentalism?" *Environmental Ethics* 22 (2000): 63–72.

42. Clarence J. Glacken, *Traces on the Rhodian Shore: Nature and Culture in Western Thought from Ancient Times to the End of the Eighteenth Century* (Berkeley: University of California Press, 1967); Madhav Gadgil and Ramachandra Guha, *This Fissured Land: An Ecological History of India* (Berkeley: University of California Press, 1993); Ramachandra Guha and J. Martinez-Alier, *Varieties of Environmentalism: Essays North and South* (London: Earthscan Publications, 1997); David E. Cooper and Simon P. James, *Buddhism, Virtue and Environment* (Aldershot: Ashgate, 2005); Martin D. Yaffe, ed., *Judaism and Environmental Ethics: A Reader* (Lanham, MD: Lexington Books, 2001). See also *The Encyclopedia of Religion and Nature* (London: Thoemmes Continuum, 2005) for a useful beginning on the relationships between various religious traditions and the environment.

43. Leila Harris, "Water and Conflict Geographies of the Southeastern Anatolia Project," *Society and Natural Resources* 15 (2002): 743–759; idem., "Postcolonialism, Postdevelopment, and Ambivalent Spaces of Difference."

44. Several scholars have noted this lack of Middle East environmental history. J. R. McNeill, "Observations on the Nature and Culture of Environmental History," *History and Theory* 42 (2003), 30; Edmund Burke III, "The Transformation of the Middle Eastern Environment, 1500 B.C.E.–2000 C.E.," in *The Environment and World History*, ed. Edmund Burke III and Kenneth Pomeranz (Berkeley: University of California Press, 2009), 81; Suraiya Faroqhi, "Introduction," in *Animals and People in the Ottoman Empire*, ed. Suraiya Faroqhi (Istanbul: Eren, 2010), 19. For useful surveys of some of the work that has been done on Middle East environmental history to date, see the following collections: Jeff Albert, Magnus Bernhardsson, and Roger Kenna, eds., *Transformations of Middle Eastern Natural Environments: Legacies and Lessons*, no. 103 of *Bulletin Series* (New Haven, CT: Yale School of Forestry and Environmental Sciences, 1998); Diana K. Davis and Edmund Burke III, eds., *Environmental Imaginaries of the Middle East and North Africa* (Athens: Ohio University Press, 2011).

45. Collins, *The Nile*; Waterbury, *Hydropolitics of the Nile Valley*.

46. Miriam R. Lowi, *Water and Power: The Politics of a Scarce Resource in the Jordan River Basin* (Cambridge: Cambridge University Press, 1993); Mark Zeitoun, *Power and Water in the Middle East: The Hidden Politics of the Palestinian-Israeli Water Conflict* (London: I. B. Tauris, 2008); Jeffrey K. Sosland, *Cooperating Rivals: The Riparian Politics of the Jordan River Basin* (Albany: State University of New York Press, 2007); Jan Selby, *Water, Power and Politics in the Middle East: The Other Israeli-Palestinian Conflict* (London: I. B. Tauris, 2003).
The reader will no doubt have noticed the absence in this book of a chapter or more devoted to the environmental history of Israel, the Palestinian Territories, and the Jordan River valley more generally. Unlike most parts of the MENA, the obvious political interests in this area and the strategic role of its environmental resources have meant that its environmental history has received substantial attention from historians and social scientists. Given the relative wealth of available materials, this book focuses on more under-researched parts of the MENA. In addition to those works already cited, see also the following for more on the environmental history of Israel and the Palestinian Territories: Stuart Schoenfeld, ed., *Palestinian and Israeli Environmental Narratives: Proceedings of a Conference Held in Association with the Middle East Environmental Futures Project* (Toronto: York University, 2005); Alon Tal, *Pollution in a Promised Land: An Environmental History of Israel* (Berkeley: University of California Press, 2002); Samer Alatout, "Towards a Bio-Territorial Conception of Power: Territory, Population, and Environmental Narratives in Palestine and Israel," *Political Geography* 25 (2006): 601–621; Sufian, *Healing the Land and the Nation*; Shaul Ephraim Cohen, *The Politics of Planting: Israeli-Palestinian Competition for Control of Land in the Jerusalem Periphery* (Chicago: University of Chicago Press, 1993).

47. Ali Ihsan Bagis, "Turkey's Hydropolitics of the Euphrates-Tigris Basin," *International Journal of Water Resources Development* 13 (1997): 567–582; Harris, "Water and Conflict Geographies."

48. For Boutros Boutros-Ghali's opinions on the politics of water in the Nile Basin and elsewhere in the Middle East, see Boutros Boutros-Ghali, *Egypt's Road to Jerusalem: A Diplomat's Story of the Struggle for Peace in the Middle East* (New York: Random House, 1997), 321–328.

49. This fact, for instance, offers Robert O. Collins the unifying theme for his book on the Nile. See Collins, *The Nile*. On pastoralism in the MENA, see Kasaba, *A Moveable Empire*; Khazeni, *Tribes and Empire*; Gould, "Pashas and Brigands."

50. This is shown quite convincingly in the following analysis of French colonial imaginaries of the North African environment: Davis, *Resurrecting the Granary of Rome*.

51. For an example of this kind of work in the Middle East, see Sharif S. Elmusa, ed., *Culture and the Natural Environment: Ancient and Modern Middle Eastern Texts*, vol. 26, no. 1 of *Cairo Papers in Social Science* (Cairo: American University in Cairo Press, 2003).

52. Bulliet, *Cotton, Climate, and Camels*.

53. White, *Climate of Rebellion*.

54. Oman, Robock, Stenchikov, Thordarson, "High-Latitude Eruptions."

55. For a general introduction to the Tanzimat, see M. Şükrü Hanioğlu, *A Brief History of the Late Ottoman Empire* (Princeton, NJ: Princeton University Press, 2008).

56. On these famines, see Erler, *Kuraklık ve Kıtlık*; Donald Quataert, *The Ottoman Empire, 1700–1922*, 2nd edn. (Cambridge: Cambridge University Press, 2005), 114–115.

57. See, for example, Farhad Kazemi and John Waterbury, eds., *Peasants and Politics in the Modern Middle East* (Miami: Florida International University Press, 1991); Joel Beinin, *Workers and Peasants in the Modern Middle East* (Cambridge: Cambridge University Press, 2001); Amy Singer, *Palestinian Peasants and Ottoman Officials: Rural Administration around Sixteenth-Century Jerusalem* (Cambridge: Cambridge University Press, 1994); Beshara Doumani, *Rediscovering Palestine: Merchants and Peasants in Jabal Nablus, 1700–1900* (Berkeley: University of California Press, 1995); Suraiya Faroqhi, "Ottoman Peasants and Rural Life: The Historiography of the Twentieth Century," *Archivum Ottomanicum* 18 (2000): 153–82; idem., "The Peasants of Saideli in the Late Sixteenth Century," *Archivum Ottomanicum* 8 (1983): 215–50.

58. See, for example, Doumani, *Rediscovering Palestine*; Leslie Peirce, *Morality Tales: Law and Gender in the Ottoman Court of Aintab* (Berkeley: University of California Press, 2003); Dror Ze'evi, *An Ottoman Century: The District of Jerusalem in the 1600s* (Albany: State University of New York Press, 1996).

59. In this regard, see Alan Mikhail, "An Irrigated Empire: The View from Ottoman Fayyum," *International Journal of Middle East Studies* 42 (2010): 569–590.

60. Analyses of the various impacts of animals on Middle Eastern history include Faroqhi, *Animals and People*; Annemarie Schimmel, *Islam and the Wonders of Creation: The Animal Kingdom* (London: al-Furqān Islamic Heritage Foundation, 2003); Mohamed Hocine Benkheira, Catherine Mayeur-Jaouen, and Jacqueline Sublet, *L'animal en islam* (Paris: Indes savantes, 2005); Basheer Ahmad Masri, *Animal Welfare in Islam* (Markfield: The Islamic Foundation, 2007); Richard C. Foltz, *Animals in Islamic Tradition and Muslim Cultures* (Oxford: Oneworld, 2006); Thomas T. Allsen, *The Royal Hunt in Eurasian History* (Philadelphia: University of Pennsylvania Press, 2006); Alan Mikhail, "Animals as Property in Early Modern Ottoman Egypt," *Journal of the Economic and Social History of the Orient* 53 (2010): 621–652; Bulliet, *The Camel and the Wheel*; idem., *Cotton, Climate, and Camels*; Suraiya Faroqhi, "Camels, Wagons, and the Ottoman State in the Sixteenth and Seventeenth Centuries," *International Journal of Middle East Studies* 14 (1982): 523–539.

61. On the political economy of animals in the early modern Ottoman Empire, see Mikhail, "Animals as Property."

62. On the Celali revolts, see Karen Barkey, *Bandits and Bureaucrats: The Ottoman Route to State Centralization* (Ithaca, NY: Cornell University Press, 1994); William Griswold, *The Great Anatolian Rebellion, 1000–1020/1591–1611* (Berlin: Klaus Schwarz Verlag, 1983).

63. White, *Climate of Rebellion*.

64. Edmund Burke III, "The Big Story: Human History, Energy Regimes, and the Environment," in *The Environment and World History*, ed. Edmund Burke III and Kenneth Pomeranz (Berkeley: University of California Press, 2009), 35.

65. For studies that make this point in some detail, see Vaclav Smil, *Energy in World History* (Boulder: Westview Press, 1994); idem., *Energy in Nature and Society: General Energetics of Complex Systems* (Cambridge: Massachusetts Institute of Technology Press, 2008); Stephen J. Pyne, *World Fire: The Culture of Fire on Earth* (Seattle: University of Washington Press, 1997); idem., *Vestal Fire: An Environmental History, Told through Fire, of Europe and Europe's Encounter with the World* (Seattle: University of Washington Press, 1997).

66. For recent work on the environmental and political histories of oil in the Middle East, see Toby Craig Jones, *Desert Kingdom: How Oil and Water Forged Modern Saudi Arabia* (Cambridge: Harvard University Press, 2010); Mitchell, *Carbon Democracy.*

67. Edmund Burke III, "The Big Story," 33–53. For another study that gives energy utilization primacy of place in a long story of human history, see J. R. McNeill, "The First Hundred Thousand Years," in *The Turning Points of Environmental History,* ed. Frank Uekoetter (Pittsburgh: University of Pittsburgh Press, 2010), 13–28. For a general discussion of energy regimes, see: idem., *Something New under the Sun,* 296–324.

68. For some of the existing literature on these topics, see Albert, Bernhardsson, and Kenna, *Transformations of Middle Eastern Natural Environments*; Davis and Burke, *Environmental Imaginaries*; Elmusa, *Culture and the Natural Environment*; Schoenfeld, *Palestinian and Israeli Environmental Narratives*; Sufian, *Healing the Land and the Nation*; Faroqhi, *Animals and People*; Foltz, "Is There an Islamic Environmentalism?"

I

The Eccentricity of the Middle East and North Africa's Environmental History

J. R. McNeill

This chapter aims to provide a global and comparative perspective on the environmental history of the MENA. It will draw attention to and briefly explore some of the eccentricities of the region as seen from an environmental history point of view. These eccentricities should not be misconstrued as exceptionalisms. Every one of them is shared with some other part or parts of the world, even if they are all eccentric in the sense that they are unusual. What is routine or commonplace in the MENA may be eccentric from the global perspective, as is true to some degree of any sizable region on the face of the Earth.[1]

In trying to adopt a global framework for contextualizing the MENA, I am consciously avoiding an East/West binary or Asia/Europe dichotomy. Herodotus, Montesquieu, Marx, Weber, and legions of lesser scholars have found that a convenient framework for their arguments, and many of them lumped the MENA together with China and India into one Asian category against which to contrast Europe's uniqueness. This approach, while not extinct, is deservedly less popular among scholars today. I aim instead for a global contextualization of the MENA's environmental history.

My vantage point is that of a generalist in environmental history with only modest acquaintance with the MENA's history and less with the debates and controversies that enliven the region's historiography. In the late 1980s and early 1990s, for a book on Mediterranean mountain landscapes, I devoted some attention to the Rif massif in Morocco and to the western stretch of the Taurus

Mountains in Anatolia. But that research was undertaken without any knowledge of Arabic or Rifian Berber, and with insufficient grounding in Turkish
(and none in Ottoman Turkish). Twenty years later, I remain an outsider to
Middle East Studies.

Within the field of environmental history, I have my own eccentricities
reflected in the pages that follow. My approach, both in general and here,
emphasizes the material components of environmental history. The genre of
environmental history, broadly speaking, includes three main (but overlapping)
approaches. One is concerned with biophysical changes to the environment,
why they occurred, and what they meant for human communities. My work
falls mainly into this category. A second approach emphasizes the conscious
and intentional regulation of the environment, mainly through state political
and legal action. One could, for example, write an interesting study of official
state responses to Tehran's or Cairo's air pollution. The third approach focuses
on cultural and intellectual perceptions of, and responses to, the environment.
This usually takes the form of examining popular environmentalism or the
writings of influential authors. One could, for example, write a useful treatise
on the environmental thought of Ibn Khaldun, whose *Muqqadimah* contains
many ideas about the relationships between power, wealth, and nature.

The eccentricities I consider below all belong in the category of comparative material environmental history. Comparative environmental history of the
MENA could be of almost any sort and need not be confined to the material
realm. One could profitably compare the record of environmentalism in MENA
societies to those elsewhere, an exercise that would surely raise interesting
questions. One could also ponder the environmental engineering ambitions
of postcolonial states in the MENA with those in South Asia and sub-Saharan
Africa, or the impact upon behavior of the MENA's religious traditions compared to those of religions elsewhere. If it is the case that the MENA's environmental history is in its infancy, then the comparative environmental history of
the MENA is a newborn.[2]

The eccentricities of MENA environmental history considered below, as a
result of my own eccentricities, are concerned with water, grass, and energy.
I do not argue that these eccentricities as a whole either favored or disadvantaged the region. Such generalizations cannot be sustained across millennia because conditions change. At certain times, such as during the late
nineteenth-century age of coal and steam, it is probably safe to say that the
MENA stood at a disadvantage with respect to many parts of the world because
most of it lacked coal. Such specific statements—anchored in particular historical moments—are plausible. But generalizations made for all time are not.
In any case, to historians for whom the past is more than a horse race among

civilizations or world regions, such statements—even when plausible—only scratch the surface. My goal, rather, is to show some of the ways in which the MENA's environmental history has been distinctive and, where I can, how those distinctive characteristics—or eccentricities, as I call them—carried broader consequences.

Eccentricities of Water

Let us begin with salt water and maritime peninsulas. The Middle East (not North Africa) has a distinctive and historically consequential pelagic geography. It features four peninsulas of salt water: the Black Sea, the easternmost Mediterranean,[3] the Red Sea, and the Persian Gulf. Unlike China or northern Europe, the MENA is not blessed with a sprawling network of easily navigable rivers, but all its seas are excellent for sailing, with comparatively reliable winds and few catastrophic storms. Add to them the Nile and the Tigris-Euphrates, and one has a navigable network equal to that of anywhere in the world. The interpenetration of land and sea and the density of bays and peninsulas of the Middle East is rivaled in Southeast Asia and perhaps the Caribbean, and on a smaller scale by the geography of the Baltic and North Seas. All three of these spaces, however, were (and are) subject to much more frequent cyclones and gales than the MENA's seas.

The antiquity of urbanization, markets, and societies in which long-distance trade played a strong role owes something to this configuration of land and sea. The unusually strong development of caravan trade draws historians' gazes away from the importance of seaborne routes to the region. Indeed, the elaboration of terrestrial trade networks to some extent resulted from the existence of maritime ones: the incentive to trek over challenging terrain from Trabzon to Kirkuk derived from the complementarity of goods stockpiled in those towns through their seaborne and riverine networks.

It could be—although this hypothesis seems less secure—that the configuration of the MENA's seas also lent them to piracy. As in the Caribbean and Southeast Asian waters, the numerous narrows and choke points, combined with countless hideouts provided by the irregularities of coastlines and their proximity to defensible crags and bluffs, encouraged the rise of piracy and seaborne protection rackets (some of which became known to history as states). The historical record is replete with stories of piracy from at least the time of Pompey, who in 67 BCE tried to exterminate it in Mediterranean waters, down to the present day, when twenty-first-century Somalis batten on ships sailing in and out of the Gulf of Aden.[4]

One further eccentricity of the MENA's seas may have also carried historical consequences: all but one are poor in fish. The Black Sea is the exception. The several large rivers that flow into it—the Danube, the Dnieper, and the Don above all—bring abundant nutrients to feed the lower trophic levels of the marine food web, which in turn feed populations of anchovies and sprat, which in turn feed top predators such as tuna, mackerel, and bonito. Until recent decades, when overfishing and pollution have undermined the catch, the Black Sea supported vibrant fisheries and fishing communities. Its surrounding peoples probably got more protein than most MENA populations and were probably a little healthier as a result. Its surrounding states enjoyed a reserve army of seafaring men who could be lured or impressed into naval duty if the occasion warranted. The famous Ottoman reconstitution of the navy in the months after the Battle of Lepanto in 1571 was a feat that rested on such a reserve army, a luxury unknown to MENA states beyond the Black Sea.[5]

The other MENA seas are warmer and carry lesser quantities of dissolved oxygen, as a result inhibiting aquatic life. All are poor in nutrients compared to the Black Sea, because they are less influenced by the influx of river water. The Mediterranean coast from Alexandria east to Gaza was an exception to this rule of impoverished seas until the Aswan High Dam in 1971 blocked the Nile's flow of nutrients into Egypt and the eastern Mediterranean. So was the head of the Persian Gulf until oil pollution soiled its waters. But none ever had the fisheries (or fishermen) of the Black Sea.

Even the Black Sea never had the bounty found in some of the world's other fisheries. Compared to the waters of the Humboldt Current off of Peru, the Japan Current in the northern Pacific, or the upwellings created by seafloor banks in the North Atlantic, the MENA's seas were fish poor. Fresh water fisheries, in the Nile Delta for example, could compensate to some small degree, but nowhere on Earth do fresh waters support the cornucopia of edible fish found in the deep oceans where cold, oxygen-rich water wells up to the surface. With less fish protein available, MENA populations, like those of the Eurasian Steppe and much of Africa, relied more heavily on domestic animals to avoid or minimize protein deficiency. Thus the prominence of livestock in MENA economic history may be connected to the biotic character of the region's seas.

The second eccentricity of water that shaped life and history in the MENA is more familiar: the sharply uneven distribution of fresh water, the prevalence of aridity, and the consequent ecological responsiveness to even modest climate change. That responsiveness took the form of florescence in times of plentiful rainfall (the mid-Holocene greening of the Sahara for example) and of crisis in times of low rainfall.

A few landscapes, however, were almost exempt from this sensitivity to fluctuations in rainfall. First, tall mountains with reliable snowpacks—the Moroccan High Atlas or the Elburz in Iran, for example—served as water towers insulating their neighborhoods from droughts lasting, as most do, only a few years. The Elburz and some mountains in eastern Anatolia have year-round glaciers that yield meltwater even in the driest months, though more in centuries past than today (because the glaciers used to be larger). Second, big river basins that drew their water from large catchments or reliable monsoon rains provided some insurance against drought in Iraq and Egypt. These areas, however, grew so dependent on these rivers that on the rare occasions when prolonged drought cut their flows sharply, the resulting human disasters were all the more complete. The best example is an old one: the prolonged drought of the twenty-second century BCE that coincided with (and likely produced) the ends of the Old Kingdom in Egypt and the Akkadian Empire in Mesopotamia.[6] The parts of the MENA with more or less reliable fresh water were the ones where people settled and built cities and states.

But most of the MENA lacked enough water for dense settlement to arise.[7] In recent millennia, more than 90 percent of the region's surface area has supported scrub vegetation and seasonal grasses, but not forest or rain-fed agriculture. No conventionally constituted world region, with the possible exception of Australia, has nearly so high a proportion of arid land. This circumstance predisposed societies to develop a particular expertise in water management, namely getting the most out of limited water rather than—as in some other settings—keeping water away from dwellings and fields. The variety of technologies and management systems that have evolved over the past 6,000 years for this purpose is impressive, as is the spread of a handful of "best practices"—such as the *qanat/khattara*—in the last 1,500 years. As arid regions go, the MENA has an abundance of fossil water in aquifers (far more than Central Asia, the Gobi Desert, Australia, or western North America), encouraging the emergence of complex and labor-intensive water management schemes, including deep wells and *qanats* in Iran and the *khattara* of Morocco.

For a thousand years or more, landscapes such as eastern China, northern India, and Western Europe (north of the Pyrenees) featured broad expanses of continual settlement. Villages dotted the land, their fields or pastures abutting one another. The MENA had no such broad expanses. It instead had small zones of continual settlement, such as western Anatolia or the river valleys, together making up an archipelago. The settlement pattern resembles that of Polynesia more than that of China or India, with larger and smaller "islands" of habitation existing where enough water could be found.

The size and borders of some of these "islands" changed with the climate. While in the Nile valley or Mesopotamia the water source remained fairly reliable and shortages were extreme anomalies, a few years of below-average rainfall on the Iranian or Anatolian plateaus or in Syria or Cyrenaica meant less to eat. A decade or two of below-average rainfall meant lower population and shrinking arable land. A century or two of below-average rainfall might have meant abandonment of villages and fields.

Large parts of the MENA had a keen sensitivity to climate change, rainfall in particular, roughly analogous to high latitudes where agriculture was marginal for reasons of temperature. In Scotland or Finland, for example, a few colder-than-average years in the seventeenth century meant starvation and population decline.[8] These lands were vulnerable because they had little farming, in contrast to agriculturally well-suited areas like France or Bengal. Much of the MENA was similarly marginal for farming, with other of its lands being vulnerable because they were marginal for pastoralism. A 20 percent reduction in average rainfall over a decade could be disastrous in Syria or Tunisia but inconsequential in Korea or Poland, just as a lowering of average temperatures by one or two degrees Celsius for a few years spelled catastrophe in Scotland or Finland but was meaningless in Portugal or Punjab.

One of the interesting avenues for future research in MENA environmental history is to integrate climate change into general social, economic, and political narratives.[9] How plausible is it to suppose that drier conditions in the seventh through ninth centuries favored the extension of pastoralism at the expense of agriculture and thereby made the Arab conquests of Syria and Persia easier? And should one believe that wetter conditions in the tenth to the thirteenth centuries helped undergird the prosperity, urbanism, and commercial efflorescence of Fatimid and Abbasid times?[10] Historians of Europe and China have begun to factor climate shifts into their analyses. By and large, they appear to have stronger documentary bases from which to work, although, as Sam White shows in Chapter 3, the Ottoman archives contain a fair share of helpful data. So far, Europe, China, and North America are much better served by proxy evidence—things such as tree rings, fossil pollen, and calcite deposits in caves that paleoclimatologists use to understand climate change—than is the MENA. This is not likely to change because climate history research is rarer in the MENA than in China, Europe, and North America, and because there are fewer old trees for dendrochronologists to study and fewer bogs and lake beds for palynology in the MENA than in those other major regions. But even with these constraints, there is ample room for historical climatology in the MENA and for historians to weave its findings into their reconstructions of the past.

The geography of sea and land, rainfall regimes, and climate change affect people and history everywhere. The MENA is not unique. But these phenomena affected the MENA differently than they affected most places. The interpenetration of land and sea is something the MENA shared with the Caribbean and Southeast Asia. Vulnerability to drought and deep social investment in water management was shared with the Amerindians of the U.S. Southwest and northwest Mexico. High sensitivity to climate change characterized Scotland and Finland as well as the MENA, albeit with respect to temperature rather than rainfall. Taken individually, these characteristics are atypicalities— at most, eccentricities. But taken together, the combination of these characteristics is distinctive and certainly eccentric. They affected the region's history in ways both obvious and subtle.

Eccentricities of Grass

To date, environmental history as a subdiscipline is most developed for North America, Europe, and India. In most cases, forests, forest management, and deforestation loom large. Although the MENA had and has its forests (and some scholars have probed their history) by and large the more important natural biome (land cover) in the MENA has been grasslands—although in many cases it is hard to know whether grasslands are natural or the result of grazing and burning. Several parts of the world feature (or featured) broad grasslands: central North America, the Pampa of South America, the West and East African savannas, the South African veld, and the steppe belt of Eurasia from southern Ukraine to Mongolia. The American grasslands did not host pastoral populations until the sixteenth century, for lack of suitable animals. When Spaniards and Portuguese brought horses, cattle, and other herbivores to the American grasslands, a form of mobile pastoralism[11] briefly flourished in South America, but in North America the presence of vast bison herds made it more attractive to hunt herbivores than to raise them. Grasslands in Africa and Asia, on the other hand, have served for several millennia as home to mobile pastoralists living off herds of sheep, goats, cattle, and horses.[12] The herds of the MENA's grasslands probably included a larger share of goats than herds elsewhere, which is one eccentricity and likely reflects the quality of forage more than cultural preferences. A second eccentricity is the often high proportion of herders within some MENA societies: in the 1520s, a quarter of the inhabitants of Anatolia and perhaps 60 percent of the population in the Ottoman Arab provinces were nomads or semi-nomads.[13] A third eccentricity is the proximity of grasslands (and pastoralists) to the MENA's great urban centers and the interdigitation of steppe and sown. This quirk of ecology had some interesting implications.

No great cities stood anywhere near the grasslands of the Americas or Africa (unless one counts Timbuktu) until the middle of the nineteenth century, when European settlement fundamentally changed the equations of energy, population, trade, and technology that determine whether and where cities might exist. In the steppelands of Eurasia the situation was sometimes broadly the same as it was throughout the MENA: steppe and city often stood not far apart. But generally in Eurasia the great grasslands lay at a distance from population centers, sharing long borderlands with arable zones, in the heart of which lay the big cities. China's cities, for example, were located mainly in the east (and after 1000 in the south), near the sea, and far from the steppe. Europe's cities too stood a long way from the grasslands. Part of the reason for this arrangement was that before the era of the railroads, cities in cool latitudes needed so much firewood that they had to be located near or downriver from forests (seaports were the only exception). According to one calculation, before fossil fuels a European or Chinese city needed wood from an expanse of forest 50–200 times its own area.[14] Cities without good waterborne transport links had to maintain woodlands all around them. At warmer latitudes this constraint was relaxed somewhat, but grasslands generally did not produce enough fuel to serve as urban hinterlands. In the MENA, unusually, no big city except Istanbul stood more than a couple days' ride from steppe or desert.

The MENA is one of the few places on Earth where grasslands and arable lands exist in a mosaic. From Balochistan to Morocco, grasslands exist interspersed with arable lands. None of these areas is comparable in extent to those in the Americas, in sub-Saharan Africa, or the Eurasian Steppe. Nor are there broad expanses of arable land of the sort seen in southern and eastern China for the last thousand years. In land use and land cover, the MENA has for millennia had more of a patchwork than elsewhere, its arable land and pastures and (comparatively modest) forests all in close proximity to different biomes. This fragmented pattern maximized the interaction between pastoralists and farmers, between tribal confederations and agrarian states. Such interaction was of course normal wherever there was pastoralism. But the MENA was eccentric in the degree to which pastoral and agrarian communities interacted, the degree to which their lands intersected. To put it in terms of geometry, the MENA's grasslands and farmlands both had longer perimeters per unit of area than was normal elsewhere on Earth.

The interaction of pastoralist and farmer over millennia became genetically inscribed among MENA populations in the form of a biological eccentricity—high rates of adult lactose tolerance. All infants are able to digest milk. Almost all people in the ancient world, however, lost the ability to metabolize milk and milk products beyond age 3 or 4 because their bodies stopped making the necessary enzyme (called lactase). Thus, nine thousand years ago, every adult was lactose intolerant. Between five and nine thousand years ago, probably in northern Europe or in the MENA, the first mutation occurred, creating lactose-tolerant adults. This must have proved a great advantage because, by the standards of genetic mutations, it spread like wildfire among populations with access to cattle, goats, camels, and other milk-giving livestock.[15] Parallel mutations took place later in Africa, where adult lactose tolerance rates are very high among two subpopulations, and perhaps in Central Asia, where most adults can digest yogurt and cheese—if not milk. Overall, about a quarter of the world's population today retains the ability to digest milk into adulthood. They are found mainly in northern Europe (and zones where northern Europeans settled), the MENA, Central Asia, and two regions in sub-Saharan Africa. These are all lands where the cultural trait of livestock-keeping and the genetic trait of adult lactose tolerance coevolved over the past few millennia. Within the MENA, different populations have different rates of adult lactose tolerance. The Bedouin of the Arabian Peninsula have the highest rates, while Lebanese have among the lowest. (Within Europe, Scandinavians and Britons have the highest rates, Sicilians the lowest; within India there is a north–south gradient from moderate to very low rates.) Ancient populations' animal husbandry practices thus made more recent populations biologically distinctive, different from people in East, Southeast, and most of South Asia and from most Africans, Amerindians, and Pacific Islanders as well. MENA peoples, like northern Europeans, are among the eccentric ones, thanks to their ancestors' easy access to grass and milk-giving herbivores.[16]

The patchwork of arable lands and grasslands in the MENA meant that the complementary economies of protein-producers and carbohydrate-producers existed in tighter harness than elsewhere. Pastoralists and farmers needed one another nutritionally and economically, and their relations included regular trade, seasonal labor, occasional intermarriage, and much else. But pastoralists usually needed the products of farm and city more than farmers and city-dwellers needed the products of the grasslands. After all, they could raise some animals on their own lands, even if they usually could not afford to devote good land to livestock. The near self-sufficiency of a farming village in medieval England or Japan was feasible in most MENA agricultural settings, but it was comparatively suboptimal, and therefore less common, because of the nearby presence of plentiful grass, livestock, and protein. Put another way, the rewards of exchange, of producing surplus for sale, were higher in the MENA than in most other places. This, combined with the good transport opportunities afforded by the interpenetration of sea and land, and with the abundance of pack animals and the related human skills of muleteers and camel-drivers, encouraged commercial development very early in the region's history.

The potential for conflict between farmer and pastoralist was also higher in the MENA than elsewhere. Such conflict is, again, routine wherever farmers and herders coexist. They could easily have incompatible ambitions for the use of land and water (although they obviously need not). More importantly, if they did not like the terms of trade, pastoralists could easily resort to violence and had little to fear in the way of consequences. For the last four thousand years in Africa and Asia, political history evidences frequent conflict between farmers and herders, just as economic history demonstrates trade and cooperation between them. Once they mastered the arts of horseback riding and archery, steppe pastoralists became a formidable foe. Beginning on the western steppe around the tenth century BCE, they inaugurated a pattern of irregular mounted raids upon farming villages. Their mobility allowed them to choose when and where they might fight settled folk, and to retreat at no cost when the odds seemed unfavorable. When attacking farmers, they could also engage in wanton destruction and human slaughter with little fear that revenge would be exacted. Steppe and desert warriors' own women, children, and herds remained safe, hundreds of kilometers away. They had no immobile property for which they had to stand and fight. Literate chroniclers typically regarded the occasional brutality of pastoralists as inherent to what these writers described as these people's savage and bellicose nature, rather than a function of impunity borne of ecological circumstance.[17]

Agrarian states spasmodically attacked pastoralists in punitive raids and genocidal campaigns. The first such efforts in the historical record are those of Darius against the Scythians as described by Herodotus in the fifth century BCE. Chinese armies ventured out onto the steppe from time to time from at least 129 BCE. But armies had very limited shelf lives on the grasslands before they starved or died of thirst. Without wagons, they could travel for five days on the steppe—assuming they could find water—before they would have to head back to farmland. With wagons, they might double their range. If they could not find water and had to bring it with them, thirst would constrict their range by about 80 percent. They could almost never inflict crushing defeat on pastoralists, but when they got the chance they often spared no one. These logistical limitations contrasted with pastoralists' mobility made life for frontier farming folk dangerous, made farmers eager for the protection of states, and made frontier zones unstable.[18]

Such conflicts existed from Manchuria to Morocco and beyond, but they led to different results in different places. In East Asia, from the time of the Qin Unification (221 BCE), big agrarian empires recurred with regularity as a means of resisting pastoralists' incursions. In turn, pastoralists often built large confederacies in order to resist the pressure from Chinese empires (or to mount credible threats through which to extort payment). The geographical segregation of arable land and grassland, of farmer and pastoralist, encouraged this scaling-up of polities in response to one another. In India, roughly the same pattern held, although on a smaller scale. As in China, all the big empires in South Asia originated in the north, in proximity to pastoralists. Big empires provoked big pastoralist confederations, although none as large as what evolved on the Chinese/Mongolian Steppe. In the MENA, the fragmented pattern of grassland and arable land encouraged the emergence of militarized states but made it hard for them to grow to the same spatial scale as occurred routinely in China and fairly frequently in northern India. Until the Ottoman Empire, the big empires usually fragmented within a century or two. Those that lasted best, like the Ottomans, incorporated pastoralist peoples and lands within their ambit, controlling both carbohydrate-producing and protein-producing zones, and the transit routes among them. The comparative rarity of huge empires in the MENA reduced the logic for pastoralists to build giant confederations of the sort more common on the larger grasslands of Central Asia and Mongolia.

All this is not to say that ecology and geography governed the political patterns of world history. Rather, it is to say that, first, the presence of grasslands and suitable livestock—especially horses for mobility—raised the probability of large agrarian empires emerging, which in turn raised the probability of large pastoralist confederations. Second, where the grasslands and arable lands existed in discrete spaces the scale of empires and confederations was likely to be larger, and where the grasslands and arable lands existed in a mosaic pattern the scale was likely to be smaller. Of course other things matter, and big empires could emerge and cohere, at least briefly, in places without pastoralist neighbors. The Inka in Peru and the Khmer in Cambodia are examples. But in the great majority of cases, big agrarian empires arose where arable lands and grasslands met; they coevolved with pastoralist polities.[19]

The eventual eclipse of pastoralist power also carried deep consequences for the MENA. From the time of mounted archery until the eighteenth century, pastoralists enjoyed some military advantages over settled populations—hence the incentives for farmers to band together under the rule of big states. But between 1700 and 1890, agrarian states around the world gained the upper hand and ultimately destroyed the political power and independence of pastoralists. This worldwide historical shift (rarely if ever recognized as such) was based on the growing logistical and technological power of agrarian states, as well as, in some cases, their epidemiological advantage over pastoralists (in the form of greater resistance to certain infectious diseases). The Qing Empire in China, which destroyed the last of the great Mongol confederacies in the middle of the eighteenth century, exhibited all of these advantages. Its generals figured out how to supply big armies on the grasslands with food and water and how to coordinate movements over vast distances so as to entrap their enemies. They increasingly exploited the potential of firearms to allow infantry to withstand and defeat cavalry. Fortuitously—and fortunately for the Chinese—smallpox epidemics battered the Mongols while leaving the Chinese unscathed (more on this follows).[20] At approximately the same time as Qing forces finally eliminated the Dzungar Mongols, the Russian state "pacified" the Pontic Steppe.[21] Soon the Chinese, Russians, Qajar dynasts in Iran, and a few like-minded agrarian states had extinguished pastoralist power in Asia.

Meanwhile, in North America a loosely parallel evolution took place. The introduction of horses by Spanish conquerors led to the rise of mobile equestrian warrior states such as the Comanche and the Sioux between 1710 and 1750. They dominated the broad grasslands from Texas to Saskatchewan and frequently raided nearby farming communities for food and slaves. Unlike the kings of the Afro-Asian grasslands, they were not pastoralists on any scale, but rather bison hunters. The farmers they traded with and raided did not have time to band together into imperial agrarian states before the U.S. Army (perhaps in this context to be understood as serving an imperial agrarian state) obliterated the equestrian empires of the Comanche and the Sioux. The last stand of autonomous semi-nomadic (not, strictly speaking, pastoralist) power in American history took place at Wounded Knee in North Dakota in 1890.[22]

The extinction of pastoralist power affected the MENA fundamentally. The Ottomans, who in their early centuries had encouraged pastoralism (especially in frontier districts), changed their policy by the 1690s. Henceforth they often tried to settle nomads and encourage farming. They wanted, as most states usually do, to be able to tax and conscript people more easily. For more than two centuries they worked at it, registering, counting, describing pastoralists in Ottoman lands, and cajoling or forcing many of them into surrendering their autonomy and mobility. Ottoman officials developed the standard prejudices against mobile pastoralists, viewing them as uncivilized, primitive, and savage—little better than animals. Their objections included the notion that pastoralists' animals damaged agriculture and degraded environments, a common refrain wherever pastoralists went.[23] The Ottomans conducted military operations against pastoral tribes. But in the absence of giant confederations of pastoral tribes, the Ottomans could not deliver a smashing blow as the Qing did to the Mongols in the 1750s. Instead, for centuries, the government in Istanbul might bribe chiefs or defeat a tribe or two, but they often would not stay bribed and defeated for very long.

But by the 1860s, the Ottomans were meeting with more success and within 30 years had settled almost all mobile pastoralists in Anatolia, their remaining Balkan territories, Syria, and Kurdistan. They never managed to achieve their goal in Libya and the Maghreb, where pastoralists retained autonomy and formed important parts of the resistance to Italian and French power into the early twentieth century (although the French had broken the big pastoralist confederacies in Algeria by the 1870s). The vastness of the Sahara made the extinction of pastoralist power there an especially difficult task.[24] Chapter 6 makes clear how easily the Turkmen tribes resisted attempts by Qajar Iran to destroy their power and autonomy into the middle of the nineteenth century. The Turkmen never formed any large confederacies; while this made it unlikely they could conquer Iran (which Arash Khazeni says they never wished to do), it also meant their fighting forces could not be trapped and crushed like the Dzungar Mongols.

In some ways, the suppression of pastoralists' autonomy in the MENA resembles a more difficult version of the suppression of piracy and privateering in the Caribbean. Privateers preyed on shipping and port cities on behalf of a monarch; pirates did it on their own account. One could be a privateer one month and a pirate the next. The line between the two was fuzzy, just as was the line between *akıncı* and *haydut* (or *armatole* and *kleft*) in the Ottoman Empire. The plantation societies of the West Indies (based on a very eccentric grass, sugarcane) proved vulnerable to pirates and privateers, whose mobility allowed them to swoop in, raid, pillage, and disappear quickly. The built environment of plantations and ports was easily destroyed, like the irrigation systems of MENA farmers. This vulnerability made it easier for pirates and privateers (like mobile pastoralists) to exact protection rent and ransoms from settled folk. In both cases it took concerted and sustained state action to suppress the threat, and that action was often as much diplomatic as military. Britain, which led the effort to end Caribbean piracy in the early eighteenth century, not only needed cooperation from other European powers such as France and Spain but from pirates who could be seduced into switching sides. The British appointed an ex-privateer and pirate, Henry Morgan, as lieutenant governor of Jamaica in 1675, just as the Ottomans appointed Yeğen Osman, a former bandit, as governor of Afyon in 1687. By 1730 piracy was on the wane in the

West Indies, although not fully suppressed. The attitude of the states involved was more uncompromising than that of the Ottomans with respect to independent pastoralists, and the scale of the project much smaller.

The interdigitation of grassland and arable land might also have affected the role of infectious disease in the political history of the MENA. The extinction of pastoralist power on the eastern steppe, accomplished by the Qing in the 1750s, was helped along by smallpox.[25] A Chinese chronicler wrote that in the final campaigns, 40 percent of the Mongols succumbed to smallpox, 30 percent were killed by Qing forces, and 20 percent escaped to Russian lands.[26] The Chinese suffered far less from smallpox, because most of them had survived it in childhood (when it is typically less lethal) and were thus immune as adults. In North America, the Comanche lost half their population to smallpox in 1780–1782 and more of their number to other epidemics in the nineteenth century.[27] Recruits in the U.S. Army and settlers on the prairies were usually immune to smallpox (and more resistant than the Comanche or Sioux to a host of other infections). The Ottomans and the French in Algeria, on the other hand, had no such help from pathogens. By the eighteenth and nineteenth centuries, if not well before, the pastoralists of the MENA had acquired approximately the same portfolio of disease immunities as settled folk. Since farmers and pastoralists interacted so regularly in the MENA, they shared infections liberally and developed antibodies against the same diseases. This might help explain why the extinction of pastoralist autonomy came later and slower in the MENA than it did in East Asia or North America.

It could well be that the origins of the Ottoman Empire also owed something to the interaction of disease and the settlement geography of the MENA. In the early fourteenth century, the Ottomans drew most of their manpower from pastoralist tribes in northwest Anatolia. They were not yet at all urbanized. But their main rivals—the Byzantines and several Turkic principalities such as Aydın, Karası, and Karaman—were either coastal settlements or closely connected to port cities. As such, they probably suffered much more acutely from the plague beginning in 1347 than did the Ottomans.[28] Unlike smallpox, plague is a vector-borne disease. The vector in question is rat fleas, which will cheerfully bite humans when rats are scarce. Thus plague is a disease confined to certain environments, especially those where grain is stored and rats congregate. Its usual pattern—noted by Procopius during the plague of Justinian in the sixth century and by many commentators since—is to hit port cities hardest, cities hard, villages less hard, and mobile populations least of all. Early Ottoman sources apparently make no mention at all of the plague—in sharp contrast to Byzantine accounts. The Byzantine Empire reeled under the impact of plague beginning from 1347. It recruited Ottoman help in its campaigns in the Balkans; from that time onward the Ottomans were a significant force in regional politics. For more than a century, their successes took place mainly in the southeast Balkans and northwest Anatolia, landscapes severely depopulated by plague. The existing political formations in these areas probably suffered much more in the way of manpower loss than did the Ottomans themselves. This suggests that the mobile pastoralism of the Ottomans shielded them from the worst of the plague, raising the odds that they would be militarily and politically successful in the fierce competitions that would occur in Anatolia and the Balkans in the late fourteenth century. In this instance (if these speculations have merit), disease helped mobile folk against settled folk, the opposite of the pattern in cases where smallpox did the deadly honors.[29]

Much of the foregoing is speculation that cannot be tested or verified. We will almost certainly never uncover data about the differential impact of the Black Death on Ottoman versus Byzantine manpower. But given what we do know about the ways of life of these populations, the habits of rats, and patterns of mortality in subsequent plague outbreaks, these are, I hope, plausible partial explanations for the lightning success of the Ottomans from 1347 onward— at least until they captured Constantinople in 1453. These suggestions might supplement traditional arguments about ghazi motivation, Beyazıt's talents, the restless state of Balkan peasantries, and so forth.

My suggestions about the eccentricities of grass in the history of the MENA flow from two geographical features. First, in the MENA there has long been more grassland and less forest than in either more temperate or more tropical zones. Second, unlike most steppe and steppe/desert zones, grassland and arable land exist, and have long existed, in a more fragmented mosaic in the MENA than is the pattern in East Asia, north India, or North America, which has resulted in a long history of pastoralist and farmer being in closer, more frequent contact. This then predisposed the MENA toward eccentric economic, political, and epidemiological tendencies that were not widely shared around the world.

Eccentricities of Energy

A third eccentricity of the MENA in global historical context is its relationship to energy. In Chapter 2, Richard W. Bulliet discusses the role of cheap animal power resulting from abundant forage, and how that nudged the MENA into an energy technology cul-de-sac. Incentives to develop new and different energy technologies were weak, because animal power was so cheap—both for transport and for milling. This idea bears a cousinly resemblance to Mark Elvin's explanation of the technological stagnation of late imperial China, his so-called high level equilibrium trap.[30] Elvin argued that by the late seventeenth century China had achieved great efficiency in its markets, had filled up all its suitable agricultural land, and had enormous reservoirs of cheap labor (human energy). This arrangement, Elvin, continues, discouraged investment in labor-saving technologies. Both arguments seek to explain why certain regions (the MENA and China) did not keep pace with Europe (mainly England) in the development of energy-intensive technologies that ultimately led to full-scale industrialization. Both also point to the importance of energy systems in world history.

The MENA's grasslands and domesticates (camels, horses, and donkeys in particular) endowed it with an eccentric history of energy, as Bulliet argues. This had implications for transport and the early and widespread development of markets, as noted above. Caravan routes, together with the Nile, the Tigris-Euphrates, a few lesser rivers, and the MENA's saltwater peninsulas knitted the region's centers together. China too showed precocity in the development of markets, based on its network of rivers gradually supplemented by its canals. In both cases, energy systems based on animal, human, and wind power undergirded regional economies as productive and commercialized as any in the world—until the emergence of fossil fuels.

In this final section I will underscore the obvious: the MENA has almost no coal but lots of oil. As the world shifted overwhelmingly to fossil fuels after 1800, these facts powerfully shaped the MENA's historical experience. I will treat this theme because it is so important, but briefly because it is nearly self-evident.

Until 1800, the entire world relied overwhelmingly upon an organic energy regime. In a few places, wind or waterpower added significantly to the energy harvest. For a century or more, Song China had a sizable iron industry based on coal. But these (and lesser) exceptions aside, usable energy for human purposes came in the form of chemical energy contained in biomass. Humans could eat some of it and thereby convert it into different forms of chemical energy, kinetic energy, and bodily heat. Their domestic animals could also eat some of it. And humans could burn some of it directly for heat. But to get things done, they had to rely on human or animal muscle.

This energy regime constrained human ambitions tightly. All energy ultimately came from the sun, which was inconvenient. Humans harvesting the energy from biomass were only able to tap—and very inefficiently—the annual flows from the sun, refracted through photosynthesis and one or two trophic levels. When burning fuelwood or charcoal, people could tap the energy stocks of a century or two. But this was the only good way to stockpile or store energy for future use, and it was good only for heat energy. It was hard to amass kinetic energy except by amassing humans and animals. The average pharaoh, emir, or sultan had less energy at his command than has a modern bulldozer operator.

After 1800, the exploitation of fossil fuels shattered the constraints of the organic energy regime. Coal and oil together contain a supply of about five hundred million years of stockpiled sunshine.[31] Since 1888 the world has burned fossil fuels at an annual rate greater than the net primary productivity of the Earth (in other words, the amount of energy that photosynthesis can store in biomass). It takes about 90 tons of ancient biomass to make the oil needed for a gallon of gasoline. So fossil fuels represent very inefficiently stored sunshine, and we have been burning up the total stock quickly. This abundance of energy, more than anything else, makes modern times different from earlier human experience.

To convert the frozen sunshine of fossil fuels into kinetic energy required an engine. The first steam engines lost 99 percent of their energy as dissipated heat. Nonetheless, by 1800 a single steam engine could do the work of two hundred men (and it still wasted 95 percent of its energy input). By 1900, an average steam engine could do the work of six thousand men. To feed steam engines, one needed coal. From 1800 onward, coal gradually replaced biomass in the world's energy system. By 1890, more energy came from coal than from biomass.

As world regions go, the MENA is thin on biomass, but it is far poorer in coal. From 1800–1950, this was one of the worst deficiencies a region could have in terms of global competition for wealth and power. In environmental terms, given the effects of coal mining and coal combustion, having little coal might be considered a blessing. In economic and political terms, it was a curse for a century and a half. In the nineteenth century, when Europe, North America, South Africa, Australia, Japan, and even some places where labor was very cheap (Bengal, for example) had begun to substitute coal and steam for biomass and muscle, the MENA remained dependent on grasses and animals for its energy harvest. Anatolia had a little coal, and so did Iran. But the quantities were tiny by the standards of other parts of the world. Since 1980, the MENA has accounted for only 0.025 percent of world coal production—one out of every four thousand tons.[32] It was not alone in this economic misfortune: tropical Africa and South America outside of Colombia lacked coal as well. But the big centers of population—China, India, Europe (and by the nineteenth century, eastern North America)—all had plenty.

Moreover, what little coal the MENA had could not be cheaply transported to cities. Moving it via animal transport was prohibitively expensive. So very little was used until the era of oil-powered machinery and transport. This put the MENA in an extremely unfavorable situation compared to Russia, where coalfields lay athwart the Don River, or Great Britain, where coal lay near the sea in South Wales, Tyneside, and elsewhere. If MENA industrialists needed coal, they had to buy it from afar and pay accordingly.

Deficiency in coal led people in some parts of the world to seek alternative bases for industrialization. When turbine technology emerged in the late nineteenth century, entrepreneurs and engineers here and there tried to exploit "white coal"—hydroelectricity—to power factories and railroads. In some locations—around the Alps or Quebec's St. Lawrence Valley, for example—this strategy worked because reliable swift-flowing water existed. Bavaria, Lombardy, and Quebec built industrial economies substantially on white coal between 1890 and 1940. But only a few places around the world could match this record. The MENA had precious little white coal. What hydroelectric potential it had generally remained unharnessed until the 1970s.

The organic energy regime had been good to the MENA in the sense that its endowments of grass and animals did not handicap its economic development or political power. Indeed, if anything, the reverse was true. But in general economic terms, the organic energy regime did not generate big differences from one region of the world to the next. Every place got some sunshine, and all but the driest deserts and coldest taiga produced biomass in abundance. In political terms, the MENA's grass and animals were an important source of military power, wielded sometimes by pastoralist tribes, but sometimes by agrarian empires employing tribesmen as fighters. But when coal became king, the MENA's economic and political position plummeted vis-à-vis other parts of the world.

Oil took away king coal's crown by about 1965 and has kept it ever since. We use about 20 percent more oil than coal (in energy content) today. In parts of the MENA the oil age dawned early in the twentieth century, but it revolutionized life only in the middle of the century and, even then, only in the oil-producing countries. But since the 1960s, the size of the oil industry and the sums of money involved have been so great as to reverberate almost everywhere throughout the MENA. Oil, as every reader of this book will know, has proved a source of both unprecedented wealth and unusual disruption in the MENA. It has empowered several MENA states politically and militarily far beyond what they could plausibly have achieved in recent decades without oil. It has opened up chasms of inequality within MENA societies and, as oil production has everywhere, fouled the environment in and around the oil fields themselves.

But its impacts will be short-lived. King coal reigned for about seventy-five years; king oil has lasted for forty-five. Soon, whether with bangs or whimpers, the world will shift away from oil as its main fuel. Perhaps coal will regain its crown. Perhaps new technologies will install solar power on the throne. No one knows. But everyone knows oil exists in finite supply and sooner or later will become too expensive for routine use. Soon the oil age will appear a fleeting aberration in the MENA's—and the world's—history.

Conclusion

Every region of the world has its peculiarities. This chapter aimed to identify some of those in the material environmental history of the MENA, specifically regarding water, grass, and energy. They are all, of course, linked. Availability of freshwater helps determine vegetation. Vegetation provides chemical energy that animals (including humans) can transform into movement and heat. But it helps to first see these eccentricities separately in order to make isolated comparisons to situations in other parts of the world.

The eccentricities of the MENA with respect to water include both the obvious constraint of limited fresh water in much of the region, and the less-noticed convenient geography of saltwater seas, gulfs, and bays. Both are shared with some other parts of the world. But no other sizable part of the world shares both. The configuration of land and sea helped in the precocious commercialization of the MENA, from Sumerian times if not before. It also helped stimulate caravan traffic where rivers did not serve as transport arteries (in other words, in lands other than Egypt and Iraq). The MENA's seas, while convenient for commerce, provided paltry quantities of fish protein. In comparison to other parts of the world, MENA populations had to raise more animals per person or suffer from protein deficiency. And MENA navies (except that of the Ottomans) had to find their crews without the useful reserve army of fishermen that typically existed where fisheries flourished. The scarcity of fresh water should be understood not merely as a background condition of life in much of the MENA, but also in relation to climate change—a concern that historians are with some hesitation beginning to take seriously. Above all else, even modest changes in rainfall regimes, whether over years or centuries, carried particularly significant consequences in the MENA, either enlarging or shrinking human possibilities. Equal changes in rainfall regimes in well-watered lands were far less consequential because none of their communities stood at the margin where rainfall agriculture, or even irrigation agriculture, was barely practical.

The eccentricity of grass in the MENA derived mainly from the fact that grasslands existed in a crazy quilt pattern rather than in huge expanses. Arable land, grass, and scrub stood side by side in (by the standards of other parts of the world) little patches. So herders and farmers lived almost cheek by jowl, raising the probability of systematic interactions between them, both warlike and peaceful. In contrast to the Pontic and the Mongolian steppelands, where big blocks of grassland encouraged the emergence of large pastoralist confederations in response to big agrarian states, the MENA's political landscape differed. Smaller polities were the rule, both in size and in population. There were of course exceptions, at least in the form of agrarian empires that lasted a century or two (and the Ottomans, who lasted roughly six). But the contrast to the situation in East Asia, where big agrarian states—in the form of successive Chinese dynasties—and big pastoralist confederacies were routine, is striking.

The eccentricities of energy in the MENA reside in its long reliance on biomass and animals, its minimal recourse to coal, and its near-total refashioning in the age of cheap oil. Most parts of the world relied on biomass and animals for most of human history; in this respect, the MENA is as normal in its energy history as it could be. Only in the nineteenth century did its energy path come to conform less fully to the global average. But until 1920 or so, the use of coal was itself an eccentricity, confined to a few parts of the world. The MENA, in relying minimally on coal, was still in the company of almost all of Africa, South America, and South and East Asia. Only a few pockets in these lands had turned to coal on any scale. So the MENA's energy pathway may appear eccentric when viewed through the lens of Europe, Russia, Japan, or North America (and those few pockets elsewhere), but if seen through the lens of India, China, Africa, or Brazil in 1920, it would look normal. That changed radically with the development of cheap crude oil beginning early in the twentieth century and exploding from the late 1940s on. From this point forward, the MENA's experience with energy looks highly eccentric compared to almost anywhere else. When the oil grows scarce, or a new energy regime displaces it, the MENA's experience will be eccentric again, in new ways and for new reasons.

Notes

1. In a parallel exercise I tried to isolate distinctive features of Chinese environmental history in J. R. McNeill, "Chinese Environmental History in World Perspective," in *Sediments of Time: Environment and Society in Chinese History*, ed. Mark Elvin and Liu Ts'ui-jung (Cambridge: Cambridge University Press, 1998), 31–52.

2. Note, however, that some of this book's chapters offer comparative perspectives from time to time.

3. By this I mean the Mediterranean east of a line between Cyrenaica and Rhodes.

4. On Pompey's campaign and Mediterranean piracy, see Philip De Souza, *Pirates in the Graeco-Roman World* (Cambridge: Cambridge University Press, 2002); Jules Sestier, *La Piraterie dans l'antiquité* (Paris: Marecq, 1880). About piracy in other regions, see Robert Anthony, *Like Froth Floating on the Sea: The World of Pirates and Seafarers in Late Imperial South China* (Berkeley: Institute of East Asian Studies, 2003); Kris Lane, *Pillaging the Empire: Piracy in the Americas, 1500–1750* (Armonk, NY: M. E. Sharpe, 1998).

5. On Lepanto and its aftermath, see Niccolò Copponi, *Victory of the West: The Great Christian-Muslim Clash at the Battle of Lepanto* (Cambridge: Da Capo Press, 2007), 299–303; Mehmet Kuru, "The Relations between Ottoman Corsairs and the Imperial Navy in the 16th Century" (Ph.D. diss., Sabancı University, 2009); İdris Bostan, *Osmanlı Bahriye Teşkilâtı: XVII. Yüzyılda Tersâne-i Âmire* (Ankara: Türk Tarih Kurumu Basımevi, 1992), especially 188; Colin Imber, "The Reconstruction of the Ottoman Fleet After the Battle of Lepanto, 1571–1572," in *Studies in Ottoman History and Law*, ed. Colin Imber (Istanbul: Isis, 1996), 85–101. The manpower of the Ottoman navy included oarsmen, who needed no knowledge of the sea and were recruited from inland districts of Rumeli and Anatolia, and true sailors, who came from the Black Sea, Sea of Marmara, and Aegean coasts. Experienced sailors were scarce and often the limiting factor in early modern navies.

6. Simone Riehl, "Climate and Agriculture in the Ancient Near East: A Synthesis of the Archaeobotanical and Stable Carbon Isotope Evidence," *Vegetation History and Archaeobotany* 17 (2008): 43–51; M. Magny, B. Vanniere, G. Zanchetta, and E. Fouache, "Possible Complexity of the Climatic Event around 4200–4000 cal. BP in the Central and Western Mediterranean," *The Holocene* 19 (2009): 823–833.

7. A recent reflection on aridity and water management is Edmund Burke III, "The Transformation of the Middle Eastern Environment, 1500 B.C.E.–2000 C.E.," in *The Environment and World History*, ed. Edmund Burke III and Kenneth Pomeranz (Berkeley: University of California Press, 2009), especially 83–88.

8. Karen J. Cullen, *Famine in Scotland: The "Ill Years" of the 1690s* (Edinburgh: Edinburgh University Press, 2010); Michael Flinn, *Scottish Population History from the Seventeenth Century to the 1930s* (Cambridge: Cambridge University Press, 1977); Eino Jutikkala, "The Great Finnish Famine in 1696–1697," *Scandinavian Economic History Review* 3 (1955): 48–63.

9. An attempt which leaves room for improvement is Arie S. Issar and Mattanyah Zohar, *Climate Change: Environment and Civilization in the Middle East* (Dordrecht: Springer, 2004). For comparison, see Emmanuel Garnier, *Les dérangements du temps: 500 ans de chaud et de froid en Europe* (Paris: Plon, 2010); Emmanuel Le Roy Ladurie, *Histoire humaine et comparée du climat*, 3 vols. (Paris: Fayard, 2004–2009); Rüdiger Glaser, *Klimageschichte Mitteleuropas: 1200 Jahre Wetter, Klima, Katastrophen* (Darmstadt: Primus, 2008); Franz Mauelshagen, *Klimageschichte der Neuzeit, 1500–1900* (Darmstadt: Wissenschaftliche Buchgesellschaft, 2010).

10. Issar and Zohar, *Climate Change*, 212–222.

11. I will use this phrase where other authors use "nomadism" because true nomadism was rare and many of the herders I write about were semi-nomadic or transhumant. Moreover, these herders may have changed their patterns of mobility from time to time.

12. A helpful introduction to early steppe pastoralism is David Anthony, *The Horse, the Wheel and Language: How Bronze-Age Riders from the Eurasian Steppes Shaped the Modern World* (Princeton, NJ: Princeton University Press, 2007). See also Nicola Di Cosmo, *Ancient China and Its Enemies: The Rise of Nomadic Power in East Asian History* (Cambridge: Cambridge University Press, 2002). And for what is now upper Egypt and northern Sudan, see Karim Sadr, *The Development of Nomadism in Ancient Northeast Africa* (Philadelphia: University of Pennsylvania Press, 1991).

13. Reşat Kasaba, *A Moveable Empire: Ottomans, Nomads, Migrants and Refugees* (Seattle: University of Washington Press, 2009), 18.

14. Vaclav Smil, *Energies: An Illustrated Guide to the Biosphere and Civilization* (Cambridge: Massachusetts Institute of Technology Press, 1999), 118.

15. The advantage consisted of supplies of high-quality nutrition in the form of milk, yogurt, and cheese. It may also have consisted of healthier liquids. Cattle and camels can produce healthful milk from polluted or brackish water and can take in moisture from the vegetation they eat. Thus, they can convert liquids that are unhealthful or unavailable to humans into helpful ones. This capacity would have been especially advantageous in places where good water was scarce—as in arid environments.

16. Clare Holden and Ruth Mace, "Phylogenetic Analysis of the Evolution of Lactose Digestion in Adults," *Human Biology* 81 (2009): 597–617; Pascale Gerbault et al., "Evolution of Lactase Persistence: An Example of Human Niche Construction," *Philosophical Transactions of the Royal Society B* 366 (2011): 863–877; S. S. Hijazi, A. Abulaban, Z. Ammarin, and G. Flatz, "Distribution of Adult Lactase Phenotypes in Bedouins and in Urban and Agricultural Populations of Jordan," *Tropical & Geographical Medicine* 35 (1983): 157–161.

17. A study of ancient Chinese texts and their conceptions of pastoralists (and other foreign peoples) is Di Cosmo, *Ancient China and Its Enemies*, 93–158. See also Romila Thapar, "The Image of the Barbarian in Early India," *Comparative Studies in Society and History* 13 (1971): 408–436.

18. Herodotus, *The Persian War* (New York: Modern Library, 1942), 4.121–128. Interesting calculations on the logistics of armies in steppe and desert terrain appear in the following: Kenneth Chase, *Firearms: A Global History to 1700* (Cambridge: Cambridge University Press, 2003), 16–18. The tendencies of pastoralists and farmers to demonize one another and exterminate one another in war are well attested. See, for example, Christopher I. Beckwith, *Empires of the Silk Road: A History of Central Eurasia from the Bronze Age to the Present* (Princeton, NJ: Princeton University Press, 2009).

19. These generalizations are based on Peter Turchin, "A Theory for Formation of Large Empires," *Journal of Global History* 4 (2009): 191–217. Turchin identifies 60 empires of larger than one million square kilometers before 1800; all but four, including the Khmer and Inka, were located in lands in regular contact with the steppe pastoralists of Morocco to Manchuria.

20. Peter C. Perdue, *China Marches West: The Qing Conquest of Central Asia* (Cambridge, MA: Harvard University Press, 2005).

21. Willard Sunderland, *Taming the Wild Field: Colonization and Empire on the Russian Steppe* (Ithaca, NY: Cornell University Press, 2004).

22. Pekka Hämäläinen, *The Comanche Empire* (New Haven, CT: Yale University Press, 2008); Robert Utley, *The Last Days of the Sioux Nation* (New Haven, CT: Yale University Press, 2004).

23. This was also the case among French officialdom in the Maghreb. See Diana K. Davis, *Resurrecting the Granary of Rome: Environmental History and French Colonial Expansion in North Africa* (Athens: Ohio University Press, 2007).

24. Kasaba, *A Moveable Empire*, 53–122; Benjamin Claude Brower, *A Desert Named Peace: The Violence of France's Empire in the Algerian Sahara, 1844–1902* (New York: Columbia University Press, 2009); Osama Abi-Mershed, *Apostles of Modernity: Saint-Simonians and the Civilizing Mission in Algeria* (Stanford, CA: Stanford University Press, 2010).

25. In addition to Perdue, *China Marches West*, see also Jiafeng Zhang, "Disease and Its Impact on Politics, Diplomacy and the Military: The Case of Smallpox and the Manchus, 1613–1795," *Journal of the History of Medicine and Allied Sciences* 57 (2002): 177–197. The Manchu feared smallpox greatly and managed to contain its ravages within their ranks while they conquered China, ruled China, and then conquered the Mongols.

26. Wei Yuan, cited in Peter C. Perdue, "Fate and Fortune in Central Eurasian Warfare: Three Qing Emperors and their Mongol Rivals," in *Warfare in Inner Asian History*, ed. Nicola Di Cosmo (Leiden: Brill, 2002), 393.

27. Hämäläinen, *Comanche Empire*, 111; Adam R. Hodge, "Pestilence and Power: The Smallpox Epidemic of 1780–1782 and Intertribal Relations on the Northern Great Plains," *The Historian* 72 (2010): 543–567.

28. Whether or not the pandemic of 1346–1352 was plague, plague and other infections, or only other infections is in dispute, although the weight of expert opinion plunks for plague. On the difficulties of retrospective diagnosis in the MENA, see Sam White, "Rethinking Disease in Ottoman History," *International Journal of Middle East Studies* 42 (2010): 549–567.

29. My prior speculations on this subject have been superseded by those of Uli Schamiloglu. See J. R. McNeill, "Ecology and Strategy in the Mediterranean," in *Naval Policy and Strategy in the Mediterranean: Past, Present, and Future*, ed. John Hattendorf (London: Frank Cass, 2000), 376–377; Uli Schamiloglu, "The Rise of the Ottoman Empire: The Black Death in Medieval Anatolia and Its Impact on Turkish Civilization," in *Views from the Edge: Essays in Honor of Richard W. Bulliet*, ed. Neguin Yavari, Lawrence G. Potter, and Jean-Marc Oppenheim (New York: Columbia University Press, 2004), 255–279. Metin Kunt suggested the Black Death might have helped the Ottomans, but on the implausible argument that their Central Asian origins provided them immunities that Greeks and Serbs did not have. Metin Kunt, "State and Sultan up to the Age of Süleyman: Frontier Principality to World Empire," in *Süleyman the Magnificent and His Age: The Ottoman Empire in the Early Modern World*, ed. Metin Kunt and Christine Woodhead (London: Longman, 1995), 3–19.

30. Mark Elvin, *The Pattern of the Chinese Past* (Stanford, CA: Stanford University Press, 1973), especially 298–315.

31. Useful perspectives can be found in the following: Charles Hall et al., "Hydrocarbons and the Evolution of Human Culture," *Nature* 426 (2003): 318–322; J. S. Dukes, "Burning Buried Sunshine: Human Consumption of Ancient Solar Energy," *Climatic Change* 61 (2003): 31–44.

32. "Middle East: Total Primary Coal Production," *Titi Tudorancea Bulletin*, October 21, 2010. http://www.tititudorancea.com/z/ies_middle_east_coal_production. htm (accessed June 18, 2011).

2

History and Animal Energy in the Arid Zone

Richard W. Bulliet

A multitude of lively articles and books have appeared in recent years debating the historical trajectory of the MENA in the late medieval to early modern periods. Did it enter into a long period of decline starting around the twelfth century? Or was decline only a relative (and pejorative) perception caused by Europe accelerating its development? Or did Europe and the Muslim world follow similar modernizing trajectories that have been identified solely (and erroneously) with respect to the former and overlooked in the latter? The keywords in this debate cluster either in the cultural/religious arena or in the economic/political/institutional realm. Those who feel that the MENA entered a period of decline following its medieval efflorescence favor the former keywords and ascribe the deterioration to intrinsic deficiencies in the Islamic religion or in faulty social and political structures that Islam made possible—or at least failed to ward off. Those who follow the second track focus on keywords that stress the undoubted malignancy of European imperialism and economic exploitation. Those who take yet a third course muster evidence that dynamism and a capacity for change persisted in the Muslim world well into recent times.

This chapter takes a wholly different tack. It proposes that geographic, macroeconomic, and technological factors that came into play after the twelfth century created—or greatly enhanced—a disequilibrium between the two geographical zones. Europe, I argue, experienced rapid population growth that increased the demand for grain. Sustaining human populations took precedence over growing animal fodder, and this inflated the costs of using draft animals—principally oxen, horses, and mules—as a power source.

These two developments intersected in an explosive growth in the number of watermills and—to a lesser extent—windmills and in the aggregate capital investment in these increasingly elaborate structures. Grain had to be ground for human consumption, and the high cost of animal labor made it uneconomical to use horses and oxen to turn millstones. This resulted in the emergence of a new social class: millers operating water- or wind-driven machinery. As comparatively wealthy commoners who did not share the aversion to commerce that was characteristic of the far wealthier land-holding aristocracy, millers provided a substantial share of the money, technical expertise, and entrepreneurship that launched Europe's prodigious economic growth.

As part of the Northern Hemisphere's arid zone, the MENA experienced a different calculus of animal labor. An abundance of cheap, and often free, grazing kept the price of animal power down regardless of fluctuations in the human population. Horses, oxen, mules, and donkeys could operate mills and other productive mechanical devices quite efficiently with minimal capitalization. As a result, the social class of millers that played such a powerful role in spurring Europe's economic growth never made an appearance in the MENA. The region therefore suffered a relative long-term lack of entrepreneurship, technical innovation, and accumulation of industrial capital.

Such is the argument. The remainder of this chapter is devoted to trying to substantiate it.

The term *arid zone* refers generally to the deserts and semi-deserts that stretch from Mauritania and Morocco on the Atlantic Coast, across North Africa and the Middle East, to Pakistan and northwest India. North of this zone, between the lower Danube valley and Manchuria, can be found a closely related geography that includes both deserts and steppe grasslands. For the purposes of this chapter, however, the arid zone and adjoining Eurasian desert and steppe are defined together according to one particular functional quality that they share rather than by purely physical aspects such as average rainfall or temperature. The arid zone I am concerned with is the region where animals whose energy is harnessed for riding; drawing vehicles and plows; and operating mechanical devices, such as mills, wells, and irrigation wheels typically grow to maturity and maintain their working energy by foraging rather than consuming purpose-grown fodder.

Foraging is not the exclusive resort of herders and animal users in this zone, nor is this the only part of the Old World where the landscape supports foraging animals. Transhumant pastoralism is common in mountainous parts of Europe, but milk, meat, and animal by-products outweigh the exploitation of animal labor in the economy of such regions. Animal labor is also of negligible importance in the northern barrens of Eurasia that support reindeer herding. As for the Southern Hemisphere at similar latitudes, animal labor does not have a long history except for the use of llamas in the Andes. Thus no other broad region matches the extensive foraging resources of the arid zone or supports such robust traditions of dispersed pastoral nomadism devoted to extracting animal sustenance from almost every seemingly desolate square mile. Natural fodder resources are so ubiquitous, despite being sparsely strewn across the landscape, that there is rarely any need for farmers to choose between growing crops for human sustenance and growing crops for working animals to consume. This consideration is not absolute, but it is quite marked by comparison with more densely populated, better watered, and more forested regions where feeding livestock and feeding human beings are often competing imperatives. Most of Europe, India, and East Asia (China, Korea, and Japan) fall outside the region where working animal foraging is both commonplace and traditionally based on extensive pastoral nomadism.

The thesis that this chapter explores arises from the observation that in ancient times, very roughly between 8000 and 1500 BCE, pastoral nomads generally lived on the margins of agricultural societies and had comparatively little impact on their cultural and economic practices. This balance in favor of agriculture over nomadism changed gradually. Many historians attribute the relative rise of nomadism to raids and general aggression first by charioteers and later by horse riders. My own book on the subject, *The Camel and the Wheel*, adduces technological influences connected with changes in the design of camel saddles.[1] Regardless of specific enabling factors, however, between roughly 300 BCE and 1400 CE the pastoral sector of the arid-zone economy generated ever stronger political and military forces that gave rise to a succession of important empires and states, including the Parthians, with a pastoral base in Turkmenistan and northeastern Iran; the Arab caliphate, with a pastoral base in the Arabian Peninsula and its northward desert extensions in Syria and Iraq; the Seljuqs, originally pastoralists from Turkmenistan and Uzbekistan; and the Mongols from the eastern end of the arid zone. Then, after about 1400, the edge that had been enjoyed for so long by pastoral nomadic political forces rapidly diminished. By the twentieth century, pastoral nomadism was more often seen as an outmoded economic form destined for extinction rather than a source of power in any sense of the word. This sequence of developments is well known, and only its latest stages will be addressed here.

The proposition that this chapter advances therefore embodies a great historical reversal. The low cost of utilizing animal energy that is intrinsic to arid-zone geography allowed pastoral nomads to achieve their great power over settled agricultural socities; later, however, this same efficiency in producing energy helped doom the region to economic stagnation. Ironically, a parallel to the last phase of this evolution can be observed at the present time—and not just in the arid zone—with low-cost petroleum taking the place of low-cost animal power as the barrier to the adoption of alternative energy sources: water-mills in the earlier case; wind, solar, and nuclear power now.

Comparing the past with the present will highlight the gravity of major transitions in energy utilization. Societies tend to take the benefits these transitions bestow for granted without giving much thought to the myriad ancillary developments that they entrain. For example, we take electricity for granted, but we rarely think about the scale of physical and human resources or the organizational and economic innovations involved in generating electrical current and wiring the planet to distribute it. Nor do we reflect on those things that passed into oblivion in the process: gaslight and its accompanying production and distribution network, the whaling industry that provided oil for lamps, and so forth.

Since we have as yet barely embarked on the transition from petroleum energy to alternative and renewable energy sources, we are in a position to see the promise of the future and guess at the economic and institutional changes that will be needed if that future is to come to pass. But can we truly grasp the structural changes involved in a world that is independent of fossil fuels? We might imagine fairly easily, for example, the universal disappearance of the gas station and the spread of electrical charging outlets and/or hydrogen-dispensing facilities. But will new energy companies supersede today's energy companies? Will overlords of wind farms and solar arrays supplant oil sheikhs and pipeline magnates? Will the first country to make the transition fully—China perhaps—thereby acquire a hegemonic position in the world economy? Will its industrialists and technicians steal such a march in terms of expertise and economies of scale that the rest of the world will have to come to them in the way non-Europeans once depended on European know-how and experience in the transition to steamboats and steam railroads?

Such future-casting aside, here is our present situation. The real cost of extracting petroleum has been extraordinarily low since at least the middle of the twentieth century. A decade ago, in a private conversation, Mobil chairman Lou Noto remarked to me that gasoline was the cheapest fluid sold at a Mobil station, cheaper than water and a lot cheaper than Coca-Cola. The price paid by end users for petroleum products varies, of course, according to governmental royalties and rents, taxation, distribution charges, and political considerations; on balance, however, energy has never been so cheap.

The problem is that the low cost of petroleum makes the cost of developing alternatives to petroleum prohibitively expensive. This has given rise to the much talked about—see virtually every *New York Times* op-ed by Thomas Friedman—conundrum of how the world can prepare for the inevitable day when petroleum extraction will fall below global needs. Access to energy will then become a function of national good fortune for lands that can continue to cheaply produce enough for themselves, of price rationing in global energy markets, or of force of arms.

Almost everyone plying the trade of economic clairvoyance believes there will be or should be a massive shift to renewable energy resources, such as wind and solar power, or to a new generation of nuclear (and/or fusion) power plants.[2] But such seers also note that the current low price of petroleum is a powerful disincentive for this development. Even if technical breakthroughs and economies of scale eventually bring the cost of energy from petroleum alternatives down to the current price range of petroleum products, the investment needed to get to that point may be prohibitive, at least until falling petroleum supplies and rising demands trigger a severe crisis. And this raises the question of who will be the winners and who the losers when that crisis arrives.

Now let us look at the parallel to this well-known and much-discussed story that began in late medieval times in Europe. Why Europe's population grew at this time has been much debated. Lynn White's 1962 classic, *Medieval Technology and Social Change*, mustered an array of technological explanations, though they spoke more to coping with population increase than to its causes.[3] Since that time climate historians have focused with greater intensity on what they call the Medieval Warm Period, an interval between roughly 950 and 1250 when Europe experienced a salubrious rise in temperature that permitted greater agricultural production and thus may have stimulated population growth. Regardless of the cause or causes, however, historians agree that the continent experienced a doubling of its population between 1000 and 1340. The population of Western Europe tripled. This created demand for food that increased the cost of maintaining the working animals that were, until that time, the mainstay of Europe's energy budget. Though animal-operated mills were common in Roman times, the degree to which animal power was used to grind grain in post-Roman Europe is unknown. References to rotary hand mills are more common than to animal mills. Be that as it may, when the new population surge called for an increased capacity to grind grain into flour, animal power could have been called upon, but it was not.

The reason the growing demand for industrial power was not met by animal power is that the cost of maintaining working animals grew along with the human population. Both humans and working animals required food.

TABLE 2.1 Costs of Maintaining a Horse and an
Ox in Medieval England

Expense	Horses	Oxen
Oats (in Winter)	8s 2d	2s 4
Pasture (in Summer)	1s	1s
Shoeing	4s 4d	—
Total (per Year)	13s 6d	3s 4d

Source: John Langdon, "The Economics of Horses and Oxen in
Medieval England," Agricultural History Review 30 (1982), 31.

And since animals were indispensable for plowing and transport, lower-cost energy alternatives for other purposes made economic sense. Evidence relating to the cost of animal power is not abundant, but one source stands out. The English writer Walter of Henley, who composed his book on the management of a manor—Le Dite de Hosebondrie—around 1280, compares the cost of maintaining an ox with that of maintaining a horse.

While table 2.1 makes apparent the relative economy of using oxen for plowing instead of horses, the salient comparison for present purposes is with the data presented in table 2.4 later in this chapter.[4] Horses may have cost more to maintain than oxen, but both were expensive in comparison with the zero cost of fodder for draft animals in the arid zone. As population growth put pressure on the production of grain for human consumption, finding the wherewithal to buy oats for a vast increase in the number of animals harnessed to millstones would have been a great economic burden.

Fortunately, water and wind power—essentially free energy—could be used for milling and related industrial purposes if the resources were available for the very substantial start-up costs in terms of plant and equipment. Aristocratic and monastic landholders possessed the necessary capital, and they could recover their investment fairly rapidly if they prohibited hand (and animal?) milling and forced their tenants to bring their grist to the new watermill. The situation is well summarized by Frances and Joseph Gies in their Cathedral, Forge, and Waterwheel: Technology and Invention in the Middle Ages:

> The tenth and following centuries witnessed steady progress in
> reclamation of unproductive areas via drainage, irrigation, and land
> clearance. Northern and western Europe, once sparsely inhabited,
> filled in.... The rapidly multiplying written records supply a wealth
> of statistics, of which perhaps the most telling is the figure given in
> Domesday Book, the survey prepared in England in 1086 at the order

of William the Conqueror. A century earlier, fewer than 100 mills are recorded in the country; Domesday Book lists 5,624 (low, since the book is incomplete) .... Continental records tell a similar story. In one district of France (Aube), 14 mills operated in the eleventh century, 60 in the twelfth, and nearly 200 in the thirteenth. In Picardy, 40 mills in 1080 grew to 245 by 1175.[5]

Though the capital for the explosive growth in mill construction undoubtedly came from wealthy landholders, the millers were the ones who acquired expertise in the design and uses of the machinery. Waterpower subsequently came to be applied to many industrial operations in Europe besides grinding grain. Here is how one historian has summarized this course of developments:

> In 1500, and for another couple of centuries after, there was for all practical purposes, still only wind, running water or muscle-power available as power to drive machines. These forces were, nevertheless, put to use much more efficiently than a few centuries earlier. Wind used to drive a windmill made an enormous difference to work, particularly in the countryside, which had previously had to be done by hand (above all grinding corn). Water-mills could more easily be used for industrial work, too, and were to provide most of Europe's machine-power (and, indeed, the world's) for centuries yet. In the thirteenth century they began to be used to drive bellows in forges, and, soon, for "fulling" (cleaning and thickening) cloth, for working hammers, and for cutting lumber in saw-mills.[6]

Intriguingly, considering the role millers and waterpower were to play in the origins of the Industrial Revolution, key water-driven mechanical processes were usually attested in the Islamic lands of the MENA before they appeared—often as borrowed technologies—in Christian Europe. Here are some examples.

Fulling mills, in which woolen cloth was cleaned, thickened, and felted by being beaten with wooden hammers instead of being trampled under foot as in Roman times, seem to have been known throughout the medieval Islamic world a century or two before the technique reached Christian Europe via Spain and Italy.[7]

Paper mills, in which water-powered trip-hammers instead of hand-wielded pestles pounded rags to pulp, may have been known, though perhaps only infrequently, in the Islamic world well before the first reference to a water-powered paper mill in the Christian Spanish kingdom of Aragon in 1282.[8] Water-powered sugar mills too came from North Africa and the

Middle East, where the Crusaders encountered them in the twelfth century.[9] These devices made crushing sugarcane more efficient and were integral to the later growth of the European sugar industry, first on Cyprus and other islands in the Mediterranean Sea and Atlantic Ocean and later in the West Indies.[10]

As for water-powered sawmills, archaeological and textual evidence attests to their existence in both the Middle East and Europe in Roman times.[11] But there was a long hiatus between these mills and the later spread of sawmills in Europe from the thirteenth century onward, a century or more after water-powered sawmills had become common in the Muslim world. A Dutchman was the first to adapt a windmill to power a saw in 1594.

These instances of the proliferation of mill technology in the late Middle Ages, whether or not they derive directly from models in Muslim lands, testify both to the importance of mill technology as a base for early modern European economic development, and to the parallel—and generally prior—knowledge of the same technologies in North Africa and the Middle East.

But why did European watermills proliferate so dramatically in the twelfth century after being only mildly popular over the preceding millennium? And why was there no parallel proliferation in the Muslim world, which had inherited the same technological traditions from the Romans? Leaving aside the history of technology in China, where watermills are well attested from the sixth century and became fairly common in the century following,[12] the watermill debuts in the west in the writings of the Roman engineer Vitruvius in the first century BCE.[13] Precursor techniques, such as rotary mills that ground grain between a stationary stone and a round stone turned by a donkey walking in a circle, had come into use a century and a half earlier and are well preserved in the ruins of Pompeii, which was destroyed in 79.[14]

Yet despite the apparent efficiency of harnessing the free energy of flowing water, watermills failed to drive animal-powered mills into oblivion. A 301 edict of the emperor Diocletian fixing prices throughout the Roman Empire lists the costs of four types of mill: a horse mill, 1,500 denarii; a donkey mill, 1,250 denarii; a watermill, 2,000 denarii; and a hand mill, 250 denarii. Times were hard and getting harder during the reign of Diocletian,[15] and a watermill was by far the most expensive device for making flour. The structure was expensive to build even if the kinetic energy of the water was free. It was no wonder that many people continued to rely on a plodding animal or their strength of arm applied to a quern or rotary hand mill.

Diocletian's edict applied to all parts of his empire, including those that came under Muslim rule during the Arab conquests of the seventh and eighth centuries: Spain, North Africa, and Syria. Though it is sometimes remarked that watermills were rare in the Muslim lands because of the paucity of rivers, waterpower was in fact extensively exploited.[16] Floating mills were anchored in the Tigris and Euphrates rivers, tidal mills served the Persian Gulf port of Basra, and seventy mills are reported to have been in use near the northeastern Iranian city of Nishapur. This last report is of particular interest because Nishapur is not on a river, or even a substantial stream. Hence we must assume that Iranian hydraulic engineers, who were known for digging thousands of underground water canals, devised a way of channeling water for this purpose, just as engineers in southern Morocco designed aqueducts to carry water to sugar mills from the Atlas Mountains. Moreover, as has already been pointed out, the Roman heritage of gear mechanisms, cranks, and other mechanical techniques that ultimately allowed European watermills to be used for industrial purposes was entirely shared by the societies of the Islamic Middle East and North Africa. Nevertheless, watermills were expensive to build, and animal-operated devices were the much cheaper and thus preferred alternative.

This comparison of post-Roman Europe and the Islamic lands breaks off in the twelfth and thirteenth centuries. Europe developed increasingly sophisticated mill technology and expanded that technology into diverse industrial uses. Ways were eventually found to replace water and wind with the energy of steam, but steam power was comparatively expensive and did not replace waterpower for all uses. While wind and water were free goods, fuel was required to produce steam. The advent of cheap coal, and then petroleum, eventually brought the age of the watermill to an end, but it had proven to be a key technology in Europe's industrial revolution. Moreover, the entrepreneurial class at the center of that revolution included innumerable millers, who brought together capital resources, mechanical expertise, and an enthusiasm for commerce and the expansion of their economic enterprises. It must be noted that the words "mill" and "factory" were more or less synonymous in the early stages of industrialization.

No comparable industrial revolution occurred in the Muslim world. When modern industrial enterprises finally began to appear, most in the nineteenth century, they were either copied from or started by Europeans. Would this economic history have been different if a class of millers had come into being in the Muslim world at the same time, or even before, it did in Europe? There is no way to be certain, but the absence of millers as a class in Muslim lands is striking. Though precise figures are lacking, it is unlikely that per-capita bread consumption differed significantly between Europe and the MENA. Hence the demand for milling must have been roughly parallel. But since harnessing a single forage-fed animal to a millstone was significantly cheaper than the apparatus needed to harness water or wind for that purpose, milling remained local and small-scale, and the miller enjoyed no more prominent social or economic role than any other artisan.

A specific comparison can be made between Cairo and Paris in the eighteenth century. André Raymond, in his magisterial study of Cairo's merchants and artisans, cites the seventeenth-century traveler Evliya Çelebi's figures for milling in that city: 1,200 mills engaging the efforts of 3,160 millers.[7] Of those millers for whom estate inventories are available in the court records, Raymond remarks that they were "rather poor."[18] While there is no specification of whether the mills were powered by water or animals, the one miller estate detailed by Raymond clearly points to an animal-operated mill. It consisted of sixteen oxen, seven camels, one mule, the apparatus of the mill, and various household goods, which altogether reflected "a relatively simple material life."[19] Figures for the city of Marrakech in Morocco likewise support the idea that most MENA mills of the pre-steam era were powered by animals. In the late nineteenth century, the city had one hundred horse mills and twelve watermills. Each of the former processed 240 kg. of wheat per day, and each of the latter 300 kg.[20] It is noteworthy that the watermills were only 20 percent more productive than the horse mills.

A detailed comparison is not necessary to demonstrate the difference in the importance and activities of millers between these large MENA cities and contemporary Paris. Steven Laurence Kaplan's book *Provisioning Paris: Merchants and Millers in the Grain and Flour Trade During the Eighteenth Century* presents an exhaustive account of that city's organizationally sophisticated grain-processing economy. By this historical juncture, on the eve of a major increase in technological and commercial efficiency known as "economic milling," most mills were owned by members of the first and second estates and leased to the millers. Of the sixty-one mills known to have been leased, thirty-one were owned by religious institutions and twenty-two by nobles. The rest were owned by millers, bakers, and other middling individuals.[21] (These data apply only to gristmills, of course. There were many other watermills devoted to operations like sawing lumber and fulling cloth.) By the century's end, however, the miller's status was on the rise. In Kaplan's words:

> Commercialization, which transformed the milling function during
> the eighteenth century imposed new responsibilities such as cleaning
> grain, bolting meal, and mixing flour. At the same time, it required—
> and enabled—the miller to abandon more of the physical labor to
> his aides and to become more of a businessman. It was not unusual
> for a miller to undertake some farming. Certain millers were also
> deeply involved in the bakery, despite the official prohibition against
> combining these trades.[22]

As one measure of the differential impact between Europe's turn to water-power and the MENA's continued use of animal power as the mainstay of its industrial processes, we can turn to the present-day distribution of American surnames. The Europe that came into being in the age of waterpower, and that—through immigration—would do the most to shape early American demographics, saw the miller elevated to high status in the society of rural commoners that accounted for most of Europe's population increase and economic production.[23] Like the smith, whose forge and skills in metalworking had for centuries garnered social esteem, the miller acquired affluence and frequently demonstrated a willingness to invest his money and technical skills in gaining yet more prestige.

Surnames tell the story more efficiently than a ream of statistics. Throughout European society, Smith and Miller became the most common trade-based surnames because of the status implied by both words. They are far more common, for example, than occupational names like Carpenter, Fuller, Fisher, Carter, Cooper, Sawyer, and so on. The vestiges of this social evolution survive in today's American name patterns, which are a semi-fossilized remnant of the social structure of the sixteenth through nineteenth centuries, when waves of European immigrants brought their talents and ambitions to the New World. According to the 2000 census, Smith is the most common surname in the United States, and Miller is ranked seventh.[24] The next most common trade-based names are Taylor (thirteenth), Wright (thirty-fourth), and Baker (thirty-eighth). As table 2.2 shows, this shared prominence for Smith and Miller is consistent across a wide variety of American ethnic surnames.

Arabic names, which in America represent the largest single-language segment of populations of MENA origin,[25] contrast starkly with the European ethnic pattern. Haddad, the Arabic version of Smith, is ranked 3,507, almost exactly as common as Kovach and Kovacs and slightly more common than

TABLE 2.2 Smith and Miller as American Surnames

Language	"Smith"	Name Rank	"Miller"	Name Rank
German	Schmidt	171	Muller	1,198
Dutch	Smit	9,754	Mulder	4,760
Italian	Ferraro	2,742	Molinaro	10,046
Italian	Ferrara	2,480	Molinari	10,384
Spanish	Herrera	175	Molina	417
French	Lefebvre	5,354	Meunier	11,301
Russian	Kuznetsov	39,885	Melnick	10,088
Hungarian	Kovach	3,570	Kovacs	3,709
Arabic	**Haddad**	3,507	Tahhan (Tahan)	50,468

Note: Names occur in many variants. Only the most common are given here.

Lefebvre. The Arabic word for Miller, however, is Tahhan, which appears only in the form Tahan (perhaps mostly Sephardic Jews) with a rank of 50,468.[26] Equivalents of Miller are similarly rare in medieval compilations of Arabic names because the miller never became an elite occupation.[27]

To return to our principal argument, the rising cost of animal power after roughly the year 1000 made possible a European exploitation of what we would today call alternative energy sources. New watermills and windmills were large, complicated, and very expensive. But the rising cost of animal energy left no cheaper alternative.

The opposite was true in the arid zone with its surplus of animal energy. Animal-operated mills, wells, and irrigation devices remained a constant of the rural landscape until the late twentieth century. Where watermills and windmills were used, they were generally of flimsier construction than their European counterparts because they had to compete with a virtually free source of energy.[28]

The questions we have been asking so far are the following: What caused Europe to fixate on harnessing the power of water in the eleventh and twelfth centuries? And why did the lands of Islam situated in the arid zone not develop a parallel fixation?

As we have seen, the number of watermills began to multiply rapidly in eleventh-century Europe giving rise over the following centuries to a wide diversification of water-powered mechanical processes and to a social class, the millers, that gradually accumulated the money, skills, and ambition to make water the dominant energy source in European manufacturing. But no parallel expansion of waterpower took place in the Islamic lands of the MENA, even though the region inherited the same technological traditions from the Roman period and—more importantly—despite the fact that watermills in the arid zone were often used in the MENA for purposes that Christian Europe hit upon, or borrowed, only at later dates.

Europe's waterpower romance, it should be noted, began well before the Renaissance, Reformation, and Scientific Revolution. In other words, the quickening of European intellectual life associated with the Renaissance, or even with the so-called twelfth-century renaissance, does not appear to have had much relevance to the strictly demographic, technological, and economic changes we are addressing. With respect to the arid zone, the most plausible explanation of the absence of a parallel turn to water or wind power is that the natural landscape produced abundant grazing for herds of work animals: camels, donkeys, horses, mules, and—to a lesser degree—oxen. These herds were managed by nomadic pastoralists who were locked into an austere and inexpensive pattern of material life by their need to frequently be on the move with their animals. As a result, animal power in the arid zone was incredibly cheap. By comparison, most of Europe was forested, and much of it had cold winters during which draft animals had to be stabled and artificially fed. Maintaining an ox or horse in England was considerably more expensive than keeping a camel or donkey in Syria. Moreover, when Europeans fed their animals grain to maximize their strength and endurance, the land that grew the animal fodder was land that was necessarily not available to grow food for human consumption.

Geographic and climatic differences thus conspired to raise the cost of animal power in Europe. European knights went to battle with two or three horses and relied on grain to sustain them during combat. With an abundance of animals at their disposal, the Mongols and other Central Asian nomads went into battle with whole strings of graze-fed horses and switched from one to another as the one they were riding flagged. Similar factors affected other uses of animal power. Since bread, made from milled grain, was a staple in both Europe and the MENA, there is no reason to suppose that MENA societies consumed less milled grain per capita than Europe did. The most important difference between the two regions was the fact that most mills in the MENA were operated not by water, as in Europe, but by a camel, ox, or donkey that had grown to working age at virtually no expense and could be replaced quite cheaply. The same applied to mechanical irrigation devices and wells operated by animals. As late as the early twentieth century, many tens of thousands of camels—most destined to operate mills and irrigation devices rather than to supply increasingly infrequent caravans—were funneled by camel merchants from the Arabian Peninsula to the camel markets of Basra, Aleppo, and Damascus, and from Sudan to the market in Cairo.

For a very long time, the difference between Europe and the arid zone in the cost of animal power was not crucial. Living at a generally simpler economic level, Europeans probably ground more grain with rotary hand mills and querns than Muslims did and relied less on animal-powered mills and mechanical devices. But technological change and population growth tipped the scale in favor of a major shift to waterpower during the eleventh and twelfth centuries. Technologically, the introduction of modern harnessing techniques, already in evidence in the Bayeux Tapestry in the late eleventh century, made it possible to plow heavier soils and haul heavier loads.[29] But these activities required an upgrade in the energy level of work animals, whether horses or oxen. And in demographic terms, rapid population growth put pressure on grazing lands and made expanding the acreage devoted to grain for human consumption both essential and more profitable than growing animal fodder. Again, as already noted, the historian of medieval Europe Josiah Russell estimated that the population of Europe almost doubled between 1000 and 1340, growing from 38.5 million to 73.5 million. In Western Europe—including France, the Low Countries, the British Isles, Germany, and Scandinavia—the population almost tripled, rising from 12 million to 35.5 million over the same period.[30]

Put another way, just as new harnessing technologies were helping European farmers bring new land into production to grow crops for a burgeoning population, the cost of feeding animals rose. Many farmers stuck with oxen and turned only slowly to using horses, which worked faster but were more expensive to maintain. But military technology was also pushing toward higher-priced animal power; heavier armor necessitated larger horses and more generous grain allowances. This process continued into the post-chivalric era due to the growing importance of horse-drawn field artillery and logistics trains.

It is not known how much milling of grain in Europe was done by hand, as opposed to by animals, before the year 1000. But even if the customers of the new watermills were peasants who had previously ground their meal by hand, it is significant that there is no indication of growth in the number of animal-powered mills alongside the new watermills. It cost less in the long run to invest a lot of money in a watermill than to regularly buy expensive new draft animals. Nevertheless, many millers using waterpower did maintain a standby capacity to harness animal power during droughts or when the millpond froze.

Ideally, we would now present an array of data showing exactly the relative costs of energy between Europe and the arid zone over a period of centuries. Alas, such data are not at hand, though they may someday be developed out of the historical record of the Ottoman Empire. In their stead, we will present the findings of a recent study of modern energy costs in Pakistan to show that cost-free animal power can compete effectively even with cheap late twentieth-century gasoline.

An article published in 1985 by scholars from the University of Pennsylvania and the Pakistan Agricultural Research Council addressed the question of whether camel carts and pack camels would long continue to be a part of Pakistan's transportation economy, and if so, why.[31] In this economic comparison, camel power was contrasted with irrigation pumps, mechanized sugarcane crushers, and small fuel-efficient vans like the Suzuki.

> The basic framework employed is traditional supply and demand
> analysis, where we essentially ask what has happened to the price of
> camels and to their number. If the Suzuki van and other machines
> have doomed the camel, we would expect the relative price of camels

to have fallen (demand decrease). If, with the demand decrease, there is even a larger supply shift to the left, the price might rise but the numbers would have to decrease... We conclude by suggesting that the camel will continue to find several ecological niches for itself in Pakistan, including head-to-bumper confrontations with the Suzuki in the crowded streets of cities like Karachi.[32]

Pakistan's camel population was not only robust and growing in 1985 when this study was published, but as table 2.3 shows, it has expanded even more since then, except in Punjab, Pakistan's most economically developed province. The study also concluded on the basis of interviews that the price of camels had been rising in the years leading up to the data analysis rather than falling.

Also listed in table 2.3 is Pakistan's donkey population, which has increased steadily in every province since the 1950s. Since donkeys are not usually used for milk and meat, as camels are in some regions, their growing numbers seem to indicate a robust market for their riding and draft services. Heston et al. remark that "the donkey is an important work animal; however, very little is written about the donkey." Nevertheless, they deduce from one study of the comparative efficiency of Pakistani draft animals that "for loads of up to 1 ton, a combination of one or two donkey carts would compete with the charge of a camel cart."[33]

TABLE 2.3 Pakistan's Camel and Donkey Population

Province/Unit	1955	1960	1972	1976	1986	1996
Pakistan						
Camels	719	490	731	789	958	816
Asses	1,038	1,474	1,901	2,157	2,998	3,559
Punjab						
Camels	333	266	365	338	321	187
Asses	610	897	1,063	1,139	1,657	1,948
Sindh						
Camels	176	62	80	144	218	225
Asses	161	159	242	373	500	694
N.W.F.P.						
Camels	140	76	101	95	70	65
Asses	196	306	408	381	446	534
Balochistan						
Camels	70	86	185	212	349	339
Asses	61	99	171	244	370	383

Source: Livestock Census, 1996.

Why are camels and donkeys able to compete against cheap, gasoline-powered vehicles and devices?

> One conclusion is that the camel can compete successfully against other work animals and/or machinery in several ecological situations. Wherever there are no direct feeding costs for camels, their number is likely to dominate that of all other work animals, except perhaps that of donkeys...It is our conjecture that if it is necessary to incur feed costs for camels, they cannot compete in traditional tasks like sugarcane crushing and drawing a Persian wheel for irrigation. In these areas the machine is likely to be more cost effective in view of the pricing of machinery and fuels.[34]

Table 2.4 lists the typical costs and income in rupees of a person renting a pack camel.

Historically, the impact of the zeroes in this budget might be compared to the zero cost of water and wind (and the zeroes here should also be compared with Walter of Henley's data on the feeding costs for oxen and horses cited in table 2.1). But the costs of a saddle and the purchase of a fresh animal every few years—a cost that might also be zero if the owner is a camel-raising nomad—cannot be compared with the vastly higher costs of building and maintaining a water- or windmill.

Obviously, no broad conclusion regarding the comparative efficiency of animal power throughout the arid zone over many centuries can rest on a single study. The value of Heston et al. is that it spells out in precise detail exactly why camels (and presumably donkeys) can maintain an edge in economic efficiency even in today's world. It costs little or nothing in monetary terms to raise them to working age, and arid zone grazing makes it possible to feed them without cost. A second study does offer a modicum of corroboration, even though it does not deal directly with the feeding costs of animals. In Old Fez in Morocco, a pre-colonial

TABLE 2.4 Costs and Incomes Associated with Renting a Pack Camel (Rupees)

Feeding Costs for 200 Working Days	0
Feeding Costs for 165 Working Days	0
Costs of Medicines	200
Costs of Harness	50
Amortization of Camel Cost	1,350
Revenues	12,000
Overall Return to Laborer/Owner	**10,400**

Source: Alan Heston, H. Hasnain, S. Z. Hussain, and R. N. Khan, "The Economics of Camel Transportation in Pakistan," *Economic Development and Cultural Change* 34 (1985), 132.

city whose narrow streets cannot accommodate automobiles, everyone depends in one way or another on equid transport, mostly by donkey or mule. People rent donkeys to carry goods and materials for about $6 per day and spend about $2.50 to feed and lodge one of the animals.[35] Where the food for the donkeys comes from is not stated, but with gasoline in Morocco costing about $1.34 a liter, it is fairly obvious that animal power is still competitive there under certain circumstances.

Though further research is needed to solidify the argument I have put forward, a conclusive finding that cheap animal power, analogous to cheap petroleum today, forestalled the development in the arid zone of the alternative energy sources that set Europe on the road to industrialization would be crucial to moving the debate over the post-1400 economic disjunction between Europe and the lands of Islam from the realm of culture, religion, and politics to the realm of the economics of natural resource management. Given the near-zero cost of animal power, the peoples of North Africa, the Middle East, and Inner Asia were no more capable of seizing the opportunities presented by Europe's ever more sophisticated utilization of water- and windpower than we are today, in the face of cheap oil, of investing vast sums on the alternative energy sources that we will one day so greatly need.

Notes

1. Richard W. Bulliet, *The Camel and the Wheel* (Cambridge, MA: Harvard University Press, 1975).

2. See, for example, Thomas L. Friedman, *The World Is Flat 3.0: A Brief History of the Twenty-First Century* (New York: Picador, 2007).

3. Lynn Townsend White, *Medieval Technology and Social Change* (Oxford: Clarendon Press, 1962).

4. Additional details on the feeding of draft animals in medieval England are found in John Langdon, "The Economics of Horses and Oxen in Medieval England," *Agricultural History Review* 30 (1982): 31–40.

5. Frances and Joseph Gies, *Cathedral, Forge, and Waterwheel: Technology and Invention in the Middle Ages* (New York: HarperCollins, 1994), 113. Detailed information on the history of watermills in Europe is superabundant. See, for example, Richard Holt, *The Mills of Medieval England* (Oxford: Blackwell, 1988); Terry S. Reynolds, *Stronger than a Hundred Men: A History of the Vertical Water Wheel* (Baltimore: Johns Hopkins University Press, 1983).

6. J. M. Roberts, *A Short History of the World* (New York: Oxford University Press, 1993), 246.

7. Adam Robert Lucas, "Industrial Milling in the Ancient and Medieval Worlds: A Survey of the Evidence for an Industrial Revolution in Medieval Europe," *Technology and Culture* 46 (2005): 1–30. For a detailed analysis of the use of watermills in fulling

see: John Munro, "Industrial Energy from Water-Mills in the European Economy, Fifth to Eighteenth Centuries: The Limitations of Power," Working Paper No. 16 (April 4, 2002), Department of Economics and Institute for Policy Analysis, University of Toronto.

8. Jonathan Bloom, *Paper Before Print: The History and Impact of Paper in the Islamic World* (New Haven, CT: Yale University Press, 2001), 50.

9. Ahmad Y. al-Hassan, *Technology Transfer in the Chemical Industries*, part 3 of *Transfer of Islamic Technology to the West*. www.history-science-technology.com/Articles/articles%2072.htm (accessed May 18, 2011).

10. "The Cairo Genizah records reveal that making and selling sugar from sugarcane was one of the most common occupations of Jews in the Middle Ages; Sukkari was a common family appellation from the beginning of the 11th until the end of the 13th centuries in Egypt and in North Africa. Sugar refineries were often in Jewish hands. Jews are mentioned as exporters of sugar from Crete in the 15th century. When sugar began to be used for everyday consumption (15th century), Marranos ['secret Jews'] played a leading role in introducing sugarcane cultivation to the Atlantic islands of Madeira, the Azores, the Cape Verde Islands, and São Tomé and Príncipe in the Gulf of Guinea, and in the 16th century to the Caribbean Islands. They also brought the cultivation of sugarcane from Madeira to America, and the first great proprietor of plantations and sugar mills, Duarte Coelho Pereira, allowed numerous Jewish experts on sugar processing to come to Brazil. Among them was one of the first important Jewish proprietors of sugar mills, Diego Fernandes." *Encyclopaedia Judaica*, 2nd edn., s.v. "Sugar Industry and Trade" (Hans Pohl, Henry Wasserman, and Zeev Barkai).

11. Lucas, "Industrial Milling in the Ancient and Medieval Worlds."

12. Joseph Needham, *Mechanical Engineering*, part 2 of *Physics and Physical Technology*, vol. 4 of *Science and Civilization in China* (Cambridge: Cambridge University Press, 1965).

13. Pollio Vitruvius, *The Ten Books on Architecture*, trans. Morris Hicky Morgan (Cambridge, MA: Harvard University Press, 1914).

14. James L. Franklin, *Pompeii: The "Casa del Marinaio" and its History* (Rome: "L'Erma" di Bretschneider, 1990), 36.

15. "For who is so hard and so devoid of human feeling that he cannot, or rather has not perceived, that in the commerce carried on in the markets or involved in the daily life of cities immoderate prices are so widespread that the unbridled passion for gain is lessened neither by abundant supplies nor by fruitful years." From Preamble of Diocletian's *Edict of Maximum Prices*. www.fordham.edu/halsall/ancient/diocletian-control.html (accessed November 6, 2010).

16. Salim T. S. al-Hassani and Mohammed A. al-Lawati, "Water Machines in the Lands of Islam," chapter 2 of "The Six-Cylinder Water Pump of Taqi al-Din: Its Mathematics, Operation and Virtual Design." www.muslimheritage.com/topics/default.cfm?ArticleID=967 (accessed November 6, 2010).

17. André Raymond, *Artisans et commerçants au Caire au XVIIIe siècle*, 2 vols. (Damascus: Institut Français de Damas, 1973–74), 1: 314.

18. Ibid., 1: 233.

19. Ibid., 1: 397.

20. Stacy E. Holden, "Famine's Fortune: The Pre-Colonial Mechanization of Moroccan Flour Production," *Journal of North African Studies* 15 (2010), 76.

21. Steven Laurence Kaplan, *Provisioning Paris: Merchants and Millers in the Grain and Flour Trade During the Eighteenth Century* (Ithaca, NY: Cornell University Press, 1984), 255.

22. Ibid., 262.

23. George Eliot's novel *The Mill on the Floss*, published in 1860 and based on the author's own family, provides a good illustration of the social and economic status associated with owning a mill for grinding grain in nineteenth-century England.

24. The census data have been taken from the following: names.mongabay.com/most_common_surnames.htm (accessed November 6, 2010).

25. Immigrants from the portion of the arid zone lying in Pakistan and northwest India may well be more numerous, but their names come from a variety of languages and cannot easily be disaggregated from Indian and Pakistani names generally.

26. For the Sephardic association of this name, see the following: *Dictionary of American Family Names*, s.v. "Tahan." The best known example of this rare name is Malba Tahan, the pen name of the popular Brazilian writer Júlio César de Mello e Souza.

27. For example, the name Tahhan never appears among the more than three thousand entries in the biographical dictionaries of medieval Nishapur compiled by al-Hakim al-Naisaburi and 'Abd al-Ghafir al-Farisi in the tenth and eleventh centuries, whereas names like Smith (Haddad), Fuller (Qassar), and Greengrocer (Baqqal) do, as does Daqqaq, meaning "flour merchant."

28. The following presents pictures of the decidedly low-cost windmills used in eastern Iran: Hans E. Wulff, *The Traditional Crafts of Persia: Their Development, Technology, and Influence on Eastern and Western Civilizations* (Cambridge: Massachusetts Institute of Technology Press, 1966), xx.

29. White, *Medieval Technology and Social Change*.

30. These data come from Josiah C. Russell, "Population in Europe," in *The Middle Ages*, vol. 1 of *The Fontana Economic History of Europe*, ed. Carlo M. Cipolla (London: Collins/Fontana, 1972), 25–71.

31. Alan Heston, H. Hasnain, S. Z. Hussain, and R. N. Khan, "The Economics of Camel Transportation in Pakistan," *Economic Development and Cultural Change* 34 (1985): 121–141.

32. Ibid., 123.

33. Ibid., 140, n. 13. The work cited is Sadaquat H. Hanjra, Bakht B. Khan, A. R. Barque, M. Tufail, and M. Aftab Khan, "Comparative Efficiency of Draught Animals," *Journal of Animal Sciences* 11 (1980): 79–84.

34. Heston et al., "The Economics of Camel Transportation in Pakistan," 130–131.

35. Diana K. Davis and Denys Frappier, "The Social Context of Working Equines in the Urban Middle East: The Example of Fez Medina," in *The Walled Arab City in Literature, Architecture and History: The Living Medina in the Maghrib*, ed. Susan Slyomovics (London: Frank Cass Publishers, 2001), 58.

3

The Little Ice Age Crisis of the Ottoman Empire: A Conjuncture in Middle East Environmental History

Sam White

In February 1621, the chronicler İbrahim Peçevi recorded "a most unusual event." The Bosphorus in Istanbul had completely frozen over, connecting the European and Asian shores:

By the will of God, the winter in Istanbul this year
has been colder than any winter since the world began.
Between Üsküdar and Istanbul it has frozen, the sea gone dry...
Who has seen so many walk over the ice on the sea
fearless as though it were dry land?[1]

Although as recently as 2007 a scholar expressed disbelief about this account,[2] what Peçevi described was almost certainly real.[3] Three more eyewitnesses, including the Venetian ambassador, confirm his account and offer further details of the thick ice and bitter cold that drove the imperial capital to starvation that winter.[4] What the Ottomans experienced formed part of a worldwide phenomenon of extreme weather events associated with the onset of the so-called Little Ice Age of the late sixteenth to the early eighteenth centuries. These were the same years in which London held fairs and festivals on the frozen Thames; early European settlers in North America starved to death in terrible winters; and lands as distant as China, Indonesia, and West Africa suffered devastating droughts and famines.[5]

The Little Ice Age also proved the underlying cause of perhaps the worst crisis in Ottoman history.[6] By the late sixteenth century, the Ottomans were at the peak of their power, controlling lands from

Hungary to the Hijaz and possessing perhaps the strongest military in the world. Yet at the same time, mounting population pressure, rising military costs, and natural disasters had begun to threaten the empire's stability. In the 1590s, a series of exceptionally severe winters and the region's longest drought in six centuries brought starvation to many parts of the empire, even as Ottoman armies were facing setbacks in their Long War (1593–1606) with the Habsburgs. This combination of pressures wreaked havoc on imperial systems for provisioning goods to the capital and the army; and high taxes, hunger, and desperation drove Anatolian subjects into a destructive rural uprising that would come to be known as the Celali Rebellion (c. 1596–1610). Over the following generation, significant parts of the empire were substantially depopulated by a synergistic mix of famine, flight, disorder, and disease driven by widespread violence and ongoing Little Ice Age climate events. Throughout the next century, nomadic intrusions into settled village land and a mass flight into already crowded cities dramatically slowed the empire's demographic recovery. The Little Ice Age crisis, therefore, represented not just an event in the life of an empire, but a critical conjuncture in Middle East environmental history that had far-reaching impacts.

This chapter argues that the Little Ice Age offers the best framework yet for understanding the Ottoman Empire's seventeenth-century crisis. While Ottomanists have come to reject the traditional historiography of "decline," which focused on the decay of classical institutions and strong sultanic authority, none has yet offered an alternative approach with the same explanatory power.[7] The revisionist historiography has instead offered new conceptualizations focusing on sociopolitical transformation and changing relationships among palace and provincial elites—developments sometimes cast in terms of "adaptability," "decentralization," "privatization," or even "proto-democratization."[8] While perhaps valuable in and of themselves, these insights do not capture the timing and suddenness of the empire's reversal of fortunes nor its profound agricultural and demographic losses. Moreover, the new historiography too often overlooks the experiences of the vast majority of the Ottoman population, who still lived from year to year on the products of the land, dependent on each season's weather to provide another harvest.

This environmental perspective places the Ottoman experience within a worldwide crisis born of ecological pressures and the global Little Ice Age. As described in the comparative work of Jack Goldstone and Geoffrey Parker, these revolutions and rebellions from England to China were neither signs of rising capitalism and modernity in Europe nor signs of stagnation and decay in Asia, but rather signs of the common vulnerabilities of pre-capitalist, pre-industrial societies around the world to population pressure and climate change.[9] In this regard, the devastation in Ottoman lands can be reasonably compared with losses in Germany in the Thirty Years' War, or Russia in its "Time of Troubles," or China in the famine and fighting that marked the fall of the Ming—cases in which a quarter or more of each society's population perished. If the Ottoman Empire proved exceptional, it was in the long duration of its crisis and the slow pace of recovery that followed.

The Buildup to Crisis

First emerging as a band of frontier warriors in western Anatolia around 1300, the Ottomans enjoyed an almost unbroken string of conquests for nearly two and a half centuries, launching their empire into the small circle of major powers that dominated the sixteenth-century world. By the reign of Süleyman I (r. 1520–1566), the Ottomans ruled a vast realm stretching over three continents and some thirty present-day countries. Its military was one of the most powerful and its capital city one of the largest in the world.[10]

As the empire expanded, successive rulers oversaw a great extension of settlement and agriculture, above all in the empire's core provinces from the southern Balkans through Anatolia and Greater Syria. Employing a variety of measures, from forced population transfers to tax incentives to strategically planned settlements, the imperial government guided sedentary cultivation back into lands abandoned in the intervening centuries of war and the Black Death.[11] At the same time, Ottoman policy gradually restricted the range of nomadic pastoralism, establishing varying degrees of control over recalcitrant tribes and paving the way for the expansion of farming.[12]

Bringing vast new lands and resources within their grasp, the Ottomans also developed varied and elaborate provisioning systems to supply their empire. Although not unique among early modern agrarian states,[13] Ottoman provisioning remained exceptional for its scope and scale, encompassing a wide range of commodities over geographically varied lands, balancing areas of surplus and deficit. By the mid-sixteenth century, grain and sheep flowed from the rich lands of the Danube down to Anatolia, timber from Anatolia was transported to the treeless river valleys of Iraq and Egypt, and rice and wheat from Egypt fed the barren but sacred lands of the Hijaz—to take only the most notable examples. Above all, these provisioning systems underpinned the continuing success of the Ottoman army and navy and the spectacular growth of its capital city, Istanbul.[14]

By the later decades of the sixteenth century, the empire was in some respects a victim of its own success. The Ottoman army numbered well over 100,000 and the navy (rebuilt almost instantly after its 1571 defeat at Lepanto) contained over 40,000 more.[15] Istanbul, virtually depopulated upon conquest in 1453, had mushroomed into a teeming metropolis, its nearly half a million people consuming endless resources from the provinces.[16] At the same time, population in the provinces swelled, nearly doubling in the Balkans, Anatolia, and Greater Syria from the late fifteenth to the late sixteenth centuries. On the eve of the Little Ice Age crisis in 1590, the empire may have numbered some thirty-five million subjects.[17]

Consequently, over the latter decades of the sixteenth century, population pressure, along with the rising demands of the military and capital city, began to take a toll. Throughout the core Mediterranean provinces, Ottoman cadastral surveys record a consistent pattern of shrinking landholdings, rising landlessness, and declining per-capita production. Diminishing marginal returns set in quickly, as agricultural technology remained simple and yield ratios low, and as a lack of rural capital and the Ottoman system of rotating military grandees discouraged major investment in irrigation. In the mountainous and semi-arid regions of Greece, Anatolia, and Greater Syria, seasonal drought and easily erodible soils hindered diversification and intensification, leaving the peasantry more and more dependent on the fragile harvest of a single crop of winter wheat or barley. Increasingly, farmers pushed into temporary fields or marginal lands and relied more on livestock to supplement their meager cultivation, putting pressure on forests and grazing lands, especially in inner Anatolia.[18]

At the same time, population pressure began to fuel inflation and social unrest. Like other early modern polities, the Ottoman Empire began to feel the effects of the "price revolution" brought on by American silver and by global and regional demographic growth. As inflation squeezed Ottoman budgets, the imperial government aggravated monetary instability by successive debasements of coinage. Meanwhile, rising numbers of landless peasants and shrinking opportunities in an age of economic instability fueled crime and disaffection; a growing underclass of desperate men turned to banditry or joined mercenary gangs in the provinces.[19]

Through the 1570s and 1580s, mounting population pressure also threatened to undermine Ottoman provisioning systems. Frequent military campaigns and worsening droughts and freezes only exacerbated the situation. Over these decades, successive waves of extreme cold and dry weather brought periodic harvest failures and even famine to wide stretches of the empire, testing imperial grain provisioning and famine management. In the worst cases, starvation drove panic and flight, spreading disorder and disease into surrounding provinces and intensifying the empire's chronic vulnerability to epidemics, including plague. In the capital and the army, these natural disasters often resulted in scarcities of basic supplies, especially wheat and sheep. These shortages became more frequent and severe as the crisis approached, and the imperial government's failure to revise official prices in line with inflation often exacerbated the situation, driving smuggling and speculation. Ongoing warfare compounded the problems, especially the major campaigns against the Safavid Empire in the mid-1580s, which drew supplies away from the cities and foisted heavy taxes and requisitions on the provinces. Moreover, as soldiers left the countryside on military campaigns, growing bandit gangs gained a free hand to extort from rural communities.[20]

The Great Drought

Although these ecological pressures had been building for decades, the tipping point that brought on the crisis came swiftly and suddenly with the arrival of the Little Ice Age. While Ottomanists have usually ignored or misunderstood this phenomenon, new research in climatology has revolutionized understandings of the Little Ice Age in general and its impact in the Middle East in particular. The Little Ice Age was not just an age of global cooling, but an era marked by multi-decadal periods of unusual climate extremes and variability, probably the effects of volcanic eruptions and variations in solar output.[21] Whereas these changes tended to bring cold wet springs and summers to Northern Europe, they often plunged the Eastern Mediterranean into freezing winters and spring droughts, while ongoing El Niño episodes may have disrupted the vital Nile floods.[22] New tree ring research demonstrates that the Eastern Mediterranean witnessed its worst drought in the past six hundred years around 1591–1596—years that coincided precisely with the onset of the empire's most destructive rebellion.[23]

Contemporary Ottoman and Venetian accounts provide ample eyewitness testimony of this Little Ice Age weather and its consequences. Starting with scattered local famines, the drought of the 1590s began to drive general shortages and then widespread starvation in regions from the Balkans to northern Syria. Prices rose inexorably despite imperial efforts to crack down on hoarding and profiteering. Peasants began to flee their villages in search of food and safety, and increasingly powerful bandit gangs pillaged the countryside for food and animals. Starving refugees brought epidemics into Istanbul, even as the capital faced mounting shortages.[24]

Worse still, starting in 1593, the empire found itself locked in another major conflict with its Habsburg rivals on the Hungarian front. Known for good reason as the Long War, the conflict would drag on for thirteen years, as once-unstoppable Ottoman advances stalled in the face of indomitable opposition from improved infantry formations and stronger fortifications.[25] As the war ground down into grueling sieges, soldiers on both sides faced a succession of frigid winters, blizzards, and floods and a severe shortage of provisions in the Balkans. Goods poured "without measure or compare" to the front, in the words of one chronicler, but the Ottoman army often remained woefully undersupplied and disaffected.[26]

In the meantime, the demands of campaigning unraveled Ottoman provisioning systems at home, draining vital resources from already famine-stricken provinces. Beginning with the onset of drought in 1591, but accelerating with the declaration of war in 1593, Ottoman imperial orders record an ever more frantic search for food and materiel. The capital faced "shortages" and then "great shortages," "extreme shortages," and even "total shortages" of goods ranging from wheat and barley to honey and grapes.[27] As the imperial government refused to raise official prices in line with market values, smuggling and speculation grew rampant. In desperation, sultans Murad III (r. 1574–1595) and then Mehmed III (r. 1595–1603) issued repeated decrees restricting consumption of various items, lashing out at "swindlers" and "hoarders," and demanding new sources of supply. In the meantime, the provinces faced rising emergency taxes and huge requisitions, which drove a downward spiral of flight and disorder, especially in central Anatolia, where according to one historian "tax collection and banditry collapsed into the same undifferentiated activity of living off the land."[28]

The tipping point for open rebellion arrived in late 1595, as the new sultan prepared to personally lead a campaign to the Hungarian front. While the Ottoman grain supply was then in a poor state, its supply of sheep—the army's crucial source of protein—fared even worse. Wheat had more than doubled in price, but meat had more than tripled by late 1595—if it could be found at all.[29] Livestock, too, had fallen victim to the Little Ice Age. Starting in 1591 or 1592, a major epizootic spread through the exposed and starving flocks and herds. Various contemporary descriptions suggest the disease may have killed off nine in ten or more of the sheep and cattle from eastern Anatolia through parts of the Balkans and the Crimea.[30] Given peasants' overwhelming reliance on livestock as insurance against failed harvests, the loss of animals must have contributed as much or more to the famine as the loss of crops.

In desperation, Mehmed III looked for every available source of sheep to feed his imperial campaign.[31] With the war disrupting the usual supplies from Hungary and Romania and the cold killing off flocks in Bulgaria, the sultan turned to the poor semi-arid province of Karaman in south-central Anatolia with a stunning demand for 200,000 head—probably more than twice as much as the region had ever provided in a single year.[32] The former home of the rival Karamanid dynasty, land-locked Karaman had long displayed a willingness to resist imperial demands and in the decades leading up to the crisis had also suffered some of the most acute population pressures in the empire.[33] The sultan's demand sparked widespread resistance, which soon snowballed into rebellion.[34]

The Celali Rebellion

Over the following year, central Anatolia's bandit gangs coalesced into a rebel army that became known as the Celalis. Most Ottomanists have interpreted this movement as part of a shift in military organization from prebendal cavalry to irregular infantry soldiers who, once dismissed from the Habsburg front, regrouped to plunder the Anatolian countryside.[35] Rather than their dismissal from the front, however, it was the *absence* of soldiers back home that left an opening for lawlessness in the provinces.[36] The Celalis were not a collection of irregular soldiers but a group of local bandits, driven by desperation and disaffection, who fell under the leadership and organization of provincial mercenary armies previously engaged in local factional fighting.[37]

The key development took place at the old Karamanid capital of Larende near the Taurus foothills in 1596–1597. As the sultan's sheep requisition engendered resistance, bandits, rebellious tribes (*aşiretler*), and unemployed madrasa students known as *sohtas* captured the town and drove out imperial officials. Even as Ottoman forces tried to put down the uprising, a local mercenary commander known as Karayazıcı seized control of the movement and forged these varied elements into a rebel army.[38] Karayazıcı and his men defied imperial forces and plundered the Anatolian countryside over the following years, forming the kernel of Celali forces that would gain in strength and number even after their leader's sudden death in 1601. Before their final defeat in 1609, the Celali rebels had attracted as many as seventy thousand men, defeated successive Ottoman armies, and ravaged the Anatolian countryside. Meanwhile, parts of Greater Syria rose in revolt under Canbuladoğlu Paşa, and Safavid Iran declared war on its distracted Ottoman rival.[39]

Although the great drought finally lifted, extreme weather continued to plague Ottoman lands, compounding the destruction of the rebellions and wars. The late 1590s brought a succession of cold, snowy winters to Anatolia. Villagers across Anatolia, Greater Syria, and the Balkans fled in search of food and safety; hundreds of thousands or perhaps millions of them perished of exposure, starvation, and disease.[40] Meanwhile, on the Hungarian front, the

Ottoman army faced ongoing hardships: rivers alternately froze and flooded; fevers broke out among the famished soldiers; and famines, murrains, and shortfalls in the imperial treasury undermined their supplies.[41]

In 1606, the Ottoman government conceded the hopelessness of the situation, brokering a peace of mutual exhaustion and restoring the *status quo ante* with their Habsburg opponents in order to focus their attention on the domestic rebellion. By this time, the Celalis—under their new general Kalenderoğlu— were preparing to sack the old imperial capital of Bursa and to come to Istanbul itself. Meanwhile, the situation in Anatolia had grown so troubled that the state could not even find supplies to direct a campaign into its own territory, leaving the army almost paralyzed by disaffection and hunger.[42] All seemed lost until 1607, when a new Ottoman commander, the veteran Kuyucu Murad Paşa, finally gathered supplies from Egypt and restored enough discipline among the troops to launch a new campaign.

The following two years marked the climax of both the Celali campaigns and the Little Ice Age famine. Following years of heavy winter and spring snows, Ottoman lands plunged into a new drought—not as long but far more severe than the great drought of the 1590s. Contemporary chronicles and reports reaching the Venetian ambassador in Istanbul describe widespread starvation and even cannibalism across Anatolia and northern Syria. Hundreds of thousands or millions more subjects fled their homes in search of food and safety, only to die of starvation and disease on the roads or in cities.[43] As Kuyucu Murad's army pushed through Anatolia to put down Canbuladoğlu's Syrian revolt, it encountered regions desolated by famine and violence.[44] Meanwhile, the intense cold drove Kalenderoğlu's rebel army back from northwest Anatolia in early 1608, as he turned his troops southward instead to take on the imperial forces. In the battle that ensued at Göksün Pass in June 1608, Ottoman forces finally scored a decisive victory over the rebels, bringing a measure of peace to the provinces.

Climate and Crisis in the Seventeenth Century and Its Aftermath

Unfortunately for the Ottomans, victory over the Celalis did not spell an end to the empire's troubles. Over the following half century, Ottoman lands would continue to witness severe climate fluctuations, natural disasters, and political and social unrest as part of the global seventeenth-century "general crisis." During the same decades that Europe descended into the Thirty Years' War, Britain faced its civil war, and China underwent the tumultuous Ming-Qing transition, the Ottoman Empire faced a series of mutinies and rebellions overlapping with the worst recurrences of Little Ice Age weather. It is likely not a coincidence that Sultan Osman II was killed (the only regicide in Ottoman history) in 1622, that Abaza Mehmed Paşa rebelled just a year after the Bosphorus froze in 1621, that the Karahaydaroğlu Mehmed Paşa uprising broke out while famine and plague in Egypt cut tribute from the Nile in the 1640s, or that the empire witnessed its worst political violence in a generation just as extreme cold and drought created a general famine in Anatolia and the Balkans in 1656–1660 and fire and plague gripped the capital a few years later.[45] Over the same period, imperial orders suggest that the provinces continued to suffer from widepread flight and banditry.[46] Ottoman tax records indicate population losses of half or more in provinces throughout Anatolia and neighboring regions of the empire by the 1640s.[47]

In the later seventeenth century, even as much of the world started to recover from the worst of the "general crisis," the Ottoman Empire underwent another period of intense climate fluctuations, violence, and unrest. By 1680, the Mediterranean entered a new phase of the Little Ice Age known as the Late Maunder Minimum.[48] The weather turned particularly severe just as the Ottomans began another war with the Habsburgs in the spring of 1683. Freezing weather and then heavy spring rains accompanied the army's advance into Austria. Following the Ottomans' historic defeat at the gates of Vienna that summer, severe cold and snow plagued their steady retreat over the Balkans for the next three years. Meanwhile, Greece and Anatolia witnessed their worst drought in a generation, culminating in a general famine from 1685–1687.[49] After a major defeat at Mohacs in August 1687, a poorly paid, poorly provisioned company of soldiers mutinied and marched on Istanbul, forcing the deposition of Sultan Mehmed IV. The following decade only brought more defeats and natural disasters, including a serious famine and plague in Egypt and more drought and freezing winters in Anatolia and the southern Balkans. Facing unrest at home and defeat on the battlefield, in 1699 the Ottomans surrendered territories in the Balkans for peace, the first of a series of military setbacks that marked the eighteenth and early nineteenth centuries.

These generations of natural and human disasters, from the 1590s to the turn of the eighteenth century, left an enduring mark on Ottoman demographics. Whereas much of the world underwent a generation of crisis in the seventeenth century, most regions actually recovered their losses within the next fifty to sixty years, and many doubled or even redoubled their populations by the middle of the nineteenth century. Ottoman lands, by contrast, suffered a succession of crises stretching over a century, and core Ottoman provinces would only just regain their peak c. 1590 populations around the 1850s. In this respect, the Ottoman experience cannot be understood simply in terms of "decline" or "adaptability." Rather, what stands out is the empire's relative experience of the Little Ice Age and the "general crisis," which must be analyzed in terms of the region's long-term environmental history.

Crisis and Ecological Transformation

Traditionally, writings on the Middle Eastern environment have taken a declensionist perspective, emphasizing the gradual decay of the landscape from a once supposedly pristine condition. From Enlightenment travelers, to early environmentalists like George Perkins Marsh, to the more recent work of environmental historians like J. V. Thirgood, observers have faulted man-made deforestation, grazing, and burning for the apparent degradation of the region—with the Turks receiving no small share of the blame.[50] More recently, however, an opposing paradigm has emerged, stressing the natural resiliency and stability of the Mediterranean and Middle Eastern environment, including the adaptation of its flora and fauna to millennia of drought, fires, and ruminant herds.[51]

However, the long-term history of the region might be read through a third emerging paradigm of Middle East environmental history: neither decline nor stability, but periodic ecological crisis and protracted recovery. From the mega-drought of 2200 BCE,[52] to the crises of the late Bronze Age and late antiquity,[53] to plagues and regional depopulation in the Middle Ages, the Middle East has long suffered more severe environmental setbacks and recovered more slowly from each than perhaps any other part of the world. This helps to explain its long-term transition from the center of ancient civilization to a relatively thinly populated and poorly developed region by the dawn of the industrial era.[54]

These crises display four key recurring patterns, emphasizing the region's underlying environmental vulnerabilities. First, the Middle East has proven especially sensitive to climate shifts, particularly in its large semi-arid tracts on the margins of rain-fed agriculture. As a growing body of archaeological evidence shows, these impacts may have begun with the Neolithic Revolution itself and continued through the early and mid-Holocene.[55] In ancient and medieval times, moreover, elaborate irrigation networks created "an ecological system sensitive to the smallest disturbance," as one author has put it, exacerbating vulnerabilities to drought and soil salination.[56] Second, throughout these successive crises, the collapse of agricultural systems and erratic climate drove nomadic movements, shifting the balance between desert and sown and hindering the subsequent recovery of population and settlement.[57] Third, at least in the Islamic period, it appears famine and insecurity regularly drove

populations from the countryside into urban areas, especially the major imperial capitals, in a region already highly urbanized by pre-industrial standards.[58]
Fourth, epidemic diseases historically afflicted the Middle East more than perhaps any other part of the world, at least since the Plague of Justinian in the
sixth century, preying especially upon the region's towns and cities.[59] Taken
together, these four factors—vulnerability to climate, nomadic invasions, flight
to urban areas, and a higher rate of epidemics—may account for the gradual
decline in the region's share of world population over thousands of years.

These patterns stand out in the Little Ice Age crisis in the Ottoman Empire.
Following the flight and disorder of the early 1600s, a wave of nomadic movements forced out settled cultivation throughout much of east and central
Anatolia, northern Iraq, eastern Syria, and Palestine. While population pressure
over the course of the late sixteenth century created rising tensions between
settled cultivators and mobile pastoralists, the imperial government intervened
to keep order even as agriculture continued to chip away at nomads' pastures.
During the Little Ice Age crisis, however, widespread disorder and depopulation
offered nomadic pastoralists an opportunity to push back. At the same time, the
aggressive resettlement policies of the Safavid leader Shah 'Abbas and erratic
rainfall may have pushed more tribes west out of the Zagros Mountains and
the Arabian Desert, leading to a cascade of nomad migrations.[60] By 1613, these
movements had coalesced into a widespread invasion of farmland, recorded
in a series of desperate but futile pleas from the imperial government to turn
the nomads back.[61] In little more than a decade, nearly a century of settlement
expansion had come undone. Over the following decades and even centuries,
European and Ottoman travelers would continue to comment on the abandonment of so much farmland to pastoralists.[62] Subsequent resettlement efforts in
the late seventeenth and eighteenth centuries failed in the face of insecurity and
erratic climate in the semi-arid provinces along the eastern frontier, leaving the
problem unresolved until later in the nineteenth century.[63]

In the two centuries following the eruption of the Celali Rebellion, the
imperial government was never able to restore full order in and control over the
provinces. During the Little Ice Age, serious shortages in the treasury forced
the imperial government into short-term expedients like tax farming and the
sale of offices, encouraging appointees to fleece the peasantry, and contribut-
ing to the downward spiral of instability and flight in the countryside.[64] Over
the longer term, the decentralization of imperial power opened the door to private commercial estates in the provinces, undermining some of the provisioning systems of the classical age and redirecting Ottoman land use toward the
production of cash crops for rising European economies, including new crops
from the Columbian Exchange such as tobacco.[65]

Peasants fleeing from nomadic invasions and provincial disorder concentrated in towns and cities, producing a demographic shift with long-term consequences. Already by the late sixteenth century, the imperial government had grown alarmed by the size and crowding of its urban areas, particularly the imperial capital of perhaps half a million inhabitants. With the sudden influx of refugees from Celali violence and Little Ice Age famines, underlying problems of housing, infrastructure, provisioning, water, and sanitation only grew worse.[66] Ottoman towns and cities thus became a severe demographic drain, given their rich microbial environment and significant excess of burials over births. Newcomers proved especially vulnerable to crowd infections, which could include smallpox, dysentery, typhus, and typhoid. Moreover, Ottoman cities had never developed adequate quarantine. Consequently, urban areas could face extraordinary death rates during epidemics of bubonic plague.[67] Such high rates of migration and urban mortality may have negated most of the empire's rural population recovery over the seventeenth and eighteenth centuries.

The Little Ice Age crisis in the Ottoman Empire thus underlines the great significance of the environment in Middle East history and the Middle East in world environmental history. The episode illustrates the powerful forces of population pressure, resource shortage, and climate fluctuation at work in the Ottoman Empire during the late sixteenth and seventeenth centuries. An environmental perspective on these events brings a new meaning to the paradigm of Ottoman "crisis and transformation," emphasizing both short-term natural disasters and long-term shifts in the region's demography and land use. Taking a global perspective, this study illustrates both the historical significance of climate change and the important role of regional impacts and adaptability. The Ottoman Empire was far from the only victim of Little Ice Age climate extremes and variability. However, the Middle East proved especially vulnerable to periods of severe cold and drought, and its agriculture and population growth proved less resilient in the wake of protracted crisis. In this way, the Little Ice Age crisis highlights key environmental forces, which have shaped the region's history and points to possible challenges in the present age of global warming.

Notes

1. İbrahim Peçevi, *Peçevî Tarihi*, ed. Murat Uraz (Istanbul: Neşriyat, 1968), 459.

2. Wolf-Dieter Hütteroth, "Ecology of the Ottoman Lands," in *The Cambridge History of Turkey, Volume 3: The Later Ottoman Empire, 1603–1839*, ed. Suraiya N. Faroqhi (Cambridge: Cambridge University Press, 2006), 18–43.

3. Similar freezing events had been recorded in the middle of the eighth century as well. See Y. Vural, N. Akçar, and C. Schlüchter, "The Frozen Bosphorus and Its Paleoclimatic Implications Based on a Summary of the Historical Data," in *The Black Sea Flood Question: Changes in Coastline, Climate, and Human Settlement*, ed. V. Yanko-Hombach et al. (Dordrecht: Springer Netherlands, 2007), 633–649.

4. The Venetian dispatch (*Archivio di Stato di Venezia, Dispacci—Costantinopoli*, filza 90) dates from February 9, 1621. The other accounts are in Hasan Bey-Zâde Ahmed, *Hasan Bey-Zâde Târîhi*, ed. Şevki Nezihi (Ankara: Türk Tarihi Kurumu Yayınevi, 2004), 928–929, and the "Tuği Tarihi" published in Fahir İz, "XVII. Yüzyılda Halk Dili ile Yazılmış bir Tarih Kitabı: Hüseyin Tuği 'Vak'a-i Sultan Osman Han,'" *Türk Dili Araştırmaları Yıllığı* (1967): 119–155.

5. For various anecdotes of the Little Ice Age, see Brian Fagan, *The Little Ice Age* (New York: Basic Books, 2000). For a review of climate studies and of advances in Little Ice Age climatology, see Rudolf Brázdil et al., "Historical Climatology in Europe—the State of the Art," *Climatic Change* 70 (2005): 363–430. On the Mediterranean in particular, see J. Luterbacher et al., "Mediterranean Climate Variability over the Last Centuries: A Review," in *The Mediterranean Climate: An Overview of the Main Characteristics and Issues*, ed. P. Lionello, P. Malanotte-Rizzoli, and R. Boscolo (Amsterdam: Elsevier, 2006), 27–148. For an overview of the Little Ice Age and its connection with the general crisis of the seventeenth century, see Geoffrey Parker, "Crisis and Catastrophe: The Global Crisis of the Seventeenth Century Reconsidered," *American Historical Review* 113 (2008): 1053–1079.

6. The narrative and arguments here are explained in more detail in the author's monograph: *The Climate of Rebellion in the Early Modern Ottoman Empire* (New York: Cambridge University Press, 2011).

7. On the historiography of decline see, for example, David Howard, "Ottoman Historiography and the Literature of 'Decline' in the Sixteenth and Seventeenth Centuries," *Journal of Asian History* 22 (1988): 52–76; Cemal Kafadar, "The Question of Ottoman Decline," *Harvard Middle Eastern and Islamic Review* 4 (1997–98): 30–75; and Dana Sajdi, "Decline, Its Discontents and Ottoman Cultural History: By Way of Introduction," in *Ottoman Tulips, Ottoman Coffee: Leisure and Lifestyle in the Eighteenth Century*, ed. Dana Sajdi (New York: Tauris Academic Studies, 2007), 1–40.

8. For examples of this historiographical development, see Linda T. Darling, *Revenue-Raising and Legitimacy: Tax Collection and Finance Administration in the Ottoman Empire, 1560–1660* (Leiden: Brill, 1996); Ariel Salzmann, "Measures of Empire: Tax Farmers and the Ottoman *Ancien Régime*, 1695–1807" (Ph.D. diss., Columbia University, 1995); Baki Tezcan, *The Second Ottoman Empire: Political and Social Transformation in the Early Modern World* (New York: Cambridge University Press, 2010).

9. Jack A. Goldstone, *Revolution and Rebellion in the Early Modern World* (Berkeley: University of California Press, 1991); Geoffrey Parker and Lesley M. Smith, eds., *The General Crisis of the Seventeenth Century*, 2nd edn. (London: Routledge, 1997); Parker, "Crisis and Catastrophe."

10. For an overview of events and institutions in the so-called classical age of Ottoman history, see Halil İnalcık, *The Ottoman Empire: The Classical Age*, trans. Norman Itzkowitz and Colin Imber (New York: Praeger Publishers, 1973); Colin Imber, *The Ottoman Empire, 1350–1650: The Structure of Power* (New York: Palgrave Macmillan, 2002).

11. On the depopulation of the empire after the Black Death, see Uli Schamiloglu, "The Rise of the Ottoman Empire: The Black Death in Medieval Anatolia and Its Impact on Turkish Civilization," in *Views from the Edge: Essays in Honor of Richard W. Bulliet*, ed. Neguin Yavari, Lawrence G. Potter, and Jean-Marc Oppenheim (New York: Columbia University Press, 2004), 255–279. For the literature on settlement and landholding policies see, for example, Ömer Lütfi Barkan, *Türkiye'de Toprak Meselesi* (Istanbul: Gözlem Yayınları, 1980); Suraiya Faroqhi, *Towns and Townsmen of Ottoman Anatolia: Trade, Crafts, and Food Production in an Urban Setting, 1520–1650* (New York: Cambridge University Press, 1984); Halil İnalcık, with Donald Quataert, *An Economic and Social History of the Ottoman Empire*, vol. 1 (New York: Cambridge University Press, 1994); Hüseyin Arslan, *16. yy. Osmanlı Toplumunda Yönetim, Nüfus, İskân, Göç ve Sürgün* (Istanbul: Kaknüs Yayınları, 2001).

12. On policies toward nomads in the Ottoman Empire see, for example, Rudi Lindner, *Nomads and Ottomans in Medieval Anatolia* (Bloomington: Indiana University Press, 1983); Reşat Kasaba, *A Moveable Empire: Ottoman Nomads, Migrants, and Refugees* (Seattle: University of Washington Press, 2009).

13. Compare, for example, J. R. McNeill, "China's Environmental History in World Perspective," in *Sediments of Time: Environment and Society in Chinese History*, ed. Mark Elvin and Liu Ts'ui-jung (New York: Cambridge University Press, 1998), 31–52. On the environmental history of other early modern empires, see John F. Richards, *The Unending Frontier: An Environmental History of the Early Modern World* (Berkeley: University of California Press, 2003).

14. For various aspects of Ottoman provisioning see, for example, Faroqhi, *Towns and Townsmen*; Gábor Ágoston, *Guns for the Sultan: Technology, Industry, and Military Power in the Ottoman Empire* (New York: Cambridge University Press, 2004); Rhoads Murphey, "Provisioning Istanbul: The State and Subsistence in the Early Modern Middle East," *Food and Foodways* 2 (1988): 217–263; Lütfi Güçer, *Osmanlı İmparatorluğunda Hububat Meselesi ve Hububattan Alınan Vergiler* (Istanbul: İstanbul Üniversitesi İktisat Fakültesi, 1964).

15. On the Ottoman military in this period see, for example, Rhoads Murphey, *Ottoman Warfare, 1500–1700* (New Brunswick, NJ: Rutgers University Press, 1999).

16. On the high consumption requirement of the city, see especially Robert Mantran, *Istanbul dans la seconde moitié du xviie siècle* (Paris: Maisonneuve, 1962), part II, chapter 1.

17. Demographic figures have been derived from cadastral surveys known as *tahrir defterleri*. For the original work in this field, see Ömer Lütfi Barkan, "Essai sur les données statistiques des registres de recensement dans l'Empire ottoman aux XVe et XVIe siècles," *Journal of the Economic and Social History of the Orient* 1 (1958): 9–36. For an overview of more recent research, see Erhan Afyoncu, "Türkiye'de Tahrir

Defterlerine Dayalı Olarak Hazırlanmış Çalışmalar Hakkında Bazı Görüşler," *Türkiye Araştırmaları Literatür Dergisi* 1 (2003): 267–286.

18. For studies of agricultural output and population levels in the Ottoman Empire, see Güçer, *Osmanlı İmparatorluğunda Hububat Meselesi*; Michael Cook, *Population Pressure in Rural Anatolia, 1450–1600* (New York: Oxford University Press, 1972); Suraiya Faroqhi and Huri İslamoğlu, "Crop Patterns and Agricultural Production Trends in Sixteenth-Century Anatolia," *Review* 2 (1979): 401–436; Margaret Venzke, "The Question of Declining Cereals' Production in the Sixteenth Century: A Sounding on the Problem-Solving Capacity of the Ottoman Cadastres," *Journal of Turkish Studies* 8 (1984): 251–264; Huri İslamoğlu-İnan, *State and Peasant in the Ottoman Empire: Agrarian Power Relations and Regional Economic Development in Ottoman Anatolia during the Sixteenth Century* (Leiden: Brill, 1994); and especially Oktay Özel, "Population Changes in Ottoman Anatolia during the 16th and 17th Centuries: The 'Demographic Crisis' Reconsidered," *International Journal of Middle East Studies* 36 (2004): 183–205. On the Karaman region of south-central Anatolia in particular, see Osman Gümüşçü, *Tarihî Coğrafya Açısından Bir Araştırma: XVI. Yüzyıl Larende (Karaman) Kazasında Yerleşme ve Nüfus* (Ankara: Türk Tarihi Kurumu Basımevi, 2001).

19. On the "price revolution" and monetary history see, for example, Ömer Lütfi Barkan, "The Price Revolution of the Sixteenth Century: A Turning Point in the Economic History of the near East," *International Journal of Middle East Studies* 6 (1975): 3–28; Şevket Pamuk, "The Price Revolution in the Ottoman Empire Reconsidered," *International Journal of Middle East Studies* 33 (2001): 69–89. The original studies of monetary instability and unrest are: Mustafa Akdağ, *Türkiye'nin İktisadi ve İçtimai Tarihi* (Ankara: Türk Tarih Kurumu Basımevi, 1971); idem., *Türk Halkının Dirlik ve Düzenlik Kavgası* (Ankara: Bilgi Yayınevi, 1975).

20. These developments, based on a correlation of events described in imperial orders (*mühimme defterleri*) and climatology data, are covered in more detail in White, *Climate of Rebellion*. For an earlier account of rising unrest in the period, see: Mustafa Akdağ, *Celâlî İsyanları (1550–1603)* (Ankara: Ankara Üniversitesi Basımevi, 1963). For an overview of late sixteenth-century famines, see: Orhan Kılıç, "Osmanlı Devleti'nde Meydana Gelen Kıtlıklar," *Türkler* 10 (2002): 718–730.

21. Brázdil, "Historical Climatology in Europe"; Luterbacher et al., "Mediterranean Climate Variability." On volcanic activity in the period and its global impacts see, for example, H. H. Lamb, "Volcanic Dust in the Atmosphere; with a Chronology and Assessment of Its Meteorological Significance," *Philosophical Transactions of the Royal Society of London (Series A., Mathematical and Physical Sciences)* 266 (1970): 425–533; William S. Atwell, "Volcanism and Short-Term Climatic Change in East Asian and World History c.1200–1699," *Journal of World History* 12 (2001): 29–98. On the varying regional impacts of volcanic weather see, for example, Drew T. Shindell et al., "Dynamic Winter Climate Response to Large Tropical Volcanic Eruptions since 1600," *Journal of Geophysical Research* 109 (2004): D05104.

22. On the Nile floods see, for example, William Popper, *The Cairo Nilometer: Studies in Ibn Taghrî Birdî's Chronicles of Egypt, I* (Berkeley: University of California Press, 1951); Peter Whetton and Ian Rutherford, "Historical ENSO Teleconnections

in the Eastern Hemisphere," *Climatic Change* 28 (1994): 221–253; Fekri Hassan, "Historical Nile Floods and Their Implications for Climatic Change," *Science* 212 (1981): 1142–1145.

23. For the original work on tree rings in Ottoman lands, see Peter Kunniholm, "Archaeological Evidence and Non-Evidence for Climate Change," *Philosophical Transactions of the Royal Society of London* 330 (1990): 645–655. Kunniholm's work also inspired the following earlier short article suggesting that climate and crisis may have been linked: William Griswold, "Climatic Change: A Possible Factor in the Social Unrest of Seventeenth Century Anatolia," in *Humanist and Scholar: Essays in Honor of Andreas Tietze*, ed. Heath W. Lowry and Donald Quataert (Istanbul: Isis Press, 1993), 37–57. Among several more recent dendroclimatology studies in the region, see Ramzi Touchan et al., "Standardized Precipitation Index Reconstructed from Turkish Tree-Ring Widths," *Climatic Change* 72 (2005): 339–353; Ünal Akkemik et al., "Tree-Ring Reconstructions of Precipitation and Streamflow for Northwestern Turkey," *International Journal of Climatology* 28 (2008): 173–183. Further regional tree-ring data is available at http://www.ncdc.noaa.gov/paleo/recons.html (accessed May 21, 2011).

24. Dispatches from the Venetian ambassador in Istanbul make frequent references to growing drought and famine. See, for example, A.S.V. *Dispacci-Costantinopoli* 43 (May 19, 1596). Numerous contemporary Ottoman chroniclers also mention the cold and famine, the most detailed and dramatic description being Mustafa Selaniki, *Tarih-i Selânikî*, ed. Mehmet İpşirli (Ankara: Türk Tarih Kurumu Basımevi, 1999), 600 and 624–635 *passim*.

25. On the logistics of the war, see Caroline Finkel, *The Administration of Warfare: The Ottoman Military Campaigns in Hungary 1593–1606* (Vienna: VWGÖ, 1988). On the wider shift in tactics and rising costs of war in the period, see Geoffrey Parker, *The Military Revolution: Military Innovation and the Rise of the West, 1500–1800* (New York: Cambridge University Press, 1996).

26. Selaniki, *Tarih-i Selânikî*, 597–598. Weather conditions and the state of the Ottoman army are also described in contemporary Venetian dispatches and in Ottoman chronicles such as the following: 'Abdülkâdir Efendi, *Topçular Kâtibi 'Abdülkâdir Efendi Tarihi*, ed. Ziya Yılmazer (Ankara: Türk Tarih Kurumu Yayınları, 2003).

27. See, for example, *Mühimme Defterleri* (hereafter MD) 71/489; MD 71/552; MD 71/413; and MD 71/440. This search for supplies occupies hundreds of imperial orders in the Ottoman archives' *Mühimme Defterleri* ("Registers of Important Affairs"), vols. 71–74. It remains unclear whether these different terms indicate precise degrees of shortages or merely the desperation of imperial authorities. They are, in any case, not often found in earlier *mühimme* registers.

28. Cook, *Population Pressure*, 40. For more on the violence of the 1590s, see: Güçer, *Celali İsyanları*.

29. See, for example, Selaniki, *Tarih-i Selânikî*, 624.

30. The most detailed description can be found in the following: Gelibolu Mustafa Âlî, *Künhü'l-Ahbâr*, ed. Faris Çerçi (Kayseri: Erciyes Üniversitesi Yayınları, 2000), 675–677. This account can be confirmed by archival documents

(such as MD 72/6) and European descriptions, including the following: *A.S.V. Dispacci—Costantinopoli* 47 (June 13, 1598).

31. For more on the breakdown of the sheep supply system at this time, see Antony Greenwood, "Istanbul's Meat Provisioning: A Study of the *Celepkeşan* System" (Ph.D. diss., University of Chicago, 1988).

32. MD 73/964. For comparisons of past requisitions, see Greenwood, "Istanbul's Meat Provisioning." Other imperial orders suggest that the sultan may have blamed central Anatolian officials for diverting previous sheep deliveries from further east, perhaps explaining the decision to single out Karaman in this case.

33. Gümüşçü, *Tarihî Coğrafya Açısından Bir Araştırma.*

34. For descriptions of the requisition and resistance, see Selaniki, *Tarih-i Selânikî*, 581; Akdağ, *Celali İsyanları*, 172–173.

35. Halil İnalcık, "The Socio-Political Effects of the Diffusion of Firearms in the Middle East," in *War, Technology and Society in the Middle East*, ed. V. J. Parry and M. E. Yapp (New York: Oxford University Press, 1975); Karen Barkey, *Bandits and Bureaucrats: The Ottoman Route to State Centralization* (Ithaca, NY: Cornell University Press, 1994).

36. This problem was most clearly seen during the Persian campaigns of 1583–1584. Perhaps dozens of orders in MD, vol. 44 cite the absence of soldiers as a cause for banditry and unrest. See, for example, MD 44/356. Small unspecified bandit gangs and groups of raiders on horseback continued to make up the largest number of bandits throughout the 1590s, and relatively few orders mention large numbers of *sekban* until the seventeenth century.

37. For more details, see White, *Climate of Rebellion*. The important role of local mercenary armies is also explained in the following: Baki Tezcan, "Searching for Osman: A Reassessment of the Deposition of the Ottoman Sultan Osman II (1618–1622)" (Ph.D. diss., Princeton University, 2001), 205–215 *et passim*.

38. The main narrative descriptions can be found in Selaniki, *Tarih-i Selânikî*, 581, 751 *et passim* and in the contemporary Venetian dispatches, including *A.S.V. Dispacci—Costantinopoli* 49 (August 7, 1599).

39. For a narrative of the campaigns, see William J. Griswold, *The Great Anatolian Rebellion, 1000–1020/1591–1611* (Berlin: Klaus Schwarz Verlag, 1983).

40. There are again dozens of reports about the freezing weather in contemporary chronicles as well as in various Ottoman official correspondence and Venetian dispatches. From the latter see, for example, *A.S.V. Dispacci—Costantinopoli* 46 (February 17, 1598). For the original study of the "great flight" see Mustafa Akdağ, "Celâli İsyanlarından Büyük Kaçgunluk," *Tarih Araştırmaları Dergisi* 2 (1964): 1–49.

41. The most detailed descriptions can be found in 'Abdülkâdir Efendi, *Topçular Katibi Tarihi*, 319–321 *et passim* and in the chronicles of İbrahim Peçevi and Hasan Bey-Zâde. Hungarian accounts have left similar descriptions. See, for example, Lajos Racz, "Variations of Climate in Hungary, 1540–1779," in *European Climate Reconstructed from Documentary Data: Methods and Results*, ed. Burkhard Frenzel, Christian Pfister, and Birgit Gläser (New York: G. Fisher, 1992), 125–136.

42. Griswold, *Great Anatolian Rebellion*, 182–191.

43. See particularly the eyewitness description in the following: George A. Bournoutian, trans., *The History of Vardapet Aṛak'el of Tabriz* (Costa Mesa: Mazda Publishers, 2005), 65–75.

44. See for example: MD 8z/479; MD 8z/434; MD 8z/438; MD 8z/442; MD 8z/73; MD 8z/286.

45. These incidents are narrated in more detail in White, *Climate of Rebellion.*

46. See MD, vols. 80–90. The volumes cover various years up to the late 1640s, after which time the *mühimmes* become increasingly less useful as a source for major events.

47. On population movements during the crisis, see especially Özel, "Population Changes in Ottoman Anatolia during the 16th and 17th Centuries"; Leila Erder and Suraiya Faroqhi, "Population Rise and Fall in Anatolia 1550–1620," *Middle East Studies* 15 (1979): 322–345.

48. Jean M. Grove and Annalisa Conterio, "Climate in the Eastern and Central Mediterranean, 1675 to 1715," in *Climatic Trends and Anomalies in Europe 1675–1715: High Resolution Spatio-Temporal Reconstructions from Direct Meteorological Observations and Proxy Data, Methods and Results*, ed. Burkhard Frenzel, Christian Pfister, and Birgit Gläser (New York: G. Fischer, 1994), 280. See also J. Luterbacher et al. "The Late Maunder Minimum—A Key Period for Studying Decadal Climate Change in Europe," *Climatic Change* 49 (2001): 441–462; E. Xoplaki et al. "Variability of Climate in Merdional Balkans during the Periods 1675–1715 and 1780–1830 and Its Impact on Human Life," *Climatic Change* 48 (2001): 581–615.

49. For a description of the famine, see Silahdar Fındıklı Mehmed Ağa, *Silahdar Tarihi*, vol. 2 (Istanbul: Devlet Matbaası, 1928), 243. The extent and depth of the drought are amply confirmed by dendroclimatology. Touchan et al., "Standardized Precipitation Index Reconstructed."

50. J. V. Thirgood, *Man and the Mediterranean Forest: A History of Resource Depletion* (London: Academic Press, 1981). See also J. Donald Hughes, *Ecology in Ancient Civilizations* (Albuquerque: University of New Mexico Press, 1975).

51. See, for example, A. T. Grove and Oliver Rackham, *The Nature of Mediterranean Europe: An Ecological History* (New Haven, CT: Yale University Press, 2001); Diana K. Davis, *Resurrecting the Granary of Rome: Environmental History and French Colonial Expansion in North Africa* (Athens: Ohio University Press, 2007).

52. In this regard, see especially Harvey Weiss, "Beyond the Younger Dryas: Collapse as Adaptation to Abrupt Climate Change in Ancient West Asia and the Ancient Eastern Mediterranean," in *Environmental Disasters and the Archaeology of Human Response*, ed. Garth Bawden and Richard Martin Reycraft (Albuquerque, NM: Maxwell Museum of Anthropology, 2000), 75–98.

53. See, for example, Barry Weiss, "The Decline of Late Bronze Age Civilization as a Possible Response to Climatic Change," *Climatic Change* 4 (1982): 173–198; Joel D. Gunn, ed., *The Years without Summer: Tracing AD 536 and Its Aftermath* (Oxford: Archaeopress, 2000).

54. See, for example, Peter Christensen, *The Decline of Iranshahr: Irrigation and Environments in the History of the Middle East, 500 B.C. to A.D. 1500*

(Copenhagen: Museum Tusculanum Press, 1993); Stuart J. Borsch, *The Black Death in Egypt and England: A Comparative Study* (Austin: University of Texas Press, 2005).

55. See, for example, Arlene M. Rosen, *Civilizing Climate: Social Responses to Climate Change in the Ancient Near East* (Lanham, MD: Altamira Press, 2007); Weiss, "Beyond the Younger Dryas."

56. Christensen, *Decline of Iranshahr*, 104.

57. In this regard, see Richard W. Bulliet, *Cotton, Climate, and Camels in Early Islamic Iran: A Moment in World History* (New York: Columbia University Press, 2009).

58. See, for example, Eliyahu Ashtor, "The Economic Decline of the Middle East During the Later Middle Ages: An Outline," *Asian and African Studies* 15 (1981): 253–286; Richard W. Bulliet, *Islam: The View from the Edge* (New York: Columbia University Press, 1994), 67–79.

59. Lawrence I. Conrad, "The Plague in the Early Medieval Near East" (Ph.D. diss., Princeton University, 1981); Michael W. Dols, *The Black Death in the Middle East* (Princeton, NJ: Princeton University Press, 1977); Daniel Panzac, *La peste dans l'Empire Ottoman, 1700–1850* (Leuven: Editions Peeters, 1985).

60. For evidence of drought on the eastern edge of the empire, see especially Ramzi Touchan et al., "A 396-Year Reconstruction of Precipitation in Southern Jordan," *Journal of the American Water Resources Association* 35 (1999): 49–59. On Persian resettlement policies, see John Perry, "Forced Migration in Iran During the Seventeenth and Eighteenth Centuries," *Iranian Studies* 8 (1975): 199–215.

61. See, for example, MD 80/259. Many of the key documents on this subject have also been reproduced in Ahmet Refik, *Anadolu'da Türk Aşiretleri (966–1200)*, 2nd edn. (Istanbul: Enderun Kitabevi, 1989).

62. For various European descriptions of Ottoman nomads and the reversion of land to pastoralism see, for example, Xavier Planhol, "Les nomades, la steppe, et la foret en Anatolie," *Geographische Zeitschrift* 52 (1965): 101–116. For patterns of land use and settlement change in the seventeenth century, see especially Wolf-Dieter Hütteroth, *Laendliche Siedlungen im südlichen Inneranatolien in den Letzen vierhundert Jahren* (Göttingen: Göttingen Universität Geographischen Institut, 1968); Wolf-Dieter Hütteroth and Kamal Abdulfattah, *Historical Geography of Palestine, Transjordan and Southern Syria in the Late Sixteenth Century* (Erlangen: Fränkische Geographische Ges., 1977).

63. On subsequent resettlement efforts, see especially Cengiz Orhonlu, *Osmanlı İmparatorluğunda Aşiretleri İskân Teşebbüsü, 1691–96* (Istanbul: Edebiyat Fakültesi Basımevi, 1963). For evidence of drought during key periods of resettlement activity, see Touchan et al., "A 396-Year Reconstruction of Precipitation in Southern Jordan"; R. Touchan and M. K. Hughes, "Dendrochronology in Jordan," *Journal of Arid Environments* 42 (1999): 291–303.

64. For a contemporary Ottoman description see, for example, Abdülkadir Özcan, ed., *Zübde-i Vekayiât* (Ankara: Türk Tarih Kurumu Basımevi, 1995), 512–513. This source also describes the introduction of lifetime tax-farming meant to remedy the situation.

65. On the Columbian Exchange and other shifts in Ottoman land use in the eighteenth century, see especially Faruk Tabak, *The Waning of the Mediterranean, 1550–1870: A Geohistorical Approach* (Baltimore: Johns Hopkins University Press, 2008).

66. For studies of Ottoman cities in the seventeenth and eighteenth centuries, see Mantran, *Istanbul*; Suraiya Faroqhi, *Men of Modest Substance: House Owners and House Property in Seventeenth-Century Ankara and Kayseri* (New York: Cambridge University Press, 1987); André Raymond, *Grandes villes arabes à l'époque ottomane* (Paris: Sindbad, 1985); Haim Gerber, *Economy and Society in an Ottoman City: Bursa 1600–1700* (Jerusalem: Hebrew University of Jerusalem, 1988); Abraham Marcus, *The Middle East on the Eve of Modernity: Aleppo in the Eighteenth Century* (New York: Columbia University Press, 1989).

67. For new views on Ottoman disease and the role of environmental factors, see Sam White, "Rethinking Disease in Ottoman History," *International Journal of Middle East Studies* 42 (2010): 549–567.

4

Fish and Fishermen in Ottoman Istanbul

Suraiya Faroqhi

Given the volume and variety of fish traveling every year between the Black Sea and the Sea of Marmara, it is no surprise that they have long played a role in the food regime of Istanbul's inhabitants. After all, during the period under discussion, the middle of the sixteenth century until the First World War and its immediate aftermath, the city consistently had a large number of Orthodox residents who, for religious reasons, abstained from meat for a large part of the year. The wealthier members of the community consumed fish instead.[1] And, needless to say, Istanbul's Muslim population was not averse to good-quality fish.

Because fish figured prominently in the diets and livelihoods of people in Istanbul, this chapter explores fish, fishing grounds, and fishermen in and around early modern Istanbul. The information contained in sources from the sixteenth and seventeenth centuries will be interpreted in the light of more comprehensive data collected in the years preceding the First World War. Did the methods of fishing change? How did fishermen go about their work, and what kinds of fish did they catch? Which fish were perhaps consumed locally, right after the nets arrived on the shores of the Bosporus and the Sea of Marmara, and which were carried to Istanbul's markets? To what extent did fish play a role in palace provisioning? This chapter provides partial answers to at least some of these questions.

However, given limited information on the population of Ottoman Istanbul and on that of the total catch as well, it is not possible to discuss the environmental impact of early modern fishing on the aquatic life of the Bosporus and the Sea of Marmara. When approaching the data from the labor historian's point of view, the researcher also comes up against major hurdles. The number of fishermen engaged in the trade in the early modern period remains unknown; it is hazardous to give much credence, for example, to the numbers provided by the seventeenth-century Ottoman travel writer Evliya Çelebi, as his figures are notoriously unreliable.[2] Given ignorance concerning the number of Istanbul's residents as a whole, the size of the Orthodox population remains unknown as well. Nor is anything known about the latter's relative propensity to consume fish. Therefore, this chapter does not go beyond questions of fishing techniques and fish consumption. Toward the end of this discussion, however, changes in these matters will become visible. Such changes were partly responses to the increase in Istanbul's population during the nineteenth century.

A comparison between the techniques employed in the sixteenth and seventeenth centuries with those in use during the First World War and its immediate aftermath shows that fishermen refined "traditional" technologies so as to increase their catch; by the early twentieth century, they may well have reached the limits of what could be achieved by these means. Nineteenth- and twentieth-century technologies imported from Europe, including canneries, only entered the picture in a significant way after our period of study, with the founding of the Republic of Turkey in 1923.

Sixteenth-Century Fishing Techniques as Observed by Pierre Belon du Mans

The earliest Ottoman records about fish concern taxes paid by fishermen plying their trade in the Sea of Marmara or in lakes near the coast. These are dated 1530 in the abridged tax register (icmal) of the province of Anadolu (western, northwestern, and southwestern Anatolia), which contained Koca-ili, among other sub-provinces. At this time Koca-ili stretched all the way to the Bosporus and even encompassed Üsküdar, an integral part of the Ottoman capital since the late 1500s.[3] A small fishing weir existed in the region of Gebze, today a satellite town of Istanbul, which paid probably modest dues to the mosque of Sultan Bayezid I in Bursa. Another favored fishing ground was located in the district of İznik, where the area's fishermen may have been active on the nearby lake or the Sea of Marmara. Some of these fishing grounds were productive enough to allow the fishermen to set up nets on poles (dalyan) and have a watchman keep an eye on them; similar installations existed on the Bosporus as well.

In his travelogue, the French naturalist Pierre Belon (1517–1564), who visited Istanbul in 1548, provides detailed information on how these weirs operated.[4] Belon arrived in Ottoman lands from France as part of the embassy of King François I (r. 1515–1547) to Sultan Süleyman (r. 1520–1566); after François's death, his successor Henri II (r. 1547–1559) sent a second embassy to Istanbul. Belon, who in the meantime had remained in Ottoman territory, accompanied the new envoy on some of his travels around the empire before returning to France alone in 1549.[5] In the 1550s, he gained a reputation for his books on fish and plants in which he made extensive use of the documentation he collected while in the Ottoman Empire. His attempts to assert the priority of direct observation, even if contradicting the claims of revered authors of antiquity, ensured him a lasting reputation.[6]

Belon has provided unique information about fish in Ottoman lands and the techniques developed by the fishermen of the Bosporus and the Sea of Marmara. He noted that the locals knew very well which kinds of fish passed through these waters at various times of the year, and he described weirs that were similar if not identical to those appearing in the Ottoman tax register of 1530. He saw two solid poles, about fifty paces apart, which the fishermen had set into the bottom of the sea in a place where the water was shallow. On top of these poles perched two little "seats" with roofs over them in which one or two people could find shelter: these structures provided the fishermen with rudimentary protection against sun and rain. Access was by means of a ladderlike contraption made by affixing small sticks or planks at right angles to the poles. From these lookouts the two fishermen watched for approaching shoals. Normally these weirs were used only in summer; it was difficult to observe fish and handle the nets when the sea was agitated by the north winds blowing over Istanbul during other times of the year.

The net used in this kind of fishing was square in shape, and the two corners not controlled by the fishermen were fixed to two further poles also rammed into the bottom of the sea. Evidently the latter were much farther away from the shore than those tenanted by the two fishermen and so short as to be barely visible on the surface of the sea. As a result, the edge of the net which was attached to the "untenanted" poles and ran parallel to the shore was immobile, while the opposite side could be moved by the two people on the lookouts pulling a set of cords attached to the net. Capturing a significant number of fish therefore depended on the skill of the fishermen to raise the net effectively and at the proper time. Once a shoal entered between the poles, the fishermen pulled ropes to raise the net—it had to be large enough to descend to a considerable depth—and then proceeded to haul it up, hopefully with most of the catch still inside. It fell to one of the men to tie his corner of the net securely to the pole before descending; then he got into a small boat and started pulling up the net so that, in the end, the fish accumulated in a single corner and could be collected in the fishing vessel.

Belon also reported that he accompanied fishermen from Pera (today's Beyoğlu) when they fished from boats, traveling as far as the Princes' Islands (Adalar) and even Mudanya, the port of Bursa. While sailing or rowing, the boats operated in pairs and the crews of any pair of vessels needed to work together very closely, the men on both boats controlling their net with the help of cordage. It was important to end up on a smooth beach rather than a rocky coast; otherwise, the fishermen risked ruining their nets. This mode of fishing also demanded a good deal of coordination. In order to retain as many fish as possible, it was necessary to pull the cords evenly and in tandem once the catch was in the net. When the boats approached the beach, the fishermen disembarked into the shallow water, checking the knots in their ropes to ensure that they pulled the net in evenly.

A less sophisticated manner of fishing involved simply trailing a net behind a single boat. In such cases the fishermen of the Sea of Marmara did not employ corks to keep the net from sinking too low, as was customary in the western Mediterranean, but instead made use of pine cones readily available in the area. A more complex variant involved the use of two nets, one on top of the other. This technique sought to ensure that large numbers of fish would get tangled up in the nets and, thus immobilized, be easier to haul into the boat.

Pierre Belon observed yet another mode of fishing that, according to the information he received, was only practiced on the European side of the Bosporus as far north as Beşiktaş. This technique involved the use of a net attached to two hoops and was supposedly peculiar to the "Spanish" prisoners of war that the sultan had liberated and encouraged to settle near the Bosporus. Many had constructed houses on piles in the water, perhaps the ancestors of the eighteenth-century seaside villas known as *yalıs*. Belon also observed fishermen who used torches made of pinewood to attract fish by night. Two people cooperated in this enterprise, one rowing a boat and the other supervising the torch. Belon did not offer any further details, but presumably the man holding the torch dropped it into the sea when the fish approached and then harpooned them, if the two fishermen did not already have a net ready to hold their catch. Belon recorded that this method was very profitable, and in fact fishing with the aid of lights has contributed greatly to the decrease of fish in the Mediterranean.

As for harpooning techniques, the French traveler described a trident with five or six points held by one of a pair of fishermen, who aimed for fish sleeping or resting in shallow water. When fishing in this manner it was essential to avoid any noise, for the fish were quick to take flight. The man holding the trident thus used sign language to communicate with his partner rowing the boat. These fishermen did best on a dark night when there was little wind and the waters were calm. They typically aimed to harpoon the spine of the fish, and their tridents were also equipped with barbs, which allowed them to lift their catch and throw it into the fishing vessel.

Evliya Çelebi's Seventeenth-Century Observations

The next author to provide a significant mass of documentation concerning the fishermen of Istanbul and their methods is Evliya Çelebi (1611–after 1683). As the son of a palace goldsmith and a former page to Sultan Murad IV (r. 1623–1640), Evliya was a member of the palace elite and thus should have been eligible for high office, particularly since on his mother's side he was a relative of Melek Ahmed Paşa, grand vizier in 1650 and 1651. However, Evliya preferred a lifetime of travel that began with an exploration of Istanbul, the subject of the first book of his ten-volume travelogue. Presumably he collected most of his information about fishing in Istanbul during his time in the city in the middle years of the seventeenth century, but it is also possible that he added information during his many stays in the Ottoman capital that punctuated his travels.

In his detailed description of Istanbul and its surroundings, Evliya had a good deal to say about the activities of local fishermen; in particular, he observed quite a few weirs (*dalyans*).[7] When describing the village of Tarabya on the European shore of the northern Bosporus, he commented that this place had been nothing but an agglomeration of fishing weirs, before Selim II (r. 1566–1574) found Tarabya to his liking as a picnic site where he could enjoy freshly caught fish. This sultan thus supposedly ordered a settlement to be established there and decided on its name.[8]

Evliya records that throughout Istanbul and its vicinity, three hundred fishing weirs were in operation, manned by seven hundred fishermen. While his figures must always be regarded with some caution, he notes that apparently five people could operate two weirs, a figure close enough to Belon's observation a century earlier of two fishermen operating one weir. Such installations existed along both the European and the Asian shores, but the most important site was in Anatolian Beykoz on the northern shore of the Bosporus, where sizable fishing weirs in three locations allowed fishermen to capture swordfish. These animals were so large that they had to be hunted individually. Evliya mentions nets fastened in the water by a mechanism that he does not specify. A man seated on a pole "as high as three ships' masts" was in charge of the whole operation. When he saw a fish he threw a stone into the water. In Evliya's words, "fearful of the net, the fish fell into the trap." When this happened, the observing fisherman shouted "ala!" and his colleagues hauled in their nets.[9] Presumably, these men moved around in boats. When they finally reached the fish they beat it to death with sticks—a scene reminiscent of twentieth-century Sicilian *matanzas*—and then sent their catch to Istanbul.

From a slightly later period, the years between 1673 and 1684, the Armenian scholar and Istanbul resident Eremya Çelebi gave a similar account. He recorded that along both sides of the landing stage of Beykoz, a number of nets were in evidence, as the site was full of fishing weirs. Swordfish, he noted, were a Beykoz specialty that the fishermen sent to the administrator of the grain distribution center in downtown Istanbul (Unkapanı). Unfortunately, Eremya Çelebi did not comment on the connection, if any, between the swordfish of Beykoz and the grain trade.[10] Perhaps there was no connection; but rather, in his time, the man who supervised fish wholesaling and collected the relevant dues (a man whose title was *balık emini*) may well have been operating somewhere in this section of the city with its busy markets.

Other fishermen worked in the Golden Horn, the maritime inlet that forms the northern border of *intra-muros* Istanbul. Evliya noted that from Seraglio Point (Sarayburnu) onward on both sides of the Golden Horn there protruded long poles from certain houses that might have served as the yardarms of ships, to which the owners—who were all Greeks—had fastened nets secured by rectangular frames. These nets, of which there were supposedly a hundred and fifty, caught a large amount of fish. Evliya called these contraptions *karitya* and the people operating the nets *karityacı* (pl. *karityacıyan*). He added that ten of these *karitya* did not pay any dues to the *balık emini*.

According to Evliya, this unique privilege was due to the fact that in 1453, during the Ottoman siege of Constantinople, certain Greeks had opened the gate known as Petrekapusu to the invaders and gained this tax exemption as their reward. However, in exchange they had to supply the palace with seals (*ayı balığı*), which they hunted near a group of islands that Evliya called the Kızıl Adalar and which seem to correspond to today's Princes' Islands (Adalar).[11] It is unclear why the Ottoman palace wanted seals, perhaps for their fur. In the early twentieth century, fishermen still hunted seals near Istanbul because these animals were effective fishers in their own right and thus unwelcome competitors. Apparently they also damaged fishing weirs.[12]

In any case, the historical component of this story about tax exemptions may well have been an "invented tradition" dating from the late sixteenth or early seventeenth century. Indeed, similar tales were on record to explain why certain Byzantine churches remained in the hands of the Orthodox, even though in a city taken by storm—as was Byzantine Constantinople—the inhabitants should have lost their places of worship according to Islamic religious law.[13] Evliya's story seems even more suspicious since there is no mention in the work of Pierre Belon of the *karityas* or of the exemption enjoyed by the ten fishermen. If the *karityas* and the story surrounding them had been current in Belon's time, it is likely that this traveler, given his interest in fishing technology and his acquaintance with Pera's fishermen, would have recorded the relevant information.

Apart from the duty that most *karityacıs* paid to the *balık emini*, these people all were subject to the jurisdiction of the sultan's head gardener (*bostancıbaşı*). This man, who also exercised extensive police powers, demanded a gold piece for every pole rammed into the sea. Perhaps the ten *karityacıs* were exempt because their nets were fastened to their houses and therefore did not involve poles being submerged in the sea. Evliya does not mention whether the *bostancıbaşı* demanded his dues every year or whether the gold piece was a one-off payment whenever fishermen built new weirs.

Perhaps the demands of the *bostancıbaşı* encouraged certain fishermen to fish with nets that were not fastened to any pole but were thrown into the sea either from boats or even from the shore. Evliya noted that the so-called *sıpkıncı* used javelins to capture large fish such as cholio, sea bass, and bonito. Furthermore, he observed a variety of other techniques including the use of large nets (*ığrıb*) that fishermen dragged through the water. When they learned of passing shoals, they went to the appropriate places, whether by boat or on foot Evliya does not say. Clearly transportation must have been a problem, for the Ottoman observer claimed that "hundreds" of men, a figure surely to be taken with a grain of salt, were required to pull the loaded nets ashore. At times the catch purportedly was so abundant that the net was impossible to lift and a diver had to be dispatched into the water to cut it to release some of the fish, thereby lightening the load.

Other fishermen, called *düzenci*, fished from boats or barges called *çırnık*.[14] Their enterprise must have been very modest, as six hundred barges supposedly served as the workplaces of a thousand fishermen; some barges therefore must have been worked by a single person and others by two. These fishermen—or anglers, to be precise—specialized in catching Mediterranean horse mackerel (*istavrid*), picarel (*izmarid*), and black goby (*kaya balığı*). Evliya mentions but says nothing specific about the technique used by another class of fishermen—the *ağcıyan* or fishermen-with-nets. Quite separately, he also admired the skill of the *saçmacı* fishermen who may well have been among the poorest of Istanbul's fishing community, as they threw out their nets from the shore, using neither boats nor *dalyans*. Yet Evliya makes clear that they were skillful at their work and managed to catch quite a lot of fish.

Other "specialty" fishermen were out for shellfish including oysters (*istridye*) and mussels. Unlike their colleagues, these men seem to have combined the roles of both fisherman and merchant, since they possessed both ships and shops. To loosen shellfish from the rocks to which they were attached, these men employed iron combs. Yet other specialist fishers used pots and baskets to catch particular kinds of sea creatures. Three hundred men, for example, apparently employed earthenware pots to catch black goby (*kaya balığı*). More remarkable were the baskets with which "two hundred" fishermen traveled up and down the Bosporus in search of places where the water was calm. In these locations, the men set out their reed baskets with bread in them. Once the desired sea creatures had been lured into the baskets, they were unable to get out. This arrangement served mainly for the capture of crabs, eels, and lobster (*ıstakoz*). Lobster in particular was reputed to be a "strong" food for men with an active libido. According to Evliya, the less than genteel reputation of this creature did not, however, prevent Istanbul salesmen from showing off their lobsters in official processions.

Early Twentieth-Century Perspectives on Belon and Evliya

When evaluating the sometimes enigmatic information conveyed by Pierre Belon and Evliya Çelebi, we need what might be called a "third opinion." Unfortunately, after the middle of the seventeenth century, few authors seem to have shown a close interest in fish and fisheries, and so the only viable standard of comparison is a study by Karekin Deveciyan (1867?–1964) published in 1915.[15] This author has produced the most comprehensive treatment of fish and fishing in and around the Ottoman capital, including not only detailed information but also a multitude of sketches. Deveciyan wrote his book with practical purposes in mind. The first was to help the owners of spaces suitable for weirs (*voli*) to construct the installations (*dalyan*) best adapted to their sites. Second, administrators collecting taxes at the Istanbul fish market could benefit from detailed information about local fisheries. After all, Deveciyan first published his book during the First World War, when food was at a premium in the city.

Deveciyan was in a superb position for collecting the relevant information, as he was a director and later a controller of the Istanbul Balıkhane, where the wholesaling of fish took place and sales dues were payable. Deveciyan spent much of his working life in the service of the Dette ottomane

(*Duyun-ı Umumiye*), a consortium of foreign bondholders that took control of the taxes on fish, in addition to many other sources of revenue, after the Ottoman state bankruptcy of 1875. When Deveciyan published the French version of his formidable study in 1926, his work for the Dette ottomane was coming to a close, since the functions of the foreign-controlled debt administration were steadily contracting with the establishment of the Republic of Turkey and the distribution of the defunct empire's debts among numerous "successor states."[16]

Although more than 350 years separate Belon and Deveciyan, and Evliya wrote more than 250 years before the early twentieth century, it is of interest to compare the information conveyed by these three authors to better understand the numerous and far-reaching changes that affected Istanbul's fisheries. The different sketches of *dalyans* that appear in Deveciyan's work are vastly more complicated than those described by either Belon or Evliya. Deveciyan's work does not show any installations closely resembling the "word-pictures" drawn by the earlier authors; most similar to the early modern models are the weirs that Deveciyan calls *çekme dalyan* and *çökertme*.[17] The former featured the shelters for fishermen that Belon had mentioned and also an area covered by a square net supported by poles that hung at some depth, roughly parallel to the bottom of the sea. At the back of the *dalyan*, as far away as possible from the entrance, there was an open area that fishermen called *havuz* (pond). Presumably this arrangement allowed the men working the *dalyan* to collect any fish that escaped the main net by holding them back by the netting delimiting the *havuz*. Furthermore there were two breakwater-like installations on both sides of the entrance to the *dalyan*, again consisting of fishing nets, which probably facilitated the entry of the fish into the area where they could be trapped.

The *çökertme* resembled the installation described by Belon in the sense that two corners of the rectangular net were immobilized, but they were attached to posts on shore rather than to an installation in the sea. This arrangement was also similar to that described by Evliya in Beykoz. In this case, however, the other two corners of the net were in the hands of men who did not sit on perches as Belon described but were rather in a boat held in place by an anchor. Still, the net was very large and the narrow sides of the rectangle needed support as well, and so two rowboats at anchor—each manned by two additional people—held up the net by means of cordages. Six people thus serviced the net instead of the two mentioned by Belon. Deveciyan reported that this arrangement was suitable only for summer fishing and that in the entire region of Istanbul, it was only used on two sites on the Golden Horn and three on the Bosporus.

Apparently the weirs of the early twentieth century were larger and more efficient than their predecessors. Perhaps even too efficient. The beginnings of standardization visible in the nets described by Deveciyan apparently served to limit the catch by letting small fish escape; this also, perhaps unwittingly, may have protected the future of the fisheries. However, apart from the use of iron instead of wooden poles and the employment of chemicals to preserve whatever wood was still in use, the technology did not differ much from what had been available in the early modern period. As yet no motorized boats or other machinery were in evidence. At the beginning of the twentieth century there was still no canning industry, so most of the fish caught in the Sea of Marmara and the Bosporus must have been for local consumption.

Unfortunately, as we lack reliable data on the population and the total size of catches, it is impossible to calculate the per-capita consumption of fish throughout the early modern period. An Ottoman register covering Istanbul from around 1500 records a number of household heads that may have corresponded to an urban population of about 100,000. By 1940, the city's population, the *intra-muros* and the suburbs combined, stood at approximately 850,000—a figure probably less than what it was around 1900, for Istanbul ceased being a capital in 1923.[18] Only very vague estimates are available for the sixteenth and seventeenth centuries. In the 1790s the French medical doctor Antoine Olivier estimated, based on the amount of grain being consumed in Istanbul, that the city numbered half a million inhabitants. A few decades later in 1829, as Betül Başaran has established, Istanbul along with its suburbs was home to 420,000 people. The loss in population these two figures imply is quite credible, as the city suffered serious food crises during the Napoleonic wars.[19] Thus, although perhaps tempting, it is impossible to prove that the people of Istanbul ate more fish per head in the early twentieth century than their ancestors had 350 years earlier.

It is also unknown when and how the owners of fishing weirs went from using the relatively simple contraptions described by Belon and Evliya Çelebi to the complicated weirs of Deveciyan's times. Certainly the consolidation of property rights following the Ottoman Land Law of 1858 must have encouraged investment in this sector. The laws of the late nineteenth and early twentieth centuries, conveniently reproduced in the appendix to Deveciyan's work, protected the rights of the owner of a fishing weir against people who might fish in the area covered by his concession without having first secured his permission, for which they normally would have needed to pay.[20] While the protection of property rights was doubtless a *conditio sine qua non*, other inducements were probably necessary as well. Growing demand must also have been an important factor. Yet there is no way of knowing whether the expansion of the fisheries began in the middle of the nineteenth century rather than earlier. Unfortunately, the sources currently available do not provide any answers to this question.

In the early twentieth century, *dalyans* were quite costly, necessitating an investment of between two hundred and one thousand Ottoman gold coins. This significant expenditure was due to the fact that most *dalyans* needed both a large boat (*mavna*) and one or two smaller ones (*kayık*). Moreover, increasingly elaborate *dalyans* needed more personnel than had been customary in the seventeenth century. For the largest installations of this kind an owner needed to employ up to twenty-three people. In addition many poles and other implements used in the construction of Istanbul's weirs were now made of iron—certainly sturdier but also significantly more expensive than the wooden contraptions described by Belon and Evliya. Deveciyan therefore warned investors in *dalyans* that they would need not only significant financial reserves but also a good deal of experience and know-how, as mistakes could quite easily result in the bankruptcy of an incautious investor.[21]

Which Fish?

Pierre Belon left behind the names of quite a few fish that were available in Istanbul and also recorded the season in which this or that kind of sea creature could be found.[22] However, mid-sixteenth-century French terminology concerning fish may not be the same as today or what was in use about a hundred years ago, when Deveciyan published his work. Nonetheless, the information furnished by Belon and Evliya offers the opportunity for a rough but still useful comparison with that given by Deveciyan.

Pierre Belon observed that Istanbul's fishermen in the middle of the sixteenth century seem to have worked mainly during the summer. This limitation obviously was due to weather conditions; yet, according to twenty-first-century information, *çipura* (gilthead sea bream, Fr. *dorade/daurade*, *Sparus Auratus*), *gümüş balığı* (atherine, Fr. *athérine*, *Atherina Presbyter*), and the popular *hamsi* (European anchovy, Fr. *anchois*, *Engraulis encrasicholus*) are only available during the cold season.[23] Did these fish travel at different times of the year during the sixteenth century, or did the fishermen of Istanbul simply neglect those varieties that were available only when stormy wintry weather made fishing dangerous? Unfortunately, Evliya did not record any information on the seasonal rounds that must also have characterized fishing in the middle of the seventeenth century. Tables 4.1 and 4.2 summarize the available information.

TABLE 4.1 Fish on Record in Pierre Belon, *Voyage au Levant* (1553), 214 and elsewhere

Term Used by Belon	Twentieth-Century Names
Athérine (found near the island of Lemnos)	*Gümüş balığı/Atherina presbyter*
Cholio	*Kolyos/Scomber colias* (Deveciyan, 550)
Dentaux	No information
Dorade	*Çipura, karagöz/Sargus rondeletii* (Deveciyan, 547)
Girole	Perhaps identical with *Girelle paon/Julis turcica* (Deveciyan, 547 and 550)
Glanis	No information
Lampugne	Perhaps identical with *Lapina balıkları/Labridai* (Deveciyan, 111–112)
Mene	No information
Mulet	*Topbaş kefal/Mugil chelo* (Deveciyan, 216)
Oblade	*Melanurya/Oblade commune/Oblada melanura* (Deveciyan, 547)
Perche	*Tatlısu levreği/perche commune/Perca fluvialis* (Deveciyan, 132–133 and 252–254)
Pélamide	*Palamud/Pelamys sarda* (Deveciyan, 547)
Rouget	*Barbunya, tekir/Mullus barbatus* (Deveciyan, 545)
Salpe	No information
Sardines, Sardelles	*Sardalya/Clupea sardine*
Sarg	*İspari/Sargus annularis* (Deveciyan, 547), *Karagöz balığı/Sargus rondeletii* (Deveciyan, 547), *Sarıgöz/Sargus salvieri, Sargus vulgaris* (Deveciyan, 548)
Sphyrène	*Iskarmoz balığı/Sphyrana sphyœna* (Deveciyan, 546)
Sur (Venetian *suro?*)	Perhaps identical with *saurel/Scomber trachurus* (Deveciyan, 73)

Note: The equivalents not referenced in Pierre Belon's text come from the following: http://www.mymerhaba.com/Poissons-des-eaux-turques-en-Turquie-1482.html (January 15, 2012). All the others are from the indispensible reference work by Deveciyan.

Evliya's list and the 1640 price list only partially coincide with the information provided by Pierre Belon. To a certain extent, this divergence is due to the fact that while both authors probably often spoke to Istanbul's fishermen, the people they consulted might well have belonged to different communities. In all likelihood Belon's contacts were mostly Christians, presumably Greeks, while Evliya must have known more Muslims. The two communities possibly used different names for certain fish. Even more confusing, people of different milieus may well have used the same term to denote different fish. Furthermore it is quite possible that Belon, with his probably limited knowledge of the dialect used by Istanbul's Greek fishermen, misunderstood some of the terms he heard.

Evliya's local knowledge was far more extensive, but given his occasional tendency toward exaggeration, it is a great boon to be able to compare his data with a more or less contemporary register of officially imposed prices (*narh*) for

TABLE 4.2 Fish on Record in Seventeenth-Century Sources (Evliya Çelebi and a Price List from 1640)

Terms Used by Evliya	Twentieth-Century Names	Terms Used in a Price List from 1640
Alakerde	Perhaps *Lakerda* (Redhouse: salted tunny)[1]	*Lâkerda*
Atrina	Perhaps *Gümüş balığı (Atherina presbyter)* quem vide	
Çiroz	*Sprats or young mackerel salted and dried* (Redhouse), *fish, especially mackerel that recently have produced eggs and are therefore thin* (Deveciyan, 73)	
Çuçurya	No information	
Fıçıda	No information	*Poçıda*, probably identical with *fıçıda*
Gelincik	*Bearded or shore rockling, sea loche (Motella vulgaris)*	*Gelin balığı*
Gümüş	*Atherine (Atherina presbyter)*	*Gümüş balıkları*
Hapsi	Probably *Hamsi: European Anchovy (Engraulis encrasicholus)*	
Horosya	Perhaps *Horosbina: Gattoruginous blenny (Blennius gattorugine)*	
İskorbid	*Sea scorpion, Black scorpionfish (Scorpaena porcus)*	*İskorpit*
İzmarid	*Mendole, Cackerell, Cockrell, Picarell (Moena vulgaris)*	
	Annular git-head (Sargis annularis)	*İspari*
İstaride	Probably identical with *istavrid*	
İstavrid	*Scad, Horse mackerel (Caranx trachurus, Scomber trachurus)*	*İstavrid*
Kalkan	*Turbot (Rhombus maximus)*	*Kalkan*
	Base (Sargus rondeletii)	*Karagöz*
Kaya (balığı)	No information	*Kaya*
Kefal	*Flathead mullet (Mugil cephalus)*	*Kefal*
Kılıçbalığı	*Swordfish (Xiphias gladius)*	*Kılıç*
Kolyoz	*Spanish mackerel (Scomber colias)*	
	Red/Grey gurnard (Trigla hirundi, trigla gurnardus)	*Kırlangıç*
Kürekbalığı	No information	
	Wrasse (Crenilabrus pavo)	*Lâpina*
Levrek	*Basse, Common basse (Labrax labrax, labrax lupus)*	*Levrek*
Lüfer	*Blue fish (Parca saltatrix)*	*Nilüfer (?)*
	Spanish sea bream (Pagellus erythrinus)	*Mercan*
	Cod (Gadus morrhica)	*Morina*
Paçoz	No information	*Paçoz*
Palamud, palamide	*Atlantic bonito (Pelamys sarda)*	*Palamud*
	Sole (Solea vulgaris) (Redhouse)	*Pisi*
Tekir	*Striped surmulet (Mullus surmuletus)*	*Tekfur/tekir*
Tirkis	No information	
Uskumru	*Mackerel (Scomber scombrus)*	*Uskumru*
Yılarya	No information	*İlarye* probably identical to *yılarya*
	Hornfish (Belone vulgaris)	*Zargana*[2]

Note: This table has been compiled from the following: Evliya Çelebi bin Derviş Mehemmed Zılli, *Evliya Çelebi Seyahatnâmesi, Topkapı Sarayı Bağdat 304 Yazmasının Transkripsyonu—Dizini*, ed. Robert Dankoff, Seyit Ali Kahraman, and Yücel Dağlı, 10 vols. (Istanbul: Yapı Kredi Yayınları, 2006), 290–291; Mübahat Kütükoğlu, *Osmanlılarda Narh Müessesesi ve 1640 Tarihli Narh Defteri* (Istanbul: Enderun Kitabevi, 1983), 92–93. Sweetwater fish and seafood other than fish have not been included here. English equivalents are from: http://forum.kusadasi.biz/threads/seafish-in-turkey.1580/ (accessed January 15, 2012); Karekin Deveciyan, *Türkiye'de Balık ve Balıkçılık*, trans. Erol Üyepazarcı (Istanbul: Aras Yayınları, 2011), 554–556. Latin equivalents are from: Deveciyan, *Türkiye'de Balık ve Balıkçılık*, 545–548, and the same Internet list.

[1] This term likely had a different meaning in the seventeenth century. Kütükoğlu refers to nets for a sea creature named *ilâkerda*. Kütükoğlu, *1640 Tarihli Narh Defteri*, 291.

[2] The compilers of the price list refer to this fish only indirectly by recording the nets needed for catching it. Kütükoğlu, *1640 Tarihli Narh Defteri*, 291.

fish sold in the markets of Istanbul. The sultan's government had the list promulgated in 1640, shortly after the accession of Sultan Ibrahim (r. 1640–1648), in order to ensure that prices would drop after debased silver coins had been withdrawn from circulation.[24] The list is important because it reflects the kinds of fish that appeared in the pots and pans of Istanbul's inhabitants, rather than rarities that people caught every once in a while and that may have entered Evliya's account precisely because they were so unusual.

Evliya's list contains twenty-six fish names (or twenty-seven if we assume that *istaride* and *istavrid* were indeed two separate species), while the price register of 1640 records twenty-five. Eight of the terms occur in Evliya's account but not in the price list, so perhaps they were not often available to buyers. Since many fish appeared only at certain times of the year, the officials may have included only those that were available in the market at the time they compiled their register. This assumption makes sense for certain varieties of fish—for example, the especially seasonal anchovy (*hamsi* or *hapsi*), well known to people living close to the Black Sea in the seventeenth century. The price list likewise also contains eight items of which Evliya apparently knew nothing. Perhaps the Ottoman traveler used a different list of administered prices as his main source of information about the fish, but we have no evidence in this regard.

Further problems occur when relating the terms used by Belon, Evliya, and the 1640 price register to the twentieth-century terminology used by Deveciyan.[25] Of course the taxonomy developed by Carl Linnaeus (1707–1778) to unambiguously identify a species of plant or animal using Greco-Latin terms did not exist in the sixteenth or seventeenth centuries. As terms designating items in common use often change over time, the fish that Belon called *athérine* and Evliya Çelebi named *atrina* may or may not be the creature known in modern Turkish as *gümüş balığı* and in Linnean zoological terms as *Atherina Presbyter*. All conclusions must therefore remain tentative. But while methods of fishing changed dramatically over the centuries, apparently the varieties of fish available in the Bosporus and Sea of Marmara—mullet, bonito, and mackerel, to name a few—remained relatively constant.

Both Belon and Evliya wrote about the swordfish that fishermen caught in the weirs of Beykoz. Although the locations in which they were found might have changed, these fish were still available in the Bosporus in the early twentieth century. Certainly Deveciyan complained that in his time swordfish were more difficult to find since they did not like the lights of houses and ships that had become more frequent as human presence along the northern Bosporus increased. Perhaps it was due to these disturbances that swordfish no longer appeared in the weirs of Beykoz. But even so, they were present in the vicinity of Anadoluhisarı, Kanlıca, Çubuklu, and Anadolukavağı, all on the less densely populated Anatolian side of the Bosporus, and on the European shore, near Büyükdere in the location known as Kalender and near Rumelikavağı.[26] In Deveciyan's time a total of six thousand swordfish appeared in the Istanbul fish market every year.

Evliya Çelebi paid quite a bit of attention to the fishermen who caught seafood such as crabs, eels, and lobster, even though in his time these were favorite foods of people drinking and spending their time in taverns, not the most "respectable" inhabitants of the city. Mehmed the Conqueror (r. 1451–1481), however, also seems to have enjoyed shellfish and made sure the palace kitchen acquired it.[27] Interestingly, the market register of 1640 did not contain any prices for such seafood, perhaps due to its—by then—low status in the food hierarchy. But the compilers acknowledged that people did indeed catch lobsters, as they recorded a price for the nets used to capture them.[28] Moreover, fish roe, caviar, and various kinds of dried or preserved fish also appeared in the register along with their officially determined prices. For his part, Pierre Belon, had no interest in such seafood. Given this limited information, fishing for lobsters, crabs, and mussels seems to have been less important in the middle of the sixteenth century than it would become in Evliya's time, a century later.

As the fragmentary kitchen registers from the period of Mehmed the Conqueror reveal, the sultan and his servitors certainly ate fish and caviar as well. Moreover, fish dishes, particularly fish soup, regularly featured in official seventeenth-century banquets.[29] Arrangements for procuring fish therefore were already in place by the late fifteenth century.[30] But if the surviving accounts of the sultanic kitchen are indeed accurate, fish was always a secondary food item in the Ottoman palace, whose denizens favored meat—especially chicken—much more than anything from the sea. Yet the registers may well record only part of the catch that entered the palace, for certain fishermen were in the regular employ of the sultan and thus did not receive payment for every individual load of fish they brought to the ruler's kitchen.[31] Thus, among the inheritances registered by the *askeri kassam*, an official in charge of recording and often confiscating the estates of deceased servitors of the sultans, fishermen working for the palace do occasionally appear.[32] Furthermore, influential servitors of the sultan might emulate their master and also commission fishermen to bring them whatever fish they preferred.[33]

Fishing in Ottoman Istanbul was clearly practiced by both Greeks and Muslims, as is readily apparent from the well-known fact that modern Turkish is replete with Greek loanwords for the names of various fish. In Evliya's time, fishermen probably were organized into guilds, although apart from his list not much information about their social organization has come to light. They seem to have been less likely to complain or litigate than tanners, silk weavers, or dyers, and in the seventeenth century most fishermen's guilds seem to have been "mixed" in terms of the members' religion, as was common throughout Istanbul at that time. Evliya's list only refers to Greeks when discussing the *karityacı*, making it appear that they predominated in this guild. Otherwise Evliya differentiates between fishermen's guilds on the basis of the techniques they used, not their social makeup or any other factor. He draws a clear line between these men and the sellers of fish in various sections of the city and the fish market near Yenikapı. Remarkably, it was these traders rather than the fishermen themselves who exhibited the largest fish—and also seals—in the parade ordered by Murad IV (r. 1623–1640) that Evliya describes.

Conclusion

As the comparison between the works of Belon, Evliya, and Deveciyan has shown, from the middle of the sixteenth to the early twentieth century there seems to have been considerable continuity in the varieties of fish available in the Bosporus and the Sea of Marmara. Much, however, remains unknown. For example, the anchovy that today are available only in winter seem to have been supplied to the markets of early modern Istanbul by fishermen whose activities were largely confined to summer. Did the shoals of these fish pass through the city at a different time of year in the sixteenth or seventeenth century than they do today? Answers to this and other similar questions require further study.

If there was continuity in the supply of mullet, bonito, and mackerel available in Istanbul over the centuries, there was also a consistency of class difference in the consumption of fish over time. While fish were of limited interest to the elite denizens of the Ottoman palace with their notable tastes for meat and chicken, commoners—Muslim and non-Muslim alike—seem to have been regular buyers of fish and other seafood. Continuity was thus a major feature of Istanbul's supply and classed demand for fish.

Although it is unknown whether fishing techniques changed from the middle of the sixteenth to the middle of the seventeenth century, assuming an immobile world of "traditional" techniques is clearly hazardous. Rather, it seems most reasonable to conclude that the fishermen of early modern Istanbul were quite flexible in their arrangements. After all, the variety of fishing techniques described by Pierre Belon and Evliya Çelebi is striking: fishing weirs, nets hung from contraptions fixed to houses on the Golden Horn, and nets fastened to poles in the sea that could be emptied either by fishermen in their boats or by pulling in the nets from the shore. Other fishermen had their nets trailed by vessels.[34] People also fished with double nets, baskets, and even earthenware pots. Anglers were also in evidence.

All these techniques were fairly simple, although Evliya thought highly of the skill and ingenuity involved in trapping the fish. Apart from pulling in a full net, an undertaking which both Belon and Evliya described as requiring a lot of manpower, most fishing techniques involved only a few people. Fishermen working alone or with one or two companions were indeed quite common, if Evliya's ratio of people to enterprise is at all realistic. This low cost of labor and equipment goes a long way in explaining why neither Belon nor Evliya mentioned the manner in which fishermen and investors divided the catch, an issue extensively discussed by Deveciyan. Fishing implements in the early modern period must thus have been accessible to people of small means, without the need to make recourse to large capital. Perhaps the wives and children of fishermen made nets at home. If families did undertake this work, the source of the necessary raw materials unfortunately remains unknown.

When comparing sixteenth- and seventeenth-century descriptions with those provided by Deveciyan, it is obvious that the amount of capital investment needed for fishing had increased and so had, at least in certain cases, the size of the enterprises involved. This observation applies especially to the weirs, which the owners could only install once they had acquired the rights to the fishing grounds they planned to use. Once they had done so, it was necessary to invest in a much greater number of large and small boats. This meant the owners also needed to hire many more fishermen. Furthermore, these weir owners in the early twentieth century were required by law to provide meals as well as protective clothing for their workers, though presumably many sought to illegally bypass these extra costs. Moreover, while the iron used in the twentieth century was certainly more durable than wood, it was also significantly more expensive. In short, fishing by weir had become much more capital intensive in the twentieth century than it had been in the early modern period.

At the same time, however, there was apparently no great qualitative change in technology. Investors attracted to the fishing sector thus needed what we might call local knowledge. It therefore makes sense to assume that the growth of Istanbul's population, and perhaps a greater propensity to consume fish, resulted in a growing demand that fishermen tried to satisfy by maximizing the yield of the techniques at their disposal. To what extent concerns about overfishing resulted in more regulation is debatable, and whether such worries were known in the early modern period remains an issue for further exploration.

Notes

1. The poor were often vegetarians by necessity.

2. Evliya Çelebi bin Derviş Mehemmed Zıllı, *Evliya Çelebi Seyahatnâmesi, Topkapı Sarayı Bağdat 304 Yazmasının Transkripsyonu—Dizini*, ed. Robert Dankoff, Seyit Ali Kahraman, and Yücel Dağlı, 10 vols. (Istanbul: Yapı Kredi Yayınları, 2006), passim.

3. Ismet Binark et al., eds., *Bolu, Kastamonu, Kengırı ve Koca-ili Livâları: Dizin ve Tıpkıbasım*, vol. 2 of *438 Numaralı Muhâsebe-i Vilâyet-i Anadolu Defteri (937/1530)* (Ankara: Başbakanlık Devlet Arşivleri Genel Müdürlüğü, 1994), 791 and 801.

4. Pierre Belon du Mans, *Voyage au Levant (1553): Les Observations de Pierre Belon du Mans*, ed. Alexandra Merle (Paris: Chandeigne, 2001), 212–221. The number "1553" refers not to the date of the journey but to the year in which this volume was first published.

5. Ibid., 14–16.

6. Frédéric Tinguely, *L'écriture du Levant à la Renaissance: Enquête sur les voyageurs français dans l'Empire de Soliman le Magnifique* (Geneva: Droz, 2000).

7. Evliya Çelebi, *Seyahatnâmesi*, 1: 229.

8. Ibid., 1: 226.

9. All quotes are from the following: ibid., 1: 290.

10. Eremya Çelebi Kömürcüyan, *İstanbul Tarihi: XVII. Asırda İstanbul*, ed. Hrand D. Andreasyan, with additions by Kevork Pamukciyan, 2nd edn. (Istanbul: Eren, 1988), 46.

11. Evliya Çelebi, *Seyahatnâmesi*, 1: 17.

12. Karekin Deveciyan, *Türkiye'de Balık ve Balıkçılık*, trans. Erol Üyepazarcı (Istanbul: Aras Yayınları, 2011), 266–267.

13. Rossitsa Gradeva, "Ottoman Policy towards Christian Church Buildings," in *Rumeli under the Ottomans, 15th–18th Centuries: Institutions and Communities*, ed. Rossitsa Gradeva (Istanbul: Isis, 2004), 339–368, especially 344.

14. James W. Redhouse, *A Turkish and English Lexicon* (Istanbul: H. Matteosian, 1921), 719. The original was published in 1890. In Redhouse's time, these boats were limited to the Danube, but in Evliya's era they were also in use in and around Istanbul.

15. Deveciyan, *Türkiye'de Balık ve Balıkçılık*.

16. Ibid., 7–14. Compared to the original 1915 Ottoman Turkish text, the author had greatly expanded the French edition, and eighty years later, in 2006, a translation of this latter text into modern Turkish appeared, along with the preface to the first Turkish edition. In this novel shape, the book has now gone through several printings, most recently in 2011.

17. Ibid., 316–333.

18. The estimate of 100,000 inhabitants at the end of the fifteenth century is contested, with estimates varying between minimally 60,000 to 70,000 and maximumly 167,000 to 175,000. *Encyclopaedia of Islam*, 2nd edn., s.v. "Istanbul" (Halil İnalcık).

19. Antonie (*sic* for Antoine) Olivier, *18. Yüzyılda Türkiye ve İstanbul*, trans. Aloda Kaplan (Istanbul: Kesit, 2007), 30; Betül Başaran, "The 1829 Census and Istanbul's Population during the late 18th and early 19th Centuries," in *Studies on İstanbul and Beyond, The Freely Papers*, ed. Robert G. Ousterhout, vol. 1 (Philadelphia: University of Pennsylvania Press, 2007), 53–71.

20. Deveciyan, *Türkiye'de Balık ve Balıkçılık*, 437–455.

21. Ibid., 316.

22. See, for example, http://www.mymerhaba.com/Poissons-des-eaux-turques-en -Turquie-1482.html (accessed June 1, 2011).

23. For the English and Latin terminology, I have used: http://forum.kusadasi. biz/threads/seafish-in-turkey.1580/ (accessed June 1, 2011).

24. Mübahat Kütükoğlu, *Osmanlılarda Narh Müessesesi ve 1640 Tarihli Narh Defteri* (Istanbul: Enderun Kitabevi, 1983), 33.

25. The same problem exists when relating the historical data to contemporary usage as reflected in my auxiliary source, the anonymous authors of word lists published on the Internet.

26. Deveciyan, *Türkiye'de Balık ve Balıkçılık*, 39–44.

27. Arif Bilgin, *Osmanlı Saray Mutfağı* (Istanbul: Kitabevi, 2004), 196–197.

28. Kütükoğlu, *1640 Tarihli Narh Defteri*, 291.

29. Ömer Lütfi Barkan, "İstanbul Saraylarına ait Muhasebe Defterleri," *Belgeler* IX/13 (1979): 1–380, especially 239–240; Hedda Reindl-Kiel, "The Chickens of Paradise: Official Meals in the Seventeenth-Century Ottoman Palace," in *The Illuminated Table, the Prosperous House: Food and Shelter in Ottoman Material Culture*, ed. Suraiya Faroqhi and Christoph Neumann (Istanbul: Orient-Institut, 2003), 59–88.

30. Bilgin, *Osmanlı Saray Mutfağı*, 196–197.

31. Ömer Lütfi Barkan, "Edirne Askeri Kassam'ına ait Tereke Defterleri (1545–1659)," *Belgeler* III/5–6 (1966): 1–479, especially 257–258. The only published example concerns a Greek fisherman who died in landlocked Edirne in 1636. The account does not tell us anything about the fishing gear of the deceased apart from the fact that he possessed three ordinary lines. Perhaps the other items belonged to the palace kitchen, and the deceased had given them up when he decided to leave the Ottoman capital. In Edirne this man had owned a fish shop and sold alcoholic drinks.

32. Bilgin, *Osmanlı Saray Mutfağı*, 196–197.

33. Ibid., 197.

34. By the early twentieth century, scraping the bottom of the sea to fish had been outlawed. Presumably the overuse of fishing grounds had become a serious danger. Deveciyan, *Türkiye'de Balık ve Balıkçılık*, 441.

5

Plague and Environment in Late Ottoman Egypt

Alan Mikhail

Many historians have documented the intimate temporal and geographic coincidences of plague, famine, drought, flood, and price inflation in Ottoman Egypt.[1] Few, though, have shown how the connections between plague and these phenomena fit into a recurring pattern of death and hardship during the sixteenth, seventeenth, and eighteenth centuries that came to inform Egyptians' experiences both of their natural environment and of plague itself. By examining one particular plague epidemic in Egypt—that of 1791—this chapter shows how plague was part and parcel of the pathology of the Egyptian environment at the end of the eighteenth century. The chapter thus highlights the multiple means through which plague functioned as a regular part of the Egyptian environment.

Plague in Egypt must be studied as one pathological element of the Egyptian environment that was known and understood by Egyptians at the end of the eighteenth century to include floods, wind, drought, and famine. Like the annual Nile flood, increases in the prices of grain, famine, or other hardships, Egyptians considered plague an accepted and expected environmental reality. Thus those who regularly dealt with plague at the end of the eighteenth century did not think of the disease as a kind of "foreign" invader, coming to Egypt in the hulls of ships from faraway ports. Indeed, although feared, plague did not cause Egyptians to flee; rather, it functioned and was thought of as a regular part of the Egyptian environment.[2]

While Ottoman armies conquered Egypt and made it a province of the Empire in 1517, all evidence for the chronology and periodicity of plague indicates that 1517 did not represent a significant turning point in the frequency, severity, or treatment of plague in Egypt. Indeed, a new plague epidemic visited Egypt once every nine years on average for the entire period from 1347–1894.[3] This high incidence of plague suggests that the disease and its epidemics were regular and expected occurrences in the lives of most Egyptians throughout this period. It also suggests that historians should pay closer attention to the role of the disease in shaping Egyptian history and Middle Eastern history more generally.[4]

Despite the artificiality of 1517 as a breaking point in the history of plague, this date proves significant for the historiography of the disease because most work on plague in the Middle East focuses on its impact during the Mamlūk or medieval periods, even though it was a consistent reality in Egypt (as elsewhere) throughout the entire Ottoman period.[5] This historiographic imbalance toward the earlier history of plague is commonly attributed to a greater number of sources for the study of the medieval disease than for comparable studies during the Ottoman period.[6] As much recent work has shown, however, such an explanation is difficult to maintain, since there are numerous kinds of sources about plague after 1517: documentary records of Ottoman governmental bureaucracies, archives of Islamic courts, manuscript sources, and records of correspondence between the Ottoman center in Istanbul and various provinces of the Empire.[7] In addition, the voluminous chronicles of the seventeenth and eighteenth centuries offer a great deal of information about the outbreaks, nature, and impact of plague during these two hundred years.[8]

The Plague of 1791

The year 1791 began on a thoroughly foreboding note for Egyptians. Based on the reports of numerous astrologers (al-falakiyyīn), they came to believe that at midnight on February 1 a great earthquake (zalzala ʿaẓīma) would strike Egypt and last for seven trembling hours.[9] Both the poor and the rich were convinced of the coming of this earthquake, and those who could manage to leave their cities and villages fled to broad open places like the desert or to one of Cairo's two main lakes—al-Azbakiyya and al-Fīl—to ready themselves for the anticipated calamity. Egyptians braced for the event all night, but the quake never came, and all found themselves safe and sound the next morning.[10] Feeling thoroughly duped, people recited the following verse about their foolish naïveté: "And how many laughable things are in Cairo / but it is laughter like crying."[11]

However unfounded their fears of a colossal earthquake might have been, Egyptians had good reason to be afraid in February 1791. Later that month, plague struck with great force. One eyewitness reports that at the start of the outbreak, a thousand people died every day. Before long, that number rose to fifteen hundred per day.[12] Elsewhere, the same chronicler estimates that the plague killed two thousand people every day.[13] The disease did not discriminate between young and old, powerful and weak, pious and heathen.[14] An "uncounted number of babies, youths, maidservants, slaves, Mamlūks, soldiers, inspectors, and amīrs" all died in the spring of 1791.[15] Before attempting his planned escape to Istanbul, the leader Ismā'īl Bey died, along with many of his followers.[16] Indeed, the disease of 1791 caused an enormous crisis of leadership in Ottoman Egypt since no appointed leader could stay alive long enough to rule effectively. An Ottoman firman sent to Egypt from Istanbul implored the surviving leadership of the province to do all it could to defend against the state of disorder (perişanlık) that was gripping Egypt due to the death of eight or nine of the province's most important beys. The decree further ordered this leadership to inform Istanbul of the names, physical characteristics, and notable attributes of those important men who had died and those who had replaced them.[17]

When the aghā[18]—a high-ranking military official—and the wālī[19]—a police administrator who served under the aghā—died, successors immediately rose to power only to die themselves three days later. And those who replaced these plague victims themselves also died in the course of a few days. The Egyptian chronicler al-Jabartī writes that "succession changed hands three times in one week."[20] Al-Khashshāb, another chronicler, describes the situation slightly differently. He writes that the appointment of an aghā and the need to replace that appointee because of his death from plague occurred three times in one day. Leaders came to power in the morning and died by late afternoon.[21] The pasha of Egypt, on the advice of his agents, left Cairo along with his amīrs to seek refuge in the region of Ṭurā.[22] Many large Cairene families (buyūt) were decimated by the plague.[23] The speed of death after the onset of the disease was something noted throughout these descriptions of the 1791 plague.

So great a number of the soldiers and marines stationed in Old Cairo, Gīza, and Būlāq died that mass graves were dug for their corpses, which were buried without ceremony or any final rites.[24] Funerals for those not connected to the military also had to be done en masse, with prayers being said for up to five people at one time. Indeed, the apparatuses charged with the management of death were stretched to their limits during this spring as the demand for undertakers (al-ḥawānīt) and corpse-washers (al-mughassilīn) far exceeded their available numbers. Most economic and social functions unrelated to the present grim circumstances ceased during the plague since "there was not left for people any work except death and its attendant matters."[25] Accounts of the spring of 1791 devote significant attention to the sick, the dead, visitors of the sick, consolers, funeral-goers, those returning from funeral or burial prayers, those busy with preparing the dead, or those weeping in anticipation of their own death.

Later, in the summer of 1791, al-Jabartī tells us of another consequence of that year's plague: an absence of male heads of households.[26] When a group of amīrs from southern Egypt attached to one Murād Bey came to Cairo in late July 1791, they found many houses without men, inhabited only by "women, maidservants, and slaves." Finding this situation agreeable, the amīrs married these women, "replaced their bedding, and prepared their wedding feasts."[27] Any amīr who did not have a house was free to enter any home he liked and to take it and everything in it without hindrance (min ghayr mānīʿ). In this way, these men took advantage of what plague had wrought to acquire land, houses, riches, and wives. From the imperial perspective in Istanbul, the outbreak of plague in Egypt created a worrisome opportunity for enemies of the state and rebellious bureaucrats to escape to Egypt undetected amidst the mayhem and general disarray. Indeed, in very strong language, Ottoman authorities ordered those still present in Egypt to prevent any fugitives from entering the province to hide.[28]

The movements of peoples and goods were of the utmost concern for Ottoman authorities during plague outbreaks. In fact, it was precisely because Egypt was so central to currents of trade and commerce within the Ottoman Empire and between the Empire and elsewhere that plague was a constant presence in Egypt. Plague is permanently maintained among rodent populations in areas of central Asia, Kurdistan, central Africa, and northwestern India.[29] Plague thus came to Egypt from these and other places through the movement of merchants' wares, rats, fleas, and people.[30] As the American missionary John Antes wrote in Cairo at the end of the eighteenth century, "I could never find sufficient ground to ascertain that the plague ever broke out in Egypt, without being brought thither from other parts of Turkey [the Ottoman Empire]."[31] Indeed, the arrival of hundreds of ships and caravans every week from places like Istanbul, India, Yemen, the Sudan, China, central Africa, and Iraq ensured a consistent flow and constantly replenished supply of goods, people, and vermin to Egypt. The two main entry points of the disease were through the major trading areas of Egypt: its Mediterranean ports and the southern route from the Sudan.[32] Thus, plague either entered Egypt through the ports of Alexandria and Rosetta before making its way farther inland, or it traveled from central Africa to the Sudan and then into Egypt. The plague of

1791 most likely entered Egypt through its Mediterranean ports, carried ashore from ships coming from Istanbul.[33]

Whether a particular epidemic's origins lay in Kurdistan or the Sudan did not affect the ultimate outcome of the disease on the Egyptian population. What mattered most for Egyptians' experiences of plague was that the disease functioned as though it were endemic to Egypt, given the consistency of its incidence and, more importantly, its regular appearance in the Egyptian environment as an ecological force alongside famine, flood, and drought.[34] For their part, many European observers considered Egypt to be the "cradle of the plague."[35] Interestingly, the Egyptian chronicler al-Jabartī seemed to concur with the notion that plague was somehow endemic to Egypt, or, more specifically, that it was present in the Nile Valley's soil. During the French occupation of Egypt between 1798–1801, al-Jabartī subscribed to French views about the etiology of plague in Egypt. "They say that putridity or rottenness (al-'ufūna) pollutes the depths of the ground. If winter arrives and the underground becomes cooled due to the flow of the Nile, rain, and humidity, the putrid vapors that had been trapped in the ground emerge and pollute the air, causing epidemic disease and the plague (al-wabā' wa al-ṭā'ūn)."[36] Al-Jabartī's explanations of the causes of plague in Egypt suggest both the persistence of the miasmatic theory of disease causation and the perception that plague was somehow natural to, endemic in, or constitutive of Egypt.

Despite the existence of multiple descriptions of the 1791 plague and of other outbreaks, determining the disease's demographic effects remains difficult.[37] There are very few studies of the Egyptian population at the end of the eighteenth century and during the period of Ottoman rule more generally.[38] Egyptian chroniclers' reports claiming that one thousand, fifteen hundred, or two thousand people died every day are therefore difficult to interpret and are in all likelihood incorrect.[39] Indeed, these descriptions seem to raise more questions than they answer.[40] How were these figures determined? For how many days did this same number of people die? What were the relative mortality rates in cities and the countryside? Were these symbolic figures meant to express the severity of a given outbreak rather than statistical numbers? Modern historian André Raymond estimates that the plague of 1623–1626 killed close to 300,000 Egyptians.[41] He states that the same number died during the 1718 epidemic as well, representing close to an eighth of the entire Egyptian population at that time. For his part, historian Daniel Panzac reports that the plague of 1784–1785 killed anywhere from 30,000 to 40,000 of Cairo's 300,000 inhabitants and that the plague of 1791 claimed the lives of a fifth of Cairo's population (still around 300,000 in that later year).[42]

The Flood before the Plague

During the fall of 1790, before the plague's attack in the spring of 1791, an unusually large amount of rainfall in Egypt caused many parts of Cairo to flood. In his characteristically hyperbolic style, al-Jabartī describes this large rainfall as the result of one enormous deluge.[43] On the night of Thursday, October 14, 1790, the skies above Cairo poured water over the city, "as if from the mouths of waterskins."[44] Accompanying this rain were continuous claps of thunder and lightning powerful enough to blind all those who saw it. The rains continued all night and for the entire next day, rushing down off the mountains and filling the desert outside the city's walls. The waters destroyed tombs and graves and caused houses to collapse, killing those trapped inside.

As if all this were not disastrous enough, this saturated Friday coincided with the return of pilgrims to Cairo from the annual pilgrimage to Mecca and Medina. Instead of enjoying the celebrations that normally marked their return, these pilgrims were cruelly welcomed back to the city with floods that carried away the pavilion of the Amīr al-Ḥajj.[45] The waters had by this time entered the city and flooded its numerous wakālas (storage facilities for grains and other foodstuffs), caravanserais, and mosques. Businesses, residences, and even entire neighborhoods were destroyed. For example, more than half of the houses in the district of al-Ḥusayniyya were swept away.[46] A huge lake was created in and around the city. Al-Jabartī ends his account of this rainfall with the simple statement that "this was a most terrible affair."[47]

Whether this flooding occurred as the result of one deafening deluge or merely a steady rain is ancillary to the result: significant destruction of the city and its resources.[48] Similar to years in which the Nile flooded far beyond its banks, the waters of 1790 likely destroyed vast areas of agricultural land in and around Cairo and its hinterland.[49] This meant a significantly lower amount of food production for the coming harvest season. And more destructive than even an excessive flood season, the torrential rains of 1790 destroyed stored grain supplies in the markets and wakālas of Cairo. This doubly magnified the danger for the coming year in and around Cairo, since both fields and grain supplies were washed away. The population thus likely experienced food shortages and even famine, weakening its resistance to and resolve against disease. That plague is preceded by famine is commonly observed throughout the history of plague epidemics in the Middle East and elsewhere.[50]

Another result of the floods in Cairo was the movement of thousands of rats seeking refuge. Although most species of rat do swim, these rodents—much like humans—seek out dry areas to escape rushing water.[51] In the fall of 1790, rats and humans were therefore competing for space in areas of Cairo not damaged by the flood waters inundating the city. Rats sought refuge in the thatched roofs of homes, in homes themselves, and in other protected areas. In the countryside along the Nile, the flood caused rats and their fleas to escape from areas near the river, like fields and embankments, in search of higher ground.[52] When the Nile flooded and then receded quickly, the Egyptian countryside was especially susceptible to the dangers of rodent and insect infestation. As John Antes notes, "Should it [the Nile] happen to rise suddenly to a very great height, but not remain long enough to soak the fields sufficiently, it will not be a fertile year, and other bad consequences may likewise follow if it leaves the fields too soon, before the air begins to cool, for many sorts of vermin will breed in the ground which are pernicious to some kinds of vegetables."[53] Although rats and humans did often share the same dry, protected spaces, rats usually hid in dry places that humans could not or did not enter, thus allowing them ample space in which to breed. As the waters receded and people returned to their dwellings, they encountered a large concentration of rats—as well as fleas and other insects—feasting on the contents of numerous unearthed graves and a great quantity of wet food left by the flood. Thus the flood caused rats, fleas, and humans to come into much closer proximity than they normally would have.[54]

This proximity is a key factor in the etiology of plague since the primary vectors of plague transmission in humans are rats and the fleas associated with them.[55] As long as the rat population in a particular area is large enough to support a sizable flea population, plague will survive.[56] Egyptian chroniclers do not directly address the epidemiological character of the 1791 plague. Nevertheless, enough evidence exists to suggest that this episode was bubonic rather than pneumonic or septicaemic plague.[57] Although al-Jabartī's account states that plague victims were not feverish and that they died in a matter of two or three days, suggestive of pneumonic plague, he goes on to describe a situation in which the sick, the dead, and the healthy were all present together in very close proximity.[58] He writes of those going to visit the sick and of those caring for the sick in their own homes. He relates also that despite the close interaction between amīrs and their wives, only the former died of plague while the women themselves did not fall ill.[59] If the 1791 plague was indeed of the pneumonic variety, then all of those not infected would have contracted the disease from their close interactions with the sick. Victims of pneumonic plague most often experience violent coughing spells, and a simple cough or sneeze would be enough to infect a healthy person in close proximity. Since al-Jabartī makes no mention of whole families and neighborhoods being wiped out by plague and because he discusses at some length how the healthy interacted with and cared for the sick, it seems likely that the 1791 plague was bubonic.

Likewise, John Antes describes a similar situation of intimate proximity and contact between plague victims and their caretakers at the end of the eighteenth century. "It [plague] perhaps takes one or two only out of twelve, fifteen, or more, and those sometimes die in the arms of others, who, with all the rest, escape unhurt. There are instances of two people sleeping in one bed, one of whom shall be carried off by it, and the other remain unaffected."[60] Antes's descriptions of the behavior of some Europeans resident in Egypt at the end of the eighteenth century also suggest that this episode of plague was less contagious. For instance, the Friars de Propaganda Fide stationed in Cairo "always appoint two of their number to visit the sick, and to administer extreme unction to those of their persuasion who are dying: and it happens but seldom, that any of these visitors die of the plague, which constantly inclines them to make a miracle of it."[61] Antes also writes of a Venetian doctor in Cairo who regularly visited plague victims but was never stricken with the disease himself.[62]

Offering even clearer evidence that Egypt's late eighteenth-century plagues were indeed of the bubonic variety, Antes writes that victims had "buboes in the arm-pits, or the soft part of the belly, with a few dark purple spots, or carbuncles, on the legs. When the buboes break, and discharge a great deal of matter, such patients may chance to recover.... The sick commonly complain of intolerable heat, and say they feel as if thrown into a fire."[63] Indeed, most historians of plague agree that bubonic plague was the most common form of the disease in Egypt.[64]

Nevertheless, one pocket of pneumonic plague was known to exist at the turn of the nineteenth century around the southern city of Asyūt.[65] Al-Jabartī includes in his chronicle a letter from his friend and associate Ḥasan al-'Aṭṭār, who was present in Asyūt in May 1801 during an especially virulent plague epidemic.[66] Al-'Aṭṭār's letter states that this unprecedented plague killed more than six hundred people every day in Asyūt alone and that it "exterminated most of the people of the region."[67] Although these statements are far from conclusive about whether or not this plague in May 1801 was pneumonic or bubonic, given the almost certain fatality of pneumonic plague and the stated severity of this outbreak, it was in all likelihood pneumonic. This seems all the more probable since even during the Black Death, "pneumonic plague was far more frequent in the south [of Egypt] than in the north."[68] Exceptionally mild temperatures in Asyūt, along with large rat and flea populations, contributed to the plague's historically endemic and likely pneumonic character in the city.[69]

The Famine after the Plague

If the causes of the 1791 plague are to be found in an excess of water, its effects in the form of continuing hardship for Egyptians at the end of the eighteenth century are to be found in a dearth of water. On August 21, 1791, the Nile crested.[70] Every year at the cresting of the river, the Ottoman government of Egypt held a series of celebrations that culminated in the breaking of a dam constructed at the mouth of an artificial canal in Cairo known as al-Khalīj.[71] This dam was erected every year to prevent water from flowing into the canal except during the flood season. When the Nile's waters crested, the dam was broken and water then rushed through the canal into Cairo, symbolically inaugurating the yearly flood and the beginning of the agricultural year.[72] Although usually an event of great joy, this ceremony could also be a harbinger of hard times ahead, for a lower or higher than expected flood would mean famine and death for many Egyptians. This was exactly the case in 1791.

Similar to instances of excessive flooding (like that experienced in 1790), a poor inundation also meant food shortages, famine, and death. These effects of a low flood were even more acute in the late summer and fall of 1791 because of the ravages of plague earlier that year. Moreover, the populations of Cairo and other parts of Egypt were already suffering from lowered resistance and hence greater susceptibility to plague because of the floods of 1790. When the disease hit in 1791, the stage was set for a very bad epidemic. And by the end of 1791, drought and famine had fully set in.[73] The combination of flood in the fall of 1790, plague in the spring of 1791, and drought in the fall of 1791 resulted in widespread famine, severe price inflations, and a massive human toll. Indeed, as al-Jabartī tells us, later in 1791 irrigation canals dried up and fields became parched because of a lack of water.[74]

With crops withering and dying, peasants became extremely anxious—"the people clamored," in al-Jabartī's words.[75] The poor harvest meant an increase in grain prices for that year and the corollary of revolts and agitations by peasants and the poor against increased food prices and against their rulers. In November and December of 1791, Egyptian authorities began seizing the property and land of merchants and peasants, ostensibly to relieve the economic pressures brought about by plague and famine.[76] Drought continued through these months and into January 1792. Al-Jabartī writes that "not one drop of water fell from heaven."[77] Some peasants did their best to farm land that seemed salvageable, but when they plowed, they found only worms and rats. These vermin competed among themselves and against their human rivals for fruits and the precious few crops that were grown in fields that year.[78] Many people had to make do with weeds, and cattle had no spring feed. There was thus an exceptionally large number of rats in rural Egypt in 1791, aiding the movement and tenacity of plague.[79]

This cycle of flood, plague, drought, famine, price inflation, and death in Egypt suggests that there was a kind of cyclical pathology to the economy that also functioned alongside the ecological pathology of plague in Egypt.[80] As foodstuffs decreased, prices and the severity of official measures to secure adequate supplies for the powerful and the military increased.[81] In response, there were numerous instances of agitations, complaints, and small-scale revolts by peasants and merchants aimed at these official actions. For example, when Murād Bey and his amīrs—the same group of soldiers who married the wives of men who died during the plague—entered Cairo in July 1791, grain prices began to soar.[82] Because of the low Nile and the weakness of the population (da'f al-nās) from plague and other hardships, the amīrs began to seize grain for themselves and their entourage. Their cruelty was proven when one of them attempted to extract an unjustly large quantity of grain from a village outside of Cairo. A revolt broke out in Cairo and in the village itself in response to this draconian move on the part of one of Murād Bey's amīrs. The villagers refused to hand over the foodstuffs, and the 'ulamā (members of the elite classes of Muslim religious scholars) denounced the amīr's illegal actions. A violent struggle between the amīr and the villagers soon ensued, but Murād Bey, fearful that unrest might spread, reined in his amīr and apologized to the 'ulamā and the villagers.

The relationship between the city of Cairo and rural villagers during plague epidemics serves to highlight yet again the importance of urban grain reserves during times of want. Rural depopulation was a common phenomenon during plague epidemics and the famines that usually accompanied them, as peasants from the Egyptian countryside fled their lands in search of food and work.[83] The flight of rural labor to cities thus exacerbated the hardships brought on by food shortages—the very problem peasants were escaping. Thus, ironically, "in time of famine, peasants actually came to Cairo in search of food rather than the reverse."[84] In addition to food, cities also offered peasants access to physicians, healers, and religious institutions.[85]

The Season of Plague

In describing the death of the amīr Riḍwān Bey, al-Jabartī writes that the man's candle was extinguished by a plague that came like "the icy gale of death" (ṣarṣar al-mawt).[86] The trope of likening plague to wind further suggests that

the disease was considered part of the natural world of Egypt. Plague is elsewhere described as something that scatters the lives and possessions of its victims.[87] In the only eyewitness account of the Black Death in the Middle East,
Ibn al-Wardī (who died of plague in Aleppo in 1349) likened plague to a cloud:
"It eclipsed totally the sun of Shemsin and sprinkled its rain upon al-Jubbah. In
al-Zababani the city foamed with coffins."[88] Later, Ibn al-Wardī writes that "the
air's corruption kills," an obvious reference to the dominant miasmatic theory
of disease causation.[89] A later plague was furthermore compared to winds that
came and scattered the dried foods of fields in southern Egypt.[90]

Perhaps the most important reason for the association of plague with wind
is that, as in 1791, plague usually afflicted Egypt contemporaneously with the
khamāsīn—the warm southerly winds that blew into Cairo every year in late
spring and early summer.[91] These winds covered the city with sand and dust
from the deserts south of Cairo. John Antes describes the khamāsīn as follows:
"In spring it [the wind] often changes to south-east, and then it is of a whirling
nature, filling the atmosphere with such quantities of sand and dust as to make
it almost totally dark. I once remember being obliged to light a candle at noon
on such a day, as the sky was at the same time covered with thick clouds."[92]
The often contemporaneousness of the khamāsīn and plague caused many
to think that plague, like dust and sand, was brought in these annual winds.
For instance, Aḥmad al-Damurdāshī Katkhudā 'Azabān draws a connection
between the plague of 1690–1691 and the khamāsīn that occurred at the same
time. He writes that the plague swept into Cairo like the winds of the khamāsīn,
filling the city's quarters and alleyways with dead bodies.[93]

The coincidence of plague and the annual khamāsīn also helps to underscore the timing of the so-called season of plague in Egypt—the period of
greatest recurrence of the disease in any given yearly cycle.[94] In 1791, plague
began in late winter, seems to have been most deadly in the spring, and
waned toward the middle of the summer. This pattern is the one most commonly observed for the course of plague in Cairo, which Panzac identifies as
beginning in February, peaking in May, and dissipating toward its eventual
end in July or August.[95] A more nuanced study of the "season of plague" in
Egypt suggests a gradual movement of the disease from the south toward the
Mediterranean.[96] Those plague epidemics that traveled from Upper Egypt
(as the south is known) to Cairo seem to have occasioned much more fear
among the population than those that originated in Egypt's Mediterranean
ports. John Antes observed, "There is a saying among the people, that the
plague, which was brought from Upper Egypt, was the most violent."[97] In
Upper Egypt, plague begins in March and ends in May. In the middle of the
country, it begins in April and ends in June. In the Egyptian Delta above

Cairo, plague begins in April and ends in July. And, in Egypt's Mediterranean ports, the disease begins in May and ends in October. The reasons for this slow movement of plague from the south to the north are climatic, since temperatures rise first in the south and then move north as the country warms through spring into summer. The ideal meteorological conditions for plague are temperatures between 20° C and 25° C with mild humidity, precisely the conditions that move up the Nile Valley from spring into summer, taking mild temperatures—along with plague—up the country. Moreover, fleas—the primary agents of plague transmission—are most active during the warm months of the spring and early summer.[98] The dry heat of the summer eventually makes Egypt unfavorable to plague and fleas, as temperatures rise well above 27° C and humidity remains lower than 40 percent.[99]

For his part, Antes also ascribes the disappearance of plague in Egypt to the increase in temperatures during the summer months.[100] To illustrate this point, he relates the following:

> In the year 1781, the plague broke out about the middle of April,
> and increased with such rapidity and virulence, that sometimes one
> thousand people died of it in one day at Grand Cairo; but, about
> the middle of May, the wind shifted to the east, which occasioned
> a few days violent heat, in consequence of which it immediately
> diminished; and though, as the weather became again cooler, the
> plague did not leave the country before the end of June, yet it never
> encreased to the same degree as before, but continued dwindling
> away, till it ceased entirely when the summer heat became regular.
> It has always been observed in Egypt, that a great degree of heat,
> if even but for a few days, had this effect; but this time it was very
> remarkable.[101]

According to Antes, heat was also effective in curing those already afflicted with plague. He often observed that the sick arriving in Egypt from other parts of the Ottoman Empire during the summer months would soon recover from the disease after only a few days in the province. In cities like Istanbul and Izmir, he continues, plague was much more virulent since heat was never as regularly intense in those places as it was in Cairo.[102]

Egyptians took June 24 as the date past which no plague could exist in Egypt due to the consistently hot summer temperatures.[103] This tradition was connected to several events celebrated annually by Egyptians, especially by Copts, at the end of June. The first was the celebration of al-Nuqṭa (the drop) on June 17, which commemorated the beginning of the yearly Nile flood season; it was also the Coptic festival of the Archangel Michael. It was believed that on this day the Archangel threw into the Nile one drop of holy water of such fermenting power that it made the river overflow its banks, thereby flooding the entire land of Egypt. On this day also, the Archangel commanded all other angels to cease striking the people of Egypt with plague, since it was believed that angels were sent by God to infect with plague those intended as sacrifice. If by June 24, the festival of Saint John, any angels were found to be still lurking about in search of humans to strike down with plague, they would have to face heavenly consequences.[104]

Conclusion

This chapter argues for the inclusion of plague in the study of the Egyptian environment alongside natural elements and forces such as flood, famine, wind, animals, and drought. In doing so, it does not ask questions about the relationships between plague and Egyptian society and culture during this same period. If plague was indeed a regular feature of the Egyptian environment, then how did this regularity affect the ways Egyptians interacted with and thought about other parts of the natural world? How, for example, did social institutions like the family and religion adapt to, explain, and employ the constant presence of plague in Egypt? How did plague enable the deployment of sets of practices and institutions related to sickness, disease, and death? To answer these and other questions, historians must consider plague in its social and cultural, as well as its natural, contexts.

Notes

Earlier versions of portions of this chapter appear in "The Nature of Plague in Late Eighteenth-Century Egypt," *Bulletin of the History of Medicine* 82 (2008): 249–275.

1. Daniel Panzac, *La Peste dans l'Empire Ottoman, 1700–1850* (Louvain: Association pour le développement des études turques, 1985), 29–57 and 381–407; Nāṣir Aḥmad Ibrāhīm, *al-Azamāt al-Ijtimāʿiyya fī Miṣr fī al-Qarn al-Sābiʿ ʿAshar* (Cairo: Dār al-Āfāq al-ʿArabiyya, 1998); André Raymond, "Les Grandes Épidémies de peste au Caire aux XVIIe et XVIIIe siècles," *Bulletin d'Études Orientales* 25 (1973): 203–210; William F. Tucker, "Natural Disasters and the Peasantry in Mamlūk Egypt," *Journal of the Economic and Social History of the Orient* 24 (1981): 215–224.

2. Many Muslim writers on the subject of plague commented on and struggled with the following three tenets derived from the teachings of the Prophet Muḥammad: plague is a mercy and a form of martyrdom from God for a pious Muslim (and a form of punishment for infidels), a Muslim should neither flee from nor enter a region

affected by plague, and plague is not contagious because it comes directly from God. Differing opinions about these principles came to constitute the religio-medico-legal underpinnings of many ideas about plague in the Muslim world. Michael W. Dols, *The Black Death in the Middle East* (Princeton, NJ: Princeton University Press, 1977), 23–25 and 109–121; idem., "Ibn al-Wardī's *Risālah al-Naba' 'an al-Waba'*, A Translation of a Major Source for the History of the Black Death in the Middle East," in *Near Eastern Numismatics, Iconography, Epigraphy and History: Studies in Honor of George C. Miles*, ed. Dickran K. Kouymjian (Beirut: American University of Beirut, 1974), 444–445; Jacqueline Sublet, "La Peste prise aux rêts de la jurisprudence: Le Traité d'Ibn Ḥaǧar al-'Asqalānī sur la peste," *Studia Islamica* 33 (1971): 141–149.

3. Micahel W. Dols, "The Second Plague Pandemic and Its Recurrences in the Middle East: 1347–1894," *Journal of the Economic and Social History of the Orient* 22 (1979), 169 and 176. For the period from 1416–1514, David Neustadt (Ayalon) reports that an outbreak of plague struck Egypt once every seven years on average. David Neustadt (Ayalon), "The Plague and Its Effects upon the Mamlûk Army," *Journal of the Royal Asiatic Society of Great Britain and Ireland* (1946), 68. See also Dols, *Black Death*, 223–224; Panzac, *Peste*, 197–207.

4. One could rightly make the point that the arbitrary geographic and political division of Egypt is just as misleading in a discussion of plague as is the identification of the year 1517 as a chronological divide.

5. Works on plague in the medieval Middle East include: Michael W. Dols, "The General Mortality of the Black Death in the Mamluk Empire," in *The Islamic Middle East, 700–1900: Studies in Social and Economic History*, ed. Abraham Udovitch (Princeton, NJ: Darwin Press, 1981), 397–428; idem., *Black Death*; idem., "Ibn al-Wardī"; idem., "al-Manbijī's 'Report of the Plague:' A Treatise on the Plague of 764–765/1362–1364 in the Middle East," in *The Black Death: The Impact of the Fourteenth-Century Plague*, ed. Daniel Williman, Papers of the Eleventh Annual Conference of the Center for Medieval and Early Renaissance Studies (Binghamton, NY: Center for Medieval and Early Renaissance Studies, 1982), 65–75; Neustadt (Ayalon), "The Plague and the Mamlûk Army"; Stuart J. Borsch, *The Black Death in Egypt and England: A Comparative Study* (Austin: University of Texas Press, 2005); Justin K. Stearns, *Infectious Ideas: Contagion in Premodern Islamic and Christian Thought in the Western Mediterranean* (Baltimore: Johns Hopkins University Press, 2011). For works on earlier plague epidemics, see Lawrence I. Conrad, "The Plague in the Early Medieval Near East" (Ph.D. diss., Princeton University, 1981); idem., "The Biblical Tradition for the Plague of the Philistines," *Journal of the American Oriental Society* 104 (1984): 281–287; Michael W. Dols, "Plague in Early Islamic History," *Journal of the American Oriental Society* 94 (1974): 371–383; Josiah C. Russell, "That Earlier Plague," *Demography* 5 (1968): 174–184. For a critical reading of many of the primary sources for the history of plague, see Lawrence I. Conrad, "Arabic Plague Chronologies and Treatises: Social and Historical Factors in the Formation of a Literary Genre," *Studia Islamica* 54 (1981): 51–93. The bibliographical information provided in these major studies points to the enormity of the primary and secondary literature on plague in the Middle East.

6. Dols, "Second Plague," 164–165.

7. For a very useful summation of the recent literature on plague in the Ottoman Empire that employs some of these sources, see Sam White, "Rethinking Disease in Ottoman History," *International Journal of Middle East Studies* 42 (2010): 549–567.

The records of Islamic courts, for example, have proven extremely useful for the study of the history of plague in the Middle East. One of the primary functions of the court was to administer the inheritance of the deceased. Thus, one regularly finds inventories of estates (*tarikāt*) in the records of local courts. By tracing vicissitudes in the number of such inventories before, during, and after known plague epidemics, one can gain an approximate idea of the scale of mortality for a given region. This is clearly not an exact science, however, as the majority of plague deaths leave no record in court registers. At present, we have no reliable way of knowing how to determine what percentage of actual deaths were recorded in a given court.

For references to some of the numerous manuscript sources available for the study of plague (many of which are copies or compilations of earlier sources), see the bibliographical information found in the following: Ibrāhīm, *al-Azamāt al-Ijtimāʻiyya*, 316–320; Dols, *Black Death*, 320–339; Mohammed Melhaoui, *Peste, contagion et martyre: Histoire du fléau en Occident musulman médiéval* (Paris: Publisud, 2005), 20–57. For a compilation that includes many translated excerpts of medieval Arabic plague treatises, see John Aberth, *The Black Death: The Great Mortality of 1348–1350, A Brief History with Documents* (New York: Palgrave Macmillan, 2005).

8. For examples of studies of plague in Egypt during this period, see Ibrāhīm, *al-Azamāt al-Ijtimāʻiyya*; André Raymond, "Grandes Épidémies," 203–210; idem., *Artisans et Commerçants au Caire au XVIIIe siècle*, 2 vols. (Damascus: Institut Français de Damas, 1974); Max Meyerhof, "La Peste en Égypte à la fin du XVIII siècle et le Mèdecin Enrico di Wolmar," *La Revue Médicale d'Égypte* 1 (1913): 1–13. Seventeenth- and eighteenth-century Arabic chronicles that contain information on various plague epidemics in Egypt during the period include—but are by no means limited to—the following: Ismāʻīl Ibn Saʻd al-Khashshāb, *Akhbār Ahl al-Qarn al-Thānī ʻAshar: Tārīkh al-Mamālīk fī al-Qāhira*, ed. ʻAbd al-ʻAzīz Jamāl al-Dīn and ʻImād Abū Ghāzī (Cairo: al-ʻArabī lil-Nashr wa al-Tawzīʻ, 1990); idem, *Khulāṣat mā Yurād min Akhbār al-Amīr Murād*, ed. and trans. Hamza ʻAbd al-ʻAzīz Badr and Daniel Crecelius (Cairo: al-ʻArabī lil-Nashr wa al-Tawzīʻ, 1992); Aḥmad al-Damurdāshī Katkhudā ʻAzabān, *Kitāb al-Durra al-Muṣāna fī Akhbār al-Kināna*, ed. ʻAbd al-Raḥīm ʻAbd al-Raḥman ʻAbd al-Raḥīm (Cairo: Institut Français d'Archéologie Orientale du Caire, 1989); Muṣṭafā Ibn al-Ḥājj Ibrāhīm Tābiʻ al-Marḥūm Ḥasan Aghā ʻAzabān al-Damurdāshī, *Tārīkh Waqāʼiʻ Miṣr al-Qāhira al-Maḥrūsa Kinānat Allah fī Arḍihi*, ed. Ṣalāḥ Aḥmad Harīdī ʻAlī, 2nd edn. (Cairo: Dār al-Kutub wa al-Wathāʼiq al-Qawmiyya, 2002); Ibrāhīm Ibn Abī Bakr al-Ṣawāliḥī al-ʻUfī al-Ḥanbalī, *Tarājim al-Sawāʼiq fī Wāqiʻat al-Ṣanājiq*, ed. ʻAbd al-Raḥīm ʻAbd al-Raḥman ʻAbd al-Raḥīm (Cairo: Institut Français d'Archéologie Orientale du Caire, 1986); ʻAbd al-Raḥman al-Jabartī, *ʻAjāʼib al-Āthār fī al-Tarājim wa al-Akhbār*, ed. Ḥasan Muḥammad Jawhar, ʻAbd al-Fattāḥ al-Saranjāwī, ʻUmar al-Dasūqī, and al-Sayyid Ibrāhīm Sālim, 7 vols. (Cairo: Lajnat al-Bayān al-ʻArabī, 1958–1967); Muḥammad Ibn Abī al-Surūr al-Bakrī, *al-Nuzha al-Zahiyya fī Dhikr Wulāt Miṣr wa al-Qāhira al-Muʻizziyya*, ed. ʻAbd al-Rāziq ʻAbd al-Rāziq ʻĪsā (Cairo: al-ʻArabī lil-Nashr wa al-Tawzīʻ, 1998); Aḥmad Shalabī Ibn ʻAbd al-Ghanī, *Awḍaḥ al-Ishārāt*

fī man Tawallā Miṣr al-Qāhira min al-Wuzarā' wa al-Bāshāt, ed. 'Abd al-Raḥīm 'Abd al-Raḥman 'Abd al-Raḥīm (Cairo: Tawzī' Maktabat al-Khānjī, 1978).

9. The story of this earthquake is reported in al-Jabartī, *'Ajā'ib al-Āthār*, 4: 132.

10. For a discussion of earthquakes and their psychological impact on the population of Mamlūk Egypt, see Tucker, "Natural Disasters," 219–220 and 222–223.

11. al-Jabartī, *'Ajā'ib al-Āthār*, 4: 132. The verse is: "*wa kam dhā bi-Miṣr min al-muḍḥikāt / wa lakinahu ḍiḥkun ka-al-bukkā'.*"

12. al-Khashshāb, *Akhbār Ahl al-Qarn al-Thānī 'Ashar*, 58.

13. Idem., *Akhbār al-Amīr Murād*, 40.

14. According to the American missionary John Antes, however, "it has been observed in Turkey, and particularly in Egypt, that persons of the age of seventy, and upwards, are not so much subject to the infection, and very old people not at all. The most vigorous and the strongest appear to be most subject to it." John Antes, *Observations on the Manners and Customs of the Egyptians, the Overflowing of the Nile and Its Effects; with Remarks on the Plague and Other Subjects. Written During a Residence of Twelve Years in Cairo and Its Vicinity* (London: Printed for J. Stockdale, 1800), 47. Born in 1740 in Frederick Township, Pennsylvania, John Antes was the first American missionary in Egypt, and during his residence there from January 13, 1771, to January 27, 1782, he witnessed three plague epidemics. As one of the few foreigners resident in Egypt at the end of the eighteenth century who wrote specifically on the subject of plague, Antes's account is particularly useful as a supplement to the various Egyptian chronicles.

15. al-Jabartī, *'Ajā'ib al-Āthār*, 4: 132.

16. Ibid., 4: 133; al-Khashshāb, *Akhbār al-Amīr Murād*, 40; idem., *Akhbār Ahl al-Qarn al-Thānī 'Ashar*, 58.

17. Başbakanlık Osmanlı Arşivi (Istanbul), Cevdet Dahiliye, 1722 (May 15–24, 1791).

18. *Aghā* was a title given to the head of each of the seven military blocs stationed in Egypt. Here the reference is most likely to the head of the Mustaḥfiẓān military bloc, who served as a kind of chief of police in Cairo. For more on the role of the *aghā*, see Stanford J. Shaw, ed. and trans., *Ottoman Egypt in the Eighteenth Century: The Niẓāmnāme-i Mıṣır of Cezzâr Aḥmed Pasha* (Cambridge, MA: Harvard University Press, 1964), 10–11; Laylā 'Abd al-Laṭīf Aḥmad, *al-Idāra fī Miṣr fī al-'Aṣr al-'Uthmānī* (Cairo: Maṭba'at Jāmi'at 'Ayn Shams, 1978), 176–177 and 229–232.

19. On the position of the *wālī*, see: Aḥmad, *al-Idāra*, 233–235.

20. al-Jabartī, *'Ajā'ib al-Āthār*, 4: 133.

21. al-Khashshāb, *Akhbār Ahl al-Qarn al-Thānī 'Ashar*, 58.

22. al-Jabartī, *'Ajā'ib al-Āthār*, 4: 138. Ṭurā is located on the east bank of the Nile south of Old Cairo in the province of Aṭfīḥ. For more on Ṭurā, see Muḥammad Ramzī, *al-Qāmūs al-Jughrāfī lil-Bilād al-Miṣriyya min 'Ahd Qudamā' al-Miṣriyyīn ilā Sanat 1945* (Cairo: al-Hay'a al-Miṣriyya al-'Āmma lil-Kitāb, 1994), pt. 2, vol. 3: 15–16.

23. al-Khashshāb, *Akhbār al-Amīr Murād*, 40.

24. al-Jabartī, *'Ajā'ib al-Āthār*, 4: 133.

25. Ibid.

26. Ibid., 4: 140.

27. Ibid.

28. Başbakanlık Osmanlı Arşivi (Istanbul), Cevdet Dahiliye, 1722 (May 15–24, 1791).

29. Because these endemic foci of plague cover huge geographical areas of sparse human and vast rodent populations, the complete eradication of plague remains unlikely. On this point, see Dols, "Second Plague," 178; Conrad, "Plague in the Early Medieval Near East," 6–7.

30. Dols, "Plague in Early Islamic History," 381; Raymond, "Grandes Épidémies," 208–209. For more general discussions of the relationships between the movements of peoples and goods and the spread of plague, see William H. McNeill, *Plagues and Peoples* (Garden City, NY: Anchor Press/Doubleday, 1976); Janet L. Abu-Lughod, *Before European Hegemony: The World System A.D. 1250–1350* (New York: Oxford University Press, 1989).

31. Antes, *Manners and Customs*, 39.

32. Raymond, "Grandes Épidémies," 208–209; Dols, "Second Plague," 179–180. On the basis of the following, Dols compiles a list of plagues that came to Egypt and North Africa from Sudan and Central Africa. Georg Sticker, *Abhandlungen aus der Seuchengeschichte und Seuchenlehre* (Giessen: A. Töpelmann, 1908–1912). For more on plague in the Sudan, see Terence Walz, *Trade between Egypt and Bilād as-Sūdān, 1700–1820* (Cairo: Institut Français d'Archéologie Orientale du Caire, 1978), 200–201.

33. Raymond, "Grandes Épidémies," 208–209.

34. On the epidemiological fact that Egypt did not house an endemic focus of plague, see Dols, "Second Plague," 183; idem., *Black Death*, 35.

35. LaVerne Kuhnke, *Lives at Risk: Public Health in Nineteenth-Century Egypt* (Berkeley: University of California Press, 1990), 70; J. Worth Estes and LaVerne Kuhnke, "French Observations of Disease and Drug Use in Late Eighteenth-Century Cairo," *Journal of the History of Medicine and Allied Sciences* 39: 2 (1984): 123. There were, of course, differing opinions on this point. John Antes writes, "I think Egypt cannot, with any truth, be called the mother of the plague." Antes, *Manners and Customs*, 41. See also ibid., 36–37. Ibn al-Wardī writes of plague beginning "in the land of darkness," which Dols (citing Alfred von Kremer) identifies as "northern Asia." Dols, "Ibn al-Wardī," 448. Ibn al-Wardī then goes on to trace the disease's course from China and India, through Sind and the land of the Uzbeks, to Persia and the Crimea, and finally into Rūm, Egypt, Syria, and Palestine. Ibid., 448–453. The Mamlūk chronicler al-Maqrīzī and others also place the origins of plague in a vague "East" or in parts of Mongolia. For more on these accounts, see Borsch, *Egypt and England*, 4–5; Dols, *Black Death*, 35–42.

36. al-Jabartī, *'Ajā'ib al-Āthār*, 4: 322.

37. In Dols's words, "What we cannot judge accurately is the severity of the major plague epidemics from the late fifteenth to the late eighteenth centuries in the Middle East…Therefore, we cannot propose, as we have done in the earlier period, a significant demographic effect of these epidemics." Dols, "Second Plague," 176–177.

38. For statistics on and discussions of the demographic effects of the Black Death in Egypt, see Dols, *Black Death*, 143–235; idem., "General Mortality"; Borsch, *Egypt and England*, 40–54. Studies of the demographic effects of plague in

nineteenth-century Egypt include: Panzac, *Peste*, 231–278 and 339–380; Kuhnke, *Lives at Risk*, 84–86. For a critical discussion of plague mortality statistics, see Conrad, "Plague in the Early Medieval Near East," 415–447.

39. By way of comparison to reported plague deaths reaching one or two thousand a day, take, for example, the records of French doctors who tracked plague deaths during the months of the French occupation of Egypt from 1798–1801. They recorded plague deaths on the order of approximately five hundred to eight hundred deaths per month, with the total number of deaths surpassing one thousand for only four out of twenty-nine reported months. Panzac, *Peste*, 346.

40. For a discussion of the veracity of mortality figures as cited in Mamlūk chronicler reports, see Neustadt (Ayalon), "The Plague and the Mamlûk Army," 68–71.

41. Raymond, "Grandes Épidémies," 209–210.

42. Panzac, *Peste*, 361. André Raymond writes that the population of Cairo when the French expedition arrived in 1798 was 260,000. André Raymond, "La population du Caire et de l'Egypte à l'époque ottomane et sous Muhammad Ali," in *Mémorial Ömer Lûtfi Barkan* (Paris: Libraire d'Amérique et d'Orient Adrien Maisonneuve, 1980), 169–178. For comparative figures of plague mortality in Milan, Aleppo, Izmir, Marseille, and other cities during the seventeenth, eighteenth, and nineteenth centuries, see Panzac, *Peste*, 353–362; idem., *Quarantaines et lazarets: l'Europe et la peste d'Orient (XVIIe-XXe siècles)* (Aix-en-Provence: Édisud, 1986), 12.

43. al-Jabartī, ʿAjāʾib al-Āthār, 4: 129–130.

44. Ibid., 4: 129.

45. Ibid., 4: 130. For more on the gathering points of pilgrims in Cairo, see Antes, *Manners and Customs*, 69.

46. A useful comparison to the rains of 1790 is Antes's description of the rains and flooding in Cairo in 1771. "It once happened, during my abode, in November 1771, that heavy showers of rain, accompanied with some thunder and lightning, followed one another for five successive nights, though it did not rain in the day time…Some houses fell down on that occasion, and several lives were lost." Antes, *Manners and Customs*, 99.

47. al-Jabartī, ʿAjāʾib al-Āthār, 4: 130.

48. For a discussion of precedents to this sort of damage from rain in Mamlūk Egypt, see Tucker, "Natural Disasters," 216–217.

49. On this point Antes writes, "Sometimes the river rises so rapidly, and to such a height, that all their [peasants'] endeavours are in vain, and all such vegetables are destroyed." Antes, *Manners and Customs*, 72.

50. For instance, in 638 or 639, the plague struck Syria killing at least 25,000 soldiers and countless others. Important for our purposes here is the observation that this instance of plague was preceded by a severe famine which likely, as in the plague of 1791 in Egypt, weakened the population, making them all the more vulnerable to infection. Dols, "Plague in Early Islamic History," 376. See also Tucker, "Natural Disasters," 217–219.

51. For more on the physical attributes and abilities of rats, see the following classic study of typhus: Hans Zinsser, *Rats, Lice and History, Being a Study in*

Biography, which, after Twelve Preliminary Chapters Indispensable for the Preparation of the Lay Reader, Deals with the Life History of Typhus Fever (London: George Routledge and Sons: 1935), 197–204.

52. Conrad, "Plague in the Early Medieval Near East," 35.

53. Antes, *Manners and Customs,* 69.

54. Dols makes a similar point about the proximity of humans and rats in the plague epidemic of 638 or 639 in Syria. Dols, "Plague in Early Islamic History," 376.

55. For more on the epidemiology, pathology, and etiology of plague, see Conrad, "Plague in the Early Medieval Near East," 4–38; Dols, *Black Death,* 68–83; Borsch, *Egypt and England,* 2–8. For the most recent works in this regard, see Hugo Kupferschmidt, *Die Epidemiologie der Pest: Der Konzeptwandel in der Erforschung der Infektionsketten seit der Entdeckung des Pesterregers im Jahre 1894* (Aarau: Sauerländer, 1993); Graham Twigg, *The Black Death: A Biological Reappraisal* (London: Batsford, 1984). The standard works are the following: Robert Pollitzer, *Plague* (Geneva: World Health Organization, 1954); L. Fabian Hirst, *The Conquest of Plague: A Study of the Evolution of Epidemiology* (Oxford: Clarendon Press, 1953).

56. The disease exists in the blood of an infected rodent, and when this rodent is bitten by a flea, the flea ingests blood infected with plague bacilli. When the rat population begins to die off from plague, fleas seek out new hosts. Very commonly, especially in situations of close proximity like the ones described above, these hosts are human.

57. When blood infected with plague enters the human bloodstream, the bacilli rapidly multiply. Lymph glands filter plague bacilli from the blood, and these bacilli accumulate in the glands where they continue to multiply, thus producing the characteristic buboes on the neck or groin associated with bubonic plague. The highly contagious pneumonic version of plague occurs when bacilli settle in the lungs rather than the lymph glands. The septicaemic variety of plague, in which bacilli attack primarily the victim's blood, is the most virulent and quickest to kill, with victims often dying within hours of the onset of the disease. On septicaemic plague, see Twigg, *Biological Reappraisal,* 19; Hirst, *Conquest of Plague,* 29; Pollitzer, *Plague,* 439–440. Other varieties of plague include the tonsillar and the vesicular. For a discussion of these and other types of plague, see Dols, *Black Death,* 73–74; Borsch, *Egypt and England,* 4; Hirst, *Conquest of Plague,* 30.

58. al-Jabartī, *'Ajā'ib al-Āthār,* 4: 133.

59. Ibid., 4: 140.

60. Antes, *Manners and Customs,* 42.

61. Ibid., 47.

62. Ibid.

63. Ibid., 42. Arab physicians and observers of plague used many terms to describe plague buboes. These included "the cucumber," "the almond," "the pustule," "grains," and "blistering." For more on the terminology used for plague buboes, see Dols, *Black Death,* 75, 77–79, 316–319.

64. Dols, "Second Plague," 176–177 and 182–189; Kuhnke, *Lives at Risk,* 72. Dols goes so far as to suggest that pneumonic plague disappeared from most parts of the Middle East in the second half of the fifteenth century. Dols, "Second Plague," 182.

65. LaVerne Kuhnke explains this singular anomaly of pneumonic plague in Asyūṭ as a product of late nineteenth-century Egyptian and British efforts to expand the irrigation network in the south of Egypt, which unwittingly created a permanent population of rats and fleas. Kuhnke, *Lives at Risk*, 72–73 and 200–201, n. 16. As I discuss here though, evidence suggests the presence of pneumonic plague in Asyūṭ long before the end of the nineteenth century.

66. al-Jabartī, *'Ajā'ib al-Āthār*, 5: 241–242.

67. Ibid., 5: 241. Mention is also made of this plague in Panzac, *Peste*, 284.

68. Dols, *Black Death*, 60, n. 92.

69. Kuhnke herself concedes as much in her discussion of plague in Asyūṭ. Kuhnke, *Lives at Risk*, 73.

70. al-Jabartī, *'Ajā'ib al-Āthār*, 4: 141–142.

71. On this waterway, Antes writes, "The remaining water is horribly corrupted, by the filth thrown in from the adjoining houses, and the great number of necessaries that empty themselves into it, which occasions a most abominable stench for several months of the year, tarnishing in a short time even gold and silver in the houses near it." Antes, *Manners and Customs*, 38.

72. For a description of the function and importance of these festivities in Fatimid Cairo, see Paula Sanders, *Ritual, Politics, and the City in Fatimid Cairo* (Albany: State University of New York Press, 1994), 99–119.

73. For a discussion of the relationship between famine and plague in Mamlūk Egypt, see Tucker, "Natural Disasters," 217–219. Instructive also on this point is Elisabeth Carpentier, "Autour de la Peste Noire: famines et épidémies dans l'histoire du XIVe siècle," *Annales* 17 (1962): 1062–1092.

74. al-Jabartī, *'Ajā'ib al-Āthār*, 4: 197.

75. Ibid., 4: 141–142.

76. Ibid., 4: 199.

77. Ibid.

78. On the absence in Arabic plague treatises of any association between the pathology of plague and rodent populations, see Dols, "al-Manbijī's 'Report of the Plague,'" 71.

79. For more on the presence of rodents in the Egyptian countryside, see Antes, *Manners and Customs*, 85–86.

80. Suggestive of a similar concept is Ira M. Lapidus's use of the phrase an "economic geography of Egypt." Ira M. Lapidus, "The Grain Economy of Mamluk Egypt," *Journal of the Economic and Social History of the Orient* 12 (1969), 13.

81. For another example of the relationship between food shortages and price increases during plague epidemics, see Dols, "al-Manbijī's 'Report of the Plague,'" 71.

82. This account is related in al-Khashshāb, *Akhbār al-Amīr Murād*, 33–34.

83. Dols, *Black Death*, 154–169. On the movement of civilian populations during plague epidemics in the Mamlūk period, see Neustadt (Ayalon), "The Plague and the Mamlûk Army," 72.

84. Lapidus, "The Grain Economy," 8, n. 2. Cited also in Borsch, *Egypt and England*, 50. On the procurement, politics, and economy of grain in Mamlūk Egypt,

see also Boaz Shoshan, "Grain Riots and the 'Moral Economy': Cairo, 1350–1517," *Journal of Interdisciplinary History* 10 (1980): 459–478.

85. Dols, *Black Death*, 163; Tucker, "Natural Disasters," 222–224; Borsch, *Egypt and England*, 49–50.

86. al-Jabartī, *'Ajā'ib al-Āthār*, 4: 188.

87. Ibid., 4: 192.

88. Dols, "Ibn al-Wardī," 450–451.

89. Ibid., 454.

90. al-Jabartī, *'Ajā'ib al-Āthār*, 5: 242.

91. Ibrāhīm, *al-Azamāt al-Ijtimā'iyya*, 72–75; Dols, "Second Plague," 181.

92. Antes, *Manners and Customs*, 94. On wind in Egypt more generally, see ibid., 93–99.

93. Aḥmad al-Damurdāshī Katkhudā 'Azabān, *al-Durra al-Muṣāna*, 29. Aḥmad al-Damurdāshī Katkhudā 'Azabān, moreover, goes on to write that during this plague, one would wake to find ten new victims every morning. As in other plague outbreaks, there were also shortages of corpse-washers, and given the great number of dead bodies, gravediggers were forced to work long into the night.

94. For a general discussion of the periodicity, timing, and seasonal incidence of plague in the Middle East, see Panzac, *Peste*, 195–227; Conrad, "Plague in the Early Medieval Near East," 323–327; Kuhnke, *Lives at Risk*, 72–78. For a more general treatment of the subject, see Hirst, *Conquest of Plague*, 254–282.

95. Panzac, *Peste*, 223.

96. Kuhnke, *Lives at Risk*, 201, n. 18.

97. Antes, *Manners and Customs*, 39.

98. Conrad, "Plague in the Early Medieval Near East," 326.

99. Panzac, *Peste*, 225; Dols, "Second Plague," 181.

100. Antes writes the following about the climate of Egypt: "There is scarcely a country on the globe, where the climate is so very regular as it is in Egypt.... The difference between the greatest degree of cold and the greatest, or, more properly, the most usual heat in summer, does not exceed thirty degrees, according to Fahrenheit's thermometer." Antes, *Manners and Customs*, 89 and 91.

101. Ibid., 44.

102. Antes uses this evidence about the effects of heat on plague to argue against the notion suggested by some that plague was a putrid fever, since heat was thought to increase, not diminish, the severity of a putrid fever. Ibid., 44–45.

103. Ibid., 43 and 67.

104. Kuhnke identifies June 26 as the date of the festival of Saint John and of "the death of the plague" in Egypt. For more on the celebrations of this day by Egyptians and Europeans, see Kuhnke, *Lives at Risk*, 73.

6

Through an Ocean of Sand: Pastoralism and the Equestrian Culture of the Eurasian Steppe

Arash Khazeni

In April 1812, an Indian horse merchant by the name of Mir Izzat Ullah (1790–1825) set off on a journey from Delhi into the Central Eurasian Steppe in search of the world's fleetest horses. He was dispatched by William Moorcroft, an explorer and veterinary surgeon of the East India Company seeking to introduce Central Eurasian horses into the British Indian cavalry. Mir Izzat Ullah was to survey the markets, trade routes, caravanserais, and settlements from Kashmir and Ladakh to Balkh and Bukhara, with an eye for the prized horses of the Turkmen tribes, considered the premier horses in Asia. In addition to finding Turkmen horses to introduce into the Indian cavalry, Mir Izzat Ullah was to gather information on the places, peoples, resources, and trade of the nomad steppes and to carry the information back to the Mughal Empire. In the narrative of his travels, *Ahval-i Safar-i Bukhara*, Mir Izzat Ullah records the itinerary of his mission in 1812–1813 and offers a valuable account of cross-ecological encounters and exchanges between Eurasian empires and the steppes.

The interface between the steppe and the sown is among the long-standing themes in the environmental history of Islamic Eurasia and North Africa. As Edmund Burke III notes, the interconnection between pastoral and urban life is a "leitmotif" in the "deep history" of the region.[1] The fourteenth-century Muslim historian, Ibn Khaldun, famously described this encounter in *The Muqaddimah*, the introduction to his universal history, penned while taking refuge among the Awlad 'Arif tribes in the Maghrib. "The desert is the basis and reservoir of civilization and cities," wrote Ibn Khaldun, adding that "the toughness of desert life precedes the softness of sedentary life."[2] Because of the hardships of life in the desert, and the desert attitude that came with it, pastoralists flourished on the arid peripheries of empires, where in Ibn Khaldun's words, they "[fed] on the desert shrubs and [drank] the salty desert water."[3] In time, pastoralists crossed from deserts to the urban spaces of river valleys to conquer cities and establish new dynasties, only to burn out and be conquered by city life in turn. The desert constituted the frontier between the pastoral and the imperial, between tent and throne, but it was a borderland full of crossings, possibilities, contacts, and exchanges between the steppe and the sown.

During the reign of the post-Timurid Turkic empires of the Ottomans, Mughals, and Safavids/Qajars, early modern Islamic Eurasia (c. 1400–1850) continued to be marked by the persistence and evolution of the interface between settled agrarian cities and the pastoral steppes. Nevertheless, the continuities of these trans-ecological exchanges are often elided and ignored in studies of encounters between Eurasian empires and pastoralists in the sixteenth through the nineteenth centuries. The narrative of this history has usually been one of the "closing" and settlement of imperial frontiers, the transformations wrought by state-driven development projects, and the eclipse of indigenous pastoral populations and "tribes."[4] In this narrative, Eurasian pastoralists have remained reactive, seemingly incapable of moving beyond resistance or compromise in response to projects of imperial conquest and assimilation. This view has been compounded by the pervasive binary framework between all powerful, hegemonic empires and relatively compliant "native" subjects in the historiography of imperialism in the MENA.[5] The existing literature on the encounter between empires and pastoralists in the Eurasian Steppe has similarly highlighted the primacy and power of the imperial over the local.[6]

Through a reading of nineteenth-century Persian natural histories and travel narratives about the Eurasian Steppe, this chapter diverges from prevailing empire-centered analyses of conquest to examine frontier exchanges and interconnections between the pastoral and the imperial. The expanse of land that forms the Eurasian Steppe extends from eastern Iran to western China, and much of this region—with the exception of the oasis cities of the Silk Road—is composed of steppelands and deserts.[7] Into the nineteenth century, this steppe landscape constituted a "middle ground," a space for a pattern of encounters and contacts that Richard White has analyzed in the context of North America. Seeking to go beyond caricatures of imperial conquest and assimilation, as well as those of cultural persistence, White defined a borderland, or contact zone, where empires and peoples met:

> The middle ground is the place in between: in between cultures, peoples, and in between empires and the non-state world of villages. It is a place where many of the North American subjects and allies of empires lived. It is the area between the historical foreground of European invasion and occupation and background of Indian defeat and retreat. On the middle ground diverse peoples adjust their differences.[8]

Into the early nineteenth century, the Eurasian Steppe comprised a similar frontier ground of encounters between the steppe and the sown.[9]

Adopting an environmental perspective offers crucial insight into the worlds of the Eurasian "middle ground." In the late sixteenth century, the Oxus River changed course, leading to the expansion of the sandy steppes of the Qara Qum or "Black Sands," the arid desert between the Caspian Sea and the Oxus River, rendering it an ecological zone beyond the full reach and control of empires. As the Oxus River changed course, no longer reaching the Caspian Sea, Turkmen pastoralists found new possibilities in the expanding arid steppes of the Qara Qum, forging a powerful and wide-reaching equestrian network in the Eurasian Steppe. In the desert, Turkmen pastoralists domesticated wild horses and the swift Akhal Tekke breed, and gained control of the oases. Between the sixteenth and nineteenth centuries, the Turkmen carved out a loose trading and raiding confederation built on the power and speed of horses capable of making seemingly impossible journeys through the steppes.[10] The Turkmen horse trade reached from the Qara Qum Desert to Eurasian cities and frontier markets. The pastoral power and equestrianism of the Turkmen frontier determined the boundaries of early modern Eurasian empires, as they sought to gain the pastoral resources of the steppes and saw the contours of their distant imperial frontiers shaped by contact with the desert. The Qara Qum became a frontier ground in between Islamic empires.[11]

Taming nature through their domestication of fast horses capable of crossing the desert, Turkmen pastoralists defined the terms by which contact with surrounding empires took place. In the expanse of the Black Sands Desert and its oases, Turkmen pastoralists found new pastures for their equestrian economy and culture and created wide networks of trading and raiding across the steppe frontier. Eurasian empires, including the Safavids and Qajars of Persia and the Mughals of India, took measure of the equestrian world of the Eurasian Steppe by commissioning the writing of natural histories and travel

books. Such were the conditions that spurred Mir Izzat Ullah on his long journey into the steppes of Central Eurasia in the spring of 1812.

A Steppe of Possibilities

In 1576, the Oxus River changed its course. The Oxus—or Amu Darya, as it
is known locally—originates in the Pamir Mountains in modern Afghanistan
and flows northward, parting the Qara Qum and the Qizil Qum, the Black and
Red Sands Deserts, before emptying into the Aral Sea, approximately 1600
miles from its source. The course of the river's flow, however, has historically
been prone to changes and fluctuations carrying environmental consequences
for the frontier between the steppe and the sown. During various periods of its
history, the river is believed to have partially flowed into the Caspian Sea by way
of the Uzboy Channel (an ancient bed of the Oxus that is dried up today) and
the Sarykamysh Depression.[12] It is unclear exactly why the Oxus River changed
its course; both climatic fluctuations occurring in the colder, drier periods of
the Little Ice Age, as well as the growth of irrigation works along the Oxus
in Khwarazm, may have affected the river's flow and which channels were
used. What is certain is that since the late sixteenth century, there occurred a
complete desiccation of the Sarykamysh Depression and hence the Oxus has
flowed north only into the Aral Sea ever since. The eastward shift of the river
is chronicled by Abu'l Ghazi Bahadur (1603–1642), Khan of Khiva, who writes
in *Shajara-yi Tarakima* (*Genealogical History of the Turkmen*) that around the
year 1576, the Oxus River changed channels, swerving toward the Aral Sea and
turning the lands between its former bed and the Caspian Sea into a waterless
desert.[13]

There remains much evidence to suggest that the Aralo-Caspian lands
saw significant environmental changes in the early modern period, including great fluctuations in water channels. These changes had consequences
for settlement patterns and irrigation canal networks.[14] The expansion of
desert sands in the alluvial plains is thought to have corresponded with an
increase in aridity, the shrinkage of seas, the contraction of streams and
deltas, the drying of channels, and the depopulation of oases.[15] The process
of desertification made the Qara Qum a distant imperial borderland and
opened new possibilities for indigenous pastoral populations. Thus, by the
eighteenth and nineteenth centuries, Turkmen peoples had transformed
the Black Sands Desert into a tribal frontier ground through their exceptional horse breeding and equestrian culture. The culture of equestrianism in the "middle ground" of the steppes was essential to the Turkmen's

MAP 6.1. Sands of the Oxus. Map of Safavid Persia with the Oxus River still depicted as flowing into the Caspian Sea (Nicolas Sanson, 1680).

pastoral economy and was the basis of their political power, as they forged an extensive trading and raiding network, reaching across the Black Sands and the Red.

The Turkmen were a conglomeration of mobile Turkish-speaking pastoralists inhabiting the steppes and oases of the Qara Qum Desert. They practiced a pastoral economy that combined both nomadism and agriculture. The tribes that adopted a nomadic life were called *chumur*, and those that were sedentary were called *charva*. The Turkmen were Muslims, specifically Hanafi Sunnis. However, they were grounded in indigenous "Inner Asian" Islam and were integrated into networks of Islamic pilgrimage to the tombs of charismatic saints belonging to the Naqshbandi Sufi brotherhood (*tariqa*). Between the sixteenth and nineteenth centuries, the Turkmen overran and conquered the unsettled steppe frontier between the Oxus and the Caspian. According to Abu'l Ghazi's seventeenth-century genealogy, *Shajara-yi Tarakima*, the Turkmen lived without fixed habitations, camping in thick tents of felt and subsisting on their numberless cattle in the green pastures found in the steppes between the Oxus River and the Caspian Sea. The many Turkmen tribes were once one, Abu'l Ghazi wrote, but over time they became divided into various branches, including the

TABLE 6.1 Turkmen Populations

Turkmen Tribe	Fraser (1825)	Burnes (1832)	SB (1844)	Vambery (1863)	Thomson (1876)	Marvin (1881)	Moser (1885)
Tekke	40,000	40,000	80,000	60,000	25,000	80,000	
Ersari		40,000	40,000	50,000	30,000	30,000	
Salor		2,000	8,000	8,000	9,000		
Sariq		20,000	12,000	10,000	14,500	13,000	
Yamut	25,000	20,000		40,000	30,000	35,000	20,000
Guklan	10,000	9,000		12,000	4,000	6,000	
Chawdur		6,000		12,000	8,000		17,000

Tekke, Salor, Sarik, Yamut, Chawdur, and Ersari. The Turkmen had entered and settled the Qara Qum Desert along the Uzboy Channel and the Üst-Yurt Plateau at a time when the Oxus still flowed into the Caspian Sea. After the sixteenth century, with the recession of the Oxus, the drying up of the Uzboy Channel, and the expansion of the Black Sands Desert, the Turkmen tribes migrated southward and eastward to the oases of Khurasan and Khwarazm.[16] Thus, the subsequent history and distribution of Turkmen peoples must be understood partly as a result of the environmental flux of the early modern period that pushed various communities in different directions as they sought out water and food.

According to nineteenth-century sources, the Turkmen tribes included the Tekke, near the oases of Marv and Akhal in the environs of the Murghab and Tejend (Hari Rud) Rivers; the Ersari, dwelling on the edge of the Upper Oxus in an area known as Lab-i Ab ("Lip of the Water"); the Salor, with their encampments in the vicinity of Marv; the Sariq, inhabiting the environs of the Panjdih Oasis; the Yamut tribes, spread from the Caspian and the eastern Iranian province of Astarabad to the banks of the Oxus River and Khwarazm; and the Chawdur, spread between Manqishlaq on the Caspian and Khiva.[17] Table 6.1 summarizes the varying estimates of Turkmen populations, approximated according to the number of tents or yurts, offered in nineteenth-century European and Persian sources.[18]

The pastoral economy of the Turkmen depended on the possession of large herds of horses and camels and flocks of sheep and goats that grazed in the dry soils and open grasses that grew under the Qara Qum sun. Horsemanship had long lay at the foundation of the Turkmen economy. According to traditional accounts, Turkmen horses were mixed with Arab breeds, introduced to the Central Eurasian Steppe perhaps at the turn of the fifteenth century during the reign of Timur Lang, who was thought to have distributed 4,200 of the best Arabian mares among the tribes. This crossing of breeds was likely revived in the eighteenth century by Nadir Shah Afshar, who offered six hundred Arabian mares to the Tekke.[19]

By the middle of the eighteenth century, the expansive pastoral economy of the Tekke tribes had allowed them to occupy the oases in the foothills of the Kopet Dagh Mountains, and to displace sedentary populations and take possession of the northern edges of Khurasan.[20] Many of the Tekke became concentrated around the oasis of Akhal, an arable strip of land bounded by the northern slopes of the Kopet Dagh and the sandy steppes of the Qara Qum. Making their encampments alongside the rivers that flowed from the Kopet Dagh, the Tekke grazed their horses in the vast steppes that reached to the north. In the ecology of Akhal, the Tekke found rich resources for their thriving equestrian culture, leading to the creation of the famed Akhal Tekke breed. As they possessed the premier breeds of Central Eurasian horses, the Tekke quickly rose to power in the sands of the Oxus and expanded their sway over the steppes and oases of the Qara Qum. In the 1830s, the Tekke began to expand outward from Akhal, seeking more pastures for their growing herds of horses and eventually occupying the oasis of Marv on the Murghab River. There they managed the city's irrigation system and absorbed thousands of Salor clans dwelling in the oasis.

Turkmen pastoralists domesticated their horses on the steppes of the Oxus River, nurturing and training them until they were able to make seemingly impossible journeys through the desert. The most renowned Turkmen horses came from the Tekke of the oasis of Akhal (see figure 6.1), with those from the Yamut being highly prized as well (see figure 6.2). By most accounts, these breeds were beautiful and much sought after. Long, slender, and sinewy in stature, Turkmen horses thrived on the dry grass of the desert and were well known for their fleetness and endurance on long marches through the steppes.[21]

Horses were revered in Turkmen culture and received careful and affectionate treatment. Horseback riding was central to the Turkmen's pastoral way of life, and the tribes were known for the great care they took of their horses, nurturing and training them to be able to undertake extremely long and arduous journeys through the steppes. Indeed, horses were the basis of mobility, independence, and power among the Turkmen tribes. This love of horses was not only practical but also ceremonial. Horse races and other equestrian feats were a main attraction at weddings and birthdays. The Turkmen adorned their horses with felts (namad) and woven textiles (jul), as well as ornaments (zayvar) of precious metals hung from their heads and necks.[22] Turkmen horses never saw a stable and were instead "picketed in the open air, and clothed with felt rugs."[23] Even during winter, it was said, they were covered with layers of thick wool felt and left out in the open air, snow piled around them. The famous and

FIGURE 6.1. The Desert Breeds. The Tekke in the steppes of the Qara Qum. Henri Moser, *À travers l'Asie Centrale: la Steppe Kirghize, le Turkestan russe, Boukhara, Khiva, le pays des Turcomans et la Perse, impressions de voyage* (Paris: Librairie Plon, 1885).

FIGURE 6.2. The Desert Breeds. The Yamut in the steppes of the Qara Qum. Henri Moser, *À travers l'Asie Centrale: la Steppe Kirghize, le Turkestan russe, Boukhara, Khiva, le pays des Turcomans et la Perse, impressions de voyage* (Paris: Librairie Plon, 1885).

still-recited Turkmen poem about the bandit-minstrel Köroğlu, recorded by the Russian Orientalist Alexander Chodzko in his 1842 *Specimens of the Popular Poetry of Persia*, contains a description of the finer points of Turkmen horses and how they were nurtured from the time they were foals (*kurra*):

> Listen and learn how a noble horse may be known. Active and brisk,
> see if his nostrils are rapidly swelling and shrinking alternately;
> whether his slender limbs are like the limbs of a gazelle, ready
> to commence its course. His haunches must resemble those of a
> chamois.... His back ought to remind you exactly of that of a hare:
> his mane is soft and silky; his neck is lofty, formed to the semblence
> of the peacock's. The best time to mount him is between his fourth
> and fifth year.... When brought out of the stable he is playful and
> prances.... A young man of a good family lends an obedient ear to
> the words of his parents: he pays the greatest attention to his horse.[24]

The pastoral economy of the Turkmen had implications beyond the local yurt; Turkmen horses were coveted across Eurasia. Merchants like Mir Izzat Ullah explored border towns and horse markets in search of Turkmen breeds. The Eurasian horse trade linked Turkmen pastoral nomads and merchants across a wide geographical terrain. The Turkmen thus proved integral to caravan networks connecting the steppe and the sown. Despite this role, they are frequently blamed for the disruption of trade and seen as a source of insecurity on roads. Writing in the 1920s, the pioneering scholar of Central Eurasia Vasily Barthold, for example, claimed to have come across only one reference to Turkmen merchants and traders, that of the history of Abu'l Ghazi.[25] Trade constituted a major element of Turkmen life, however, and there are ample traces of Turkmen involvement in Eurasian networks of caravan traffic.

The early nineteenth-century chronicle of Mir 'Abd al-Karim Bukhari gives a view of the overland caravan trade, "distance by distance" (*farsakh* by *farsakh*), and the civilization of Central Eurasia, a region defined by contacts and exchanges between different cultures, encompassing both the sedentary and the nomadic.

> In the environs of Bukhara there are many nomadic tribes (*ahsham nishinan*): Arabs, Turkmen, Uzbak, Qaraqalpaq. The Turkmen are
> settled along the banks (*lab-i ab*) of the Amu Darya River for four
> or five days march and are made up of the following tribes (*tayifa*):
> Ersari, Sariq, Baqah, Salor, Tekke, Amir 'Ali, and Chawdur.... The
> number of nomads is equal to the number of townsfolk and from
> Bukhara to Samarqand one passes a succession of villages, towns,
> and nomadic encampments.[26]

Through the horse trade, the Turkmen were significantly involved in the wider economic networks that integrated the steppes and oases of Central Eurasia. The Turkmen's trade in horses—as well as in camels, sheep, wool, and textiles—linked them to the markets of Iran, India, and Central Eurasia, where they found such goods as rice, tobacco, tea, and sugar.[27] In the nineteenth-century geographical chronicle *Mir'at al-Buldan*, I'timad al-Saltana described the Turkmen horse trade in the provinces of Qajar Iran: "The horses of the Turkmen, which are from the Turkmen tribes of Tekke at Marv, are taken to Mazandaran, Tehran, and other lands in Iran, and are sold and traded there."[28] In the early nineteenth century, British cavalry regiments wrote reports on the Turkmen horses in their ranks, often including descriptions of individual horses, and sought to acquire more of them for their stables.[29]

The taming of horses, and the mobility these animals provided, also had political implications. The culture of equestrianism lay behind the lightning-fast Turkmen slave raids that set the steppelands apart from surrounding empires in the eighteenth and nineteenth centuries. The possession of a fast horse was of prime importance among the Turkmen, allowing them to make long-distance raids and journeys across the Black Sands Desert.[30] Beginning in the middle of the eighteenth century, the Turkmen increasingly raided the borderlands of Iran for Shi'i slaves. By the turn of the nineteenth century, it was customary for the amir and 'ulamā of Bukhara "to issue a decree ordering the Turcomauns in the desert to march to Khorassan and Persia, to make 'tchapow,' *i.e.* foray; which order those tribes obey, capturing whole caravans, burning down villages, and selling the inhabitants as slaves in the cities of Turkestan."[31]

Turkmen slave raids (*alaman*) and surprise attacks (*chapu*) were assertions of tribal power and politics that would not have been possible if not for the speed and endurance of Turkmen horses. In the mid-nineteenth-century Qajar chronicle *Nasikh al-Tavarikh*, Mirza Muhammad Taqi Lisan al-Mulk Sipihr describes the raids of mounted Turkmen tribesmen in the Shia shrine city of Mashhad on the eastern borderlands of Iran in 1852, highlighting the centrality of horses in the exploits of the Akhal Tekke. "A thousand Turkmen *savar* (horsemen) from the Turkmen tribes of Akhal and Tekke from Tejend entered the mosque of light and took thirty women as slaves."[32] According to contemporary sources, the mounted raids of the Turkmen occurred rapidly; they took victims by surprise and returned to the steppes with their new captives. In the early Qajar chronicle *Tarikh-i Zu al-Qarnayn*, Mirza Fazl Allah Shirazi Khavari conveys the suddenness of a Turkmen raid on Khurasan in 1830:

In the winter of this year, about two thousand people strayed from the heads of the Turkmen, and as is characteristic of the Tekke, they were led astray by Allah Quli Tura, the *vali* of Khvarazm, and came to raid (*takht va taz*) the holy land of Mashhad, galloping (*asbandaz*) and raiding (*turktaz*) the land.... Finding well fortified positions and lurking in corners, they stalked the pilgrimage routes and the stages of the roads.[33]

The success of Turkmen excursions depended on the ability of their horses to make long voyages across the Black Sands Desert. The best horses were able to make the journey of "six hundred miles in six, or even five days," surviving on the scant grains carried in woven bags of wool plus what the desert supplied.[34]

Conditioning horses for the extraordinary hardships of distance and desert was crucial to ensuring the success of Turkmen raids. Those who committed to making the raid put their horses through a rigorous, monthlong training regimen in preparation for the impending journey. In this manner, the Turkmen "cooled their horses." Horses were fed lean amounts of hay and barley to reduce their weight and improve their pace. They were ridden at full gallop for half an hour every day, kept sweating until their fat entirely disappeared, and given little water. In his *Narrative of a Journey into Khorasān*, James Baillie Fraser even suggested that it was "a common practice for the Turkmen to teach their horses to fight with their heels, and thus assist their master in the time of action, and, at the will of their rider, to run at, and lay hold of with their teeth whatever men or animals may be before them."[35] After thirty days of such training, the Turkmen would then begin the raiding expedition, each man taking two horses into the field—one a beast of burden to wait with supplies at the Iranian borderlands and the other a charger to be used in the raid. "When a Turcoman horse has given great proofs of strength and endurance in a *chap-auol* [*chapu* being the name of the Turkmen's sudden raids]," wrote J. P. Ferrier, a French soldier in the service of the Qajar Dynasty, "he never leaves the tribe except by force of arms."[36] According to the Swiss explorer and travel writer Henri Moser, raiding expeditions were led by a commander and involved anywhere from three to one thousand horsemen, depending on the number of men in an encampment with a good horse, some arms, and some courage.[37]

Eurasian empires tried to no avail to put an end to the raids and thus reclaim the desert. The Safavids of Persia were vexed in their efforts to establish state authority on the Turkmen frontier and the steppes of the Oxus. In the 1590s, for instance, the Turkmen overran and destroyed the fortress of Mubarakabad (Aq Qal'a) near Astarabad, which had been designed as a bulwark against their incursions. Shah Abbas I (1588–1629) was perennially forced to take measures to bring order to the Turkmen frontiers of the empire. In 1598, he marched from Isfahan to Khurasan, restoring Mubarakabad, retaking Mashhad and Herat, and attempting to reclaim parts of the empire's Central Asian fringes. The shah sought to drive the Turkmen from the valleys of the Kopet Dagh Mountains. He ordered a deep trench to be built from the foot of the mountains to the shores of the Caspian Sea, and resettled thousands of Kurds and other tribal groups to guard the eastern frontier.[38] The Safavids pursued a cautious and defensive policy toward their Eurasian Steppe frontier. The various Turkmen tribes—the Tekke foremost among them—established their control over the Qara Qum after the fall of the Safavid Dynasty in 1722. Thus, from the dynastic instability of the eighteenth century, the Turkmen came away with the upper hand in the steppe world of the Eurasian frontier.

During the first half of the nineteenth century, the Qajar Dynasty undertook repeated military expeditions to assert its authority on the eastern edges of "the guarded domains of Iran." During the seventeenth century, much of the region from Astarabad to the oasis of Marv had been the pastoral yurt territory of the Qajar tribes, who were then among the elite Safavid cavalry known as the Qizilbash. As the Qajars moved from the tent to the throne in Tehran, they carried out a series of largely unsuccessful expeditions to reclaim the Eurasian Steppe. In 1832–1833, the crown prince 'Abbas Mirza (1789–1833) and his new modern army, the Nizam-i Jadid, campaigned against Turkmen raiders and their fortresses in the eastern borderlands of Qajar Iran, pacifying the Salor Turkmen at the frontier post of Sarakhs. But 'Abbas Mirza died in 1833, while preparing to march on the oasis of Marv.[39] Later in 1851, when Iran's reform-minded minister Mirza Taqi Khan Amir Kabir (1807–1852) sent a mission to the Oxus River and the oasis of Khwarazm in order to project Qajar power into the steppes and halt the Central Eurasian slave trade in Persian Shi'i subjects, members of the mission were themselves almost taken captive.[40] In what would effectively be the last Qajar campaign to lay claim to their steppe borderlands, the Turkmen routed thousands of Qajar infantry and cavalry that had marched on Marv in 1860, leading many of them into captivity. [41] Thus, the expanse of the Black Sands Desert was left virtually independent of all imperial authority, while the Turkmen boasted of resting "neither under the shade of a tree or a king."[42] The equestrianism of the Turkmen peoples involved them in networks of commerce and cross-ecological exchanges between the desert and the sown, and enabled pastoralists to craft an indigenous steppe world where power pushed outward against empires. Thus, through their horsemanship, the Turkmen established trade networks, made territorial claims, and maintained an enduring steppe landscape in the Qara Qum.

From Delhi to Bukhara: The Travels of Mir Izzat Ullah in the Steppes of the Turkmen

Surrounding Eurasian empires, from the Safavids and Qajars in Iran to the East India Company in Mughal South Asia, surveyed the equestrian culture and economy of the steppes and sought to acquire swift Turkmen horses for their own stables. As empires pursued steppe horses, they also sought to understand equestrian breeds, an endeavor that produced a great many written texts. A rich tradition of Persianate literature revolves around the taming of horses—aside from the well-known tales of the exploits of Rustam and his fabled piebald colt, Rakhsh, in Firdawsi's *Shahnama*, and the wondrous horses of the pastoral steppes painted by Muhammad Siyah Qalam ("Black Pen") and Kamal al-Din Bihzad to illustrate stories. Horse books and travel accounts are other types of works that survey and leave a record of equestrianism in the Eurasian Steppe.

The writing and compilation of knowledge about the horse in the genre of the *farasnama* or "horse book" (from *faras*, Arabic for horse) was encouraged during the reign of the Mughal, Safavid, and Qajar dynasties, suggesting the importance of horses for empires into the early modern "gunpowder" period and the nineteenth century. The genre of the *farasnama* provided these Islamic empires, which relied on horsemen (*savaran*) for their expansion and longevity, with much needed information about the Central Eurasian horse trade and its culture. Bringing together knowledge from Sanskrit, Arabic, and Persian texts, *farasnamas* were books solely about horses. Variously titled *farasnama* (horse book), *khaylnama* (book of horsemen), or *baytarnama* (veterinary book), they collected mythical accounts and other information about the taming of horses, the colors of horses, the classification of different breeds, and the diseases of horses and their cures.[43] As such, horse books belong to the tradition of Perso-Arabic wonders and marvels literature (*'aja'ib al-makhluqat*), which classifies the flora, fauna, and minerals of the sub-lunar world.[44]

Horse books drew their lessons about the rearing and taming of horses from eclectic sources and a variety of literary traditions. Thus, for example, one book included Hindu legends attributed to the sage Salihotra. The book relates that the horse was created a winged animal that no human could tame until Salihotra deprived it of its wings. No longer able to visit distant lands in search of medicinal herbs, horses approached Salihotra and requested that he write a book about the treatment of their diseases.[45] A Persian *farasnama* written by 'Abdallah Khan Firuz Jang during the reign of the Mughal emperor Shah Jahan (1628–1658) states that according to Muslim traditions attributed to 'Ali ibn Abu Talib, the horse was created of wind as man was created of dust.[46] *Farasnamas* also contain facts about different types and colors of horses and offer detailed treatments for their various diseases. The finest and swiftest breeds of horses described in the *farasnama* literature are the Arab (*asb-i tazi*) and the Turkmen (*asb-i Turkmani*). According to these works, the Turkmen horse was bred from the Arabian. As written in an early nineteenth-century Qajar *farasnama* by Asadallah Khan Khvansari, the horses of the Turkmen had become mixed with Arab breeds and contained the two bloodlines (*du rag*).[47]

In the genre of the *farasnama*, Turkmen horses were famed for their gallop (*duidan, randan, takhtan*). The horses of the Turkmen were patient and well suited for fast riding (*savari*). Most of all, they were thought to be the best breeds for charging across long distances. An anonymous *farasnama*, written in the early nineteenth century during the reign of Fath 'Ali Shah Qajar (1797–1834), provides details about what was known of wild Turkic horses (*m'arifat-i asban-i Turki-yi vahshi*) that grazed in the desert steppes (*sahra*) and wilderness (*biyaban*). Exposed to rain and snow, they were quite hardy and were livelier (*bi nishat tar*) than other horses.[48] Some of these horses were known to abandon their packs and run to the hills, becoming wild again. Wild horses and their foals, known as *asb-i daghi* or "mountain horses," could only be caught, made tame (*ram kardan*), and saddled (*zin bastan*) after being chased down by several horsemen.[49]

Nineteenth-century natural histories on the subject of Turkmen horses and equestrianism were complemented by surveys of the Eurasian horse trade in the genre of travel writing (*safarnama*). Mir Izzat Ullah's early nineteenth-century travel account, written by a "native" explorer sent by the East India Company from Mughal India to penetrate the Central Eurasian horse economy and find Turkmen horses for the company stables, exemplifies this form of the travel survey. According to William Moorcroft of the East India Company, the purpose of Mir Izzat Ullah's *Ahval-i Safar-i Bukhara* was to gather information useful in the "infusion of the blood and the bone" of Turkmen horses into the Indian cavalry.[50]

Mir Izzat Ullah's journal surveys the Central Eurasian horse trade and delineates the interdependence between the landscape of the steppes or "wildlands" (*biyaban*) and the sown or cultivated (*abadan*).[51] Mir Izzat Ullah's travel account was, as Mohamad Tavakoli-Targhi has suggested, a "homeless text"—a work commissioned by the British and written in Persian—that surveyed the frontiers and borderlands of Central Eurasia. Mir Izzat Ullah was an "in-between" writer. That is, his narrative conformed to certain Western conventions of information gathering while still adhering to standards of the Persian *safarnama*. It was a text produced under the influence of both indigenous and European conventions, part road book (in the genre of the Perso-Arabic book of routes and provinces, *kitab al-masalik al-mamalik*) and part colonial gazetteer.[52]

Mir Izzat Ullah departed for his journey on April 20, 1812. From Delhi, he traveled to Kashmir and Ladakh, and then to Kashghar, Khoqand, Samarqand, Bukhara, and Balkh, returning to India by way of Bamiyan and Kabul.[53] His journal serves as a kind of road guide to the horse markets of early nineteenth-century Central Eurasia. In its opening, he acknowledges his objective, noting that he has "written down by pen all that he has himself seen and heard."[54] Mir Izzat Ullah's *Ahval-i Safar-i Bukhara* exemplifies the nineteenth-century British imperial policy of relying on a "native explorer"—in this case a traveler from the Persophone world—as a go-between to survey and penetrate Central Eurasia and its horse trade.[55]

During the course of his 1813 journey, Mir Izzat Ullah recorded information about the environs, markets, resources, and curiosities of the places he visited. He subsequently traced the stages (*manazil*) of overland Eurasian trade routes from Kashmir, Tibet, and the Farghana Valley in the east to Samarqand, Bukhara, and Balkh in the west, entering into the heartland of Turkmen horse country. Reaching Bukhara in April 1813, Mir Izzat Ullah surveyed and collected information on the geography, architecture (most notably, the many tombs of saints), and commerce of the city; he also drew a map detailing the city's eleven gates (*darvaza*).[56] But Mir Izzat Ullah's interest in trade and commerce seems to have always been at the forefront of his preoccupations. He, for instance, observed that Bukhara was a market for the sale of black sheep skins (*qaraqul*), cotton, silk, turquoise, and horses.[57] These goods could be found at bazaars in front of the Rigistan, the Charsu, and the shrine of Baha al-Din Naqshband, among other markets held throughout the city.[58] Mir Izzat Ullah compiled lists of the markets, caravanserais, merchants, commodity prices, and currencies—gold (*tala*) and silver (*tanga*) coins were the principle units of exchange—to be found in Bukhara.[59] He also described the state of the caravan roads that led to and from the city, in certain instances offering practical warnings to other travelers.[60] In his description of the overland caravan trade of Central Eurasia, Mir Izzat Ullah also took measure of the Turkmen nomads of the steppes and their important seasonal contacts with cities and markets, where they traded and sold their horses. The Central Eurasian horse trade enmeshed the Turkmen and other pastoral nomadic tribes into wider economic networks.

The Turkmen's integration into the trade and traffic of Central Eurasia is conveyed by Mir Izzat Ullah's description of the frequent horse fairs of Bukhara.

He writes of the weekly horse market (*bazaar-i asb*) held at the tomb (*mazar*) of Khvaja Baha al-Din Naqshband on the outskirts of Bukhara. There, beautiful and lively (*khush shikl va chalak*) Turkmen horses (*asb-i Turkoman*) could be found:

> Horse fairs are held four times weekly: on Mondays, Thursdays, and Saturdays within the walls at the Rigistan Gate near the citadel, and on Wednesdays outside the city, near the tomb of Khvaja Baha al-Din Naqshband. Horses, of low price only, fetching from ten to fifteen *tala*, are exposed for sale at the fairs: the more valuable animals are bought and sold through merchants. Mares are difficult to procure, and fetch as much as twenty *tala*: they are generally kept by the wealthy for breeding purposes, but are occasionally sold when, owing to a scarcity of horses, the market value reaches a high figure.... The best class of horses fetch one hundred to one hundred and fifty *tala*, but few animals of this description are to be had, perhaps not more than fifteen or twenty in the whole city. Horses ranging from fifty to sixty *tala* could be procured in large numbers, if sufficient notice were given. Some fifty or sixty horses are exposed for sale on every fair day, and probably from five to ten of these are sold.... Those of the Turkmen breed are handsome and active.... They come from the neighborhood of Marv and the country on both sides of the Jayhun [Oxus] and are consequently rare in Bukhara.[61]

It was known that every Wednesday morning, the amir of Bukhara, Mir Haydar Khan, walked on foot to the hallowed Naqshbandi tomb, offered prayers and alms, and then returned to the city mounted on a horse he had acquired at the fair.[62] Along with the Kurabheer or Uzbaki breed, Turkmen horses were among the most widely coveted varieties in Bukhara. In the markets, "tall, striking Turkmen horses with active dispositions" (*asb-i Turkman-i qamat buland, khubsurat va khushraftar*) were sold for between twenty and one hundred *tala*.[63] Mir Izzat Ullah reported that Turkmen horses were domesticated and bred in the steppes on both banks of the Oxus River, in proximity to horse markets and fairs in the frontier cities of Transoxiana and Khurasan. Turkmen breeds could be found in the markets of Samarqand, Bukhara, Balkh, Herat, Marv, Mashhad, and Nayshapur, among other cities.[64] Mir Izzat Ullah found that Balkh, a historic Central Asian oasis and trading entrepot known as *Umm al-Buldan* or "The Mother of Cities," still possessed a spacious bazaar in the early nineteenth century and still served as a market for the Eurasian horse trade. In Balkh, bazaars were held every Tuesday and Saturday, and Turkmen horses were readily available for sale.[65]

In his search for Turkmen horses, Mir Izzat Ullah continued his travels to Bamiyan and Kabul, finally returning to Delhi on December 16, 1813. His pioneering twenty-one-month journey provided invaluable information about the routes, distances, and markets of the Eurasian caravan trade. In his journal, he surveyed the Oxus region in 1812–1813 on behalf of the East India Company. His *Ahval-i Safar-i Bukhara* offers a point of entry into the networks of Central Eurasian horse traffic and the ecology of the steppes. In the so-called badlands between Islamic empires, Turkmen pastoralists and steppe peoples forged a trading and raiding equestrian world. As Turkmen horses became coveted across Eurasia, merchants and travelers entered the steppes and trekked through border towns and horse markets in search of them, taking measure along the way of the equestrianism of the Black Sands Desert.

The equestrian culture of the Turkmen and their horse economy in the Qara Qum Desert was thus etched on the edges of early modern Eurasian empires and on the pages of Persianate travel narratives. Mir Izzat Ullah would return to the steppes as part of the landmark, though ill-fated, Moorcroft Expedition of 1819. Most of the members of the mission, including Mir Izzat Ullah and Moorcroft, died of cholera in Afghanistan in 1825.[66] The peregrinations of Mir Izzat Ullah took him through a borderland space where contacts and encounters between empires and steppe peoples were commensurate and symmetrical. His journeys betray the strength and resilience of a pastoral world on empire's edge. Even more, Mir Izzat Ullah's journeys demonstrate the interaction and encounter between different "regimes of circulation," between the pastoral societies of the Eurasian Steppe and a merchant traveler from the imperial setting of the East India Company.[67]

Conclusion

This chapter argues for a steppe of possibilities rather than a steppe of closures. The standard narrative of imperial contacts with indigenous populations is one of empires closing and settling frontiers and pacifying "tribal" peoples. But this narrative is hard to plant in the sands of the Qara Qum even into the early nineteenth century. Far from attempting to impose settlements and agriculture in the steppes, sedentary empires actually sought to enter the mobile networks of the desert to gain from pastoral resources. An environmental perspective demonstrates the transfrontier exchanges, trade, and traffic that linked pastoralists and empires on the "middle ground" of the Eurasian Steppe.

The shifting of the Oxus River and the expansion of the Black Sands Desert in the early modern period created new possibilities for indigenous pastoral populations that lasted well into the nineteenth century. Through their skilled horsemanship, the Turkmen carved out a loose trading and raiding confederation built on the power and speed of horses capable of making long and arduous journeys across the steppes. The steppes exerted their pull on surrounding empires and shaped the nature of imperial frontiers. The steppe landscape could destabilize, even conquer, surrounding imperial regimes, just as it could supply empires with much sought-after pastoral resources. As Turkmen horses gained fame and became coveted commodities, early modern Eurasian empires set out to find Central Eurasian breeds and hence harness the pastoral resources of the steppes. The steppe world of the Turkmen was inscribed in nineteenth-century travel narratives such as Mir Izzat Ullah's *Ahval-i Safar-i Bukhara*. In traversing the steppes, Mir Izzat Ullah entered an environment thoroughly outside of imperial control and encountered a landscape of immeasurable ecological transformations, a desert frontier under the sway of Turkic horsemen.

During the age of the early modern Islamic empires of the Ottomans, Mughals, and Safavids/Qajars, the deserts of the MENA and Central Asia remained a reservoir of cities and the desert steppes a vast borderland space of multilayered contacts and encounters. The pastoral and urban—the steppe and the sown—were interdependent, and the interface between the two was defined by modes of traffic, circulation, and exchange.

Acknowledgment

I thank Alan Mikhail, Sanjay Subrahmanyam, Nile Green, and Harriet Ritvo for their comments and suggestions on earlier versions of this chapter.

Notes

1. Edmund Burke III, "The Transformation of the Middle Eastern Environment, 1500 B.C.E.–2000 C.E.," *The Environment and World History*, ed. Edmund Burke III and Kenneth Pomeranz (Berkeley: University of California Press, 2009), 82.

2. Ibn Khaldun, *The Muqaddimah: An Introduction to History*, trans. Franz Rosenthal (Princeton, NJ: Princeton University Press, 1967), 93.

3. Ibid.

4. For a classic Marxist perspective on the conquest of indigenous peoples by empires on a global scale, see Eric Wolf, *Europe and the People Without History* (Berkeley: University of California Press, 1982). On the closing of Eurasian

frontiers, and the eclipse of indigenous pastoral populations and "tribal" power, see John F. Richards, *The Unending Frontier: An Environmental History of the Early Modern World* (Berkeley: University of California Press, 2003); Peter C. Perdue, *China Marches West: The Qing Conquest of Central Eurasia* (Cambridge, MA: Harvard University Press, 2005); Arash Khazeni, *Tribes and Empire on the Margins of Nineteenth-Century Iran* (Seattle: University of Washington Press, 2010).

5. Such studies include two seminal works on empire in the MENA: Edward W. Said, *Orientalism* (New York: Pantheon Books, 1978); Timothy Mitchell, *Colonising Egypt* (Berkeley: University of California Press, 1991). For recent correctives to the prevailing dualistic view of encounters between empires and Muslim subjects, see Julia A. Clancy-Smith, *Mediterraneans: North Africa and Europe in an Age of Migration, c. 1800–1900* (Berkeley: University of California Press, 2011); Nile Green, *Bombay Islam: The Religious Economy of the West Indian Ocean, 1840–1915* (Cambridge: Cambridge University Press, 2011).

6. For some examples of the prevailing empire-centered perspective on Central Eurasian history, see Perdue, *China Marches West*; Christopher I. Beckwith, *Empires of the Silk Road: A History of Central Eurasia from the Bronze Age to the Present* (Princeton, NJ: Princeton University Press, 2009); Mark C. Elliott, *The Manchu Way: The Eight Banners and Ethnic Identity in Late Imperial China* (Stanford, CA: Stanford University Press, 2001); Pamela Kyle Crossley, *Orphan Warriors: Three Manchu Generations and the End of the Qing World* (Princeton, NJ: Princeton University Press, 1990); Thomas J. Barfield, *The Perilous Frontier: Nomadic Empires and China* (Cambridge, UK: Basil Blackwell, 1989); René Grousset, *The Empire of the Steppes: A History of Central Asia*, trans. Naomi Walford (New Brunswick, NJ: Rutgers University Press, 1970).

7. Nicola Di Cosmo, Allen J. Frank, and Peter B. Golden, "Introduction," in *The Cambridge History of Inner Asia: The Chinggisid Age*, ed. Nicola Di Cosmo, Allen J. Frank, and Peter B. Golden (Cambridge: Cambridge University Press, 2009), 1.

8. Richard White, *The Middle Ground: Indians, Empires, and Republics in the Great Lakes Region, 1650–1815* (Cambridge: Cambridge University Press, 1991), x.

9. On trans-ecological exchanges in Central Eurasian history, see David Christian, "Silk Roads or Steppe Roads? The Silk Roads in World History," *Journal of World History* 11 (2000): 1–26. On the links between Indian "trade diasporas" and Central Eurasia, see for example Stephen Frederic Dale, *Indian Merchants and Eurasian Trade, 1600–1750* (Cambridge: Cambridge University Press, 1994); Scott C. Levi, *The Indian Diaspora in Central Asia and its Trade, 1550–1900* (Leiden: Brill, 2002).

10. While there is a relatively abundant literature on the Central Eurasian horse trade, much of it focuses either on the role of horses in the building of medieval and early modern empires or on the Eurasian horse in the context of world history. There has yet to be a concerted examination of the subject of equestrianism in Central Eurasia from the perspective of environmental history. For some examples from the existing literature, see Jos Gommans, "The Horse Trade in Eighteenth-Century South Asia," *Journal of the Economic and Social History of the Orient* 37 (1994): 228–250; idem., "Warhorse and Post-Nomadic Empire in Asia, c. 1000–1800," *Journal of Global History* 2 (2007): 1–21; idem., *The Rise of the Indo-Afghan Empire, c. 1710–1780* (Leiden: Brill, 1995); idem., *Mughal Warfare: Indian Frontiers and Highroads to Empire*

(New York: Routledge, 2002). Also see J. Masson Smith Jr., "Nomads and Ponies vs. Slaves on Horses," *Journal of the American Oriental Society* 118 (1998): 54–63; Rudi Paul Lindner, "Nomadism, Horses, and Huns," *Past and Present* 92 (1981): 3–20; G. Rex Smith, *Medieval Muslim Horsemanship: A Fourteenth-Century Arabic Cavalry Manual* (London: British Library, 1979); Garry Alder, "The Origins of the 'Pusa experiment': The East India Company and Horse-Breeding in Bengal, 1793–1808," *Bengal Past and Present* 98 (1979): 10–12; Denis Sinor, "Horse and Pasture in Inner Asian History," *Oriens Extremus* 19 (1972): 171–183; Simon Digby, *War-Horse and Elephant in the Delhi Sultanate: A Study of Military Supplies* (Oxford: Orient Monographs, 1971). Among the features of this research on Eurasian equine traffic has been the designation of the period after 1500 as signaling the arrival of a "post-nomadic" world. Yet if we were to change our vantage point from the imperial center to the frontiers, it would hardly seem possible to view early modern Eurasia as post-nomadic. For some recent works on horses in world history, see David W. Anthony, *The Horse, the Wheel, and Language: How Bronze-Age Riders from the Eurasian Steppes Shaped the Modern World* (Princeton, NJ: Princeton University Press, 2007); Pita Kelekna, *The Horse in Human History* (Cambridge: Cambridge University Press, 2009).

11. While one is somewhat hard-pressed to call the Turkmen's realm in the Qara Qum an informal "empire," such as seemingly was the case with the Comanche in the Southern Plains of North America, these different borderlands cases share some striking similarities. However, the Turkmen of the Qara Qum, despite the rising ascendancy of the Tekke, were divided into different tribes (*tavayif*) and never established a confederacy (*il*), let alone an empire. On the Comanche, see Pekka Hämäläinen, *The Comanche Empire* (New Haven, CT: Yale University Press, 2008). I thank Helena Wall for bringing Hämäläinen's outstanding work to my attention.

12. René Létolle, "Histoire de l'Ouzboi, cours fossil de l'Amou Darya: synthese et elements nouveaux," *Studia Iranica* 29 (2002): 195–240; René Létolle, Philip Micklin, Nikolay Aladin, and Igor Plotnikov, "Uzboy and the Aral Regressions: Hydrological Approach," *Quaternary International* 173–174 (2007): 125–136; M. Konchine, "La Question de l'Oxus," *Annales de Géographie* 5 (1895–1896): 496–504. Some references to the flow of the Oxus may be gathered from contemporary written sources, which offer scattered descriptions of the river's changing course. The tenth-century Arab geographer al-Muqaddasi reported that the Oxus River flowed into the Aral Sea but identified an old bed that led to the Caspian. Writing in the fifteenth century, following the destruction of dams and irrigation works on the Oxus that diverted the river's flow toward the Caspian Sea, the Timurid geographer, Hafiz-i Abru, claimed that the Aral Sea had nearly disappeared. G. Le Strange, *The Lands of the Eastern Caliphate: Mesopotamia, Persia, and Central Asia from the Moslem Conquest to the Time of Timur* (Cambridge: University Press, 1905), 455–457.

13. Abu'l Ghazi Bahadur Khan's *Shajara-yi Tarakima* was translated into French as: Ebülgâzî Bahadir Han, *Histoire des Mogols et des Tatares*, trans. Le Baron Desmaisons, 2 vols. (St. Pétersbourg: Imprimerie de l'Académie Impériale des sciences, 1871–1874). For the references to the changing course of the Oxus, see: ibid., 1: 221, 312; 2: 207, 291.

14. Konchine, "La Question de l'Oxus," 496–504.

15. Raphael Pumpelly, ed., *Explorations in Turkestan, Expedition of 1904: Prehistoric Civilizations of Anau: Origins, Growth, and Influence of Environment*, 2 vols. (Washington, DC: Carnegie Institution of Washington, 1908), 2: 295.

16. Yuri Bregel, "Uzbeks, Qazaqs, and Turkmen," in *The Cambridge History of Inner Asia: The Chinggisid Age*, ed. Nicola Di Cosmo, Allen J. Frank, and Peter B. Golden (Cambridge: Cambridge University Press, 2009), 233.

17. Anonymous, "Safarnama-yi Bukhara," [1259–1260/1844], Kitabkhana-yi Majlis, Tehran, mss. 2860, published as Husayn Zamani, ed., *Safarnamah-i Bukhara* (Tehran: Pizhūhishgāh-i 'Ulūm-i Insani va Mu'tala'at-i Farhangi vā'bastah bih Vizarat-i Farhang va Āmūzish-i 'Ālī, 1994), 72–73.

18. Population figures are based on the following: James Baillie Fraser, *Narrative of a Journey into Khorasān, in the years 1821 and 1822* (London: Longman, Hurst, Rees, Orme, Brown, and Green, 1825); Alexander Burnes, *Travels into Bokhara; Being the Account of a Journey from India to Cabool, Tartary, and Persia*, 3 vols. (London: J. Murray, 1834); Zamani, ed., *Safarnamah-i Bukhara*; Arminius Vambery, *Travels in Central Asia: Being the Account of a Journey from Teheran across the Turkoman Desert on the Eastern Shore of the Caspian to Khiva, Bokhara, and Samarcand Performed in the year 1863* (London: J. Murray, 1864); The National Archives of the United Kingdom, Foreign Office 60/379, "Report by Ronald Thomson on the Toorkoman tribes occupying districts between the Caspian and the Oxus," Tehran, February 29, 1876; Charles Marvin, *Merv, the Queen of the World; and the Scourge of the Man-Stealing Turcomans* (London: W. H. Allen, 1881); Henri Moser, *À travers l'Asie Centrale: la Steppe Kirghize, le Turkestan russe, Boukhara, Khiva, le pays des Turcomans et la Perse, impressions de voyage* (Paris: Librairie Plon, 1885).

19. J.P. Ferrier, *Caravan Journeys and Wanderings in Persia, Afghanistan, Turkistan, and Beloochistan* (London: J. Murray, 1856), 94–95.

20. Bregel, "Uzbeks, Qazaqs, and Turkmen," 233.

21. The Yamut breed was generally smaller than and not as fast as the Akhal Tekke, but was known to be stronger-boned and more durable. Amin Guli, *Tarikh-i Siyasi va Ijtima'i-i Turkman'ha* (Tehran: Nashr-i 'Ilm, 1987), 308–314. See also Marvin, *Merv, the Queen of the World*, 166.

22. Guli, *Tarikh-i Siyasi va Ijtima'i-i Turkman'ha*, 312–314. Charles Metcalfe MacGregor presents a passage intricately illustrating the saddle equipment attached to the Turkmen horse, including the "saddle tree" (*zin*); the saddle pad (*tukultu*) of felt (*namad*), leather, and cloth (often embroidered silk); and felts (*jul*) laid on the horses' backs. Each horseman also carried a woven saddlebag (*khurjin; juval*), which contained all the provisions necessary during marches. Charles Metcalfe MacGregor, *Narrative of a Journey Through the Province of Khorassan and on the N.W. Frontier of Afghanistan in 1875*, 2 vols. (London: W. H. Allen, 1879), 2: 16–18.

23. Ferrier, *Caravan Journeys and Wanderings*, 97.

24. John Malcolm, *The History of Persia, from the Most Early Period to the Present Time*, 2 vols. (London: J. Murray, 1815), 2: 157–158.

25. V. V. Bartol'd, *Mīr 'alī-Shīr: A History of the Turkman People*, vol. 3 of *Four Studies on the History of Central Asia*, trans. V. and T. Minorsky (Leiden: Brill, 1962), 154.

26. Mir 'Abd al-Karim Bukhari, *Histoire de l'Asie Centrale (Afghanistan, Boukhara, Khiva, Khoqand) depuis les dernières années de règne de Nadir Chah (1153)*, ed. Charles Schefer (Paris: E. Leroux, 1876), 77.

27. Hajji 'Abdullah Khan Qaragazlu, *Majmu'a-yi Athar*, ed. 'Inayatullah Majidi (Tehran: Miras Makub, 2003), 129 and 133–134. For an account of the bazaar of Marv in the late nineteenth century, see Edmund O'Donovan, *The Merv Oasis: Travels and Adventures East of the Caspian during the Years 1879–80–81, Including Five Months' Residence among the Tekkés of Merv*, 2 vols. (London: Smith Elder, 1882), 2: 321–337.

28. Muhammad Hasan Khan I'timad al-Saltana, *Mir'at al-Buldan*, ed. 'Abdul Husayn Nava'i (Tehran: University of Tehran, 1988), 1: 352.

29. Some of these reports are now archived at the India Office Library in London. See, for instance, India Office Records F/4/1533 (February 1834–April 1835), "Reports from the Commanding Officers of various Madras Cavalry Regiments on the Turkoman and Khorasan horses in their regiments, includes descriptions of individual horses."

30. Pumpelly, *Explorations in Turkestan*, 2: 434.

31. Joseph Wolff, *Travels and Adventures of the Reverend Joseph Wolff* (London: Saunders, Otley, and Co., 1861), 285.

32. Lisan al-Mulk Sipihr, *Nasikh al-Tavarikh*, ed. Jamshid Kiyanfar, 6 vols. (Tehran: Asatir, 1998), 3: 1193–1194.

33. Fazl Allah Shirazi, *Tarikh-i Zu al-Qarnayn*, ed. Nasir Afsharfar, 2 vols. (Tehran: Vizarat-i Farhang va Irshad-i Islami, 2001), 2: 764.

34. Ferrier, *Caravan Journeys and Wanderings*, 95.

35. Fraser, *Narrative of a Journey into Khorasān*, 271.

36. Ferrier, *Caravan Journeys and Wanderings*, 95.

37. Moser, *À travers l'Asie Centrale*, 324.

38. Riza Quli Khan Hidayat, *Sifaratnama-yi Khvarazm*, ed. Charles Schefer (Paris: Ernest Leroux, 1876), 27. In the early seventeenth century, the task of administering and preserving order on the Turkmen frontier was entrusted to Faridun Khan (d. 1621), the governor of Astarabad, whose campaigns against the Turkmen are related in a *fathnama* (book of victory) penned by Muhammad Tahir Bistami. See Muhammad Tahir Bistami, *Futuhat-i Fariduniyah: Sharh-i Jang'ha-yi Faridun Khan Charkas Amir al-Umara-yi Shah 'Abbas-i Avval*, ed. Mir Muhammad Sadiq and Muhammad Nadir Nasiri Muqaddam (Tehran: Nuqtah, 2001).

39. On the campaigns of 'Abbas Mirza against the Turkmen in Khurasan as recorded in Qajar chronicles, see Fazl Allah Shirazi, *Tarikh-i Zu al-Qarnayn*, 2: 819–836 and 872–887; Lisan al-Mulk Sipihr, *Nasikh al-Tavarikh*, 1: 457, 483–488, 500–505; Riza Quli Khan Hidayat, *Rawzat al-Safa-yi Nasiri*, vol. 9, ed. Jamshid Kiyanfar (Tehran: Asatir, 2001), 7948 and 8022–8029. See also Gavin R. G. Hambly, "Iran during the Reigns of Fath 'Ali Shah and Muhammad Shah," in *From Nadir Shah to the Islamic Republic*, vol. 7 of *The Cambridge History of Iran*, ed. Peter Avery, Gavin Hambly, and Charles Melville (Cambridge: Cambridge University Press, 1991), 166; Nasir Najmi, *Iran dar Miyan-i Tufan ya Zindigani-yi 'Abbas Mirza* (Tehran: Kanun-i

Ma'rifat, 1957); Homa Nategh, "'Abbas Mirza va Turkamanan-i Khurasan," *Nigin* 10, no. 112 (September 22, 1974): 13–17; Emineh Pakravan, *Abbas Mirza* (Paris: Buchet/ Chastel, 1973).

40. For an account of the 1851 Khwarazm Mission, see Hidayat, *Sifaratnama-yi Khvarazm*. The following French translation of this text was published in 1879: Charles Schefer, *Relation de l'Ambassade au Kharezm* (Paris: Ernest Leroux, 1879). All citations refer to the Paris 1876 edition of the *nasta'liq* Persian text originally published by Bulaq.

41. For a narrative of the disastrous Persian campaign to Marv in 1861, see Henri de Couliboeuf de Blocqueville, "Quatorze mois de captivite, chez les Turcomans aux frontieres du Turkestan et de la Perse, 1860–1861 (Frontières du Turkestan et de la Perse)," in *Le Tour du Monde* (Paris: Libraire de L. Hachette, 1866), 225–272. For nineteenth-century Persian captivity narratives in Central Eurasia, see Mirza Mahmud Taqi Ashtiyani, *'Ibratnama. Khatirati az Dawran-i Pas az Jangha-yi Herat va Marv [c. 1278–1288/1860–1870]*, ed. Husayn 'Imadi Ashtiyani (Tehran: Nashr-i Markaz, 2003); Sarhang Isma'il Mirpanja, *Khatirat-i Asarat: Ruznama-yi Safar-i Khvarazm va Khiva [1280/1862]*, ed. Safa al-Din Tabarrayan (Tehran: Mu'assasa-yi Pajhuhish va Mutala'at-i Farhangi, 1991).

42. Marvin, *Merv, the Queen of the World*, 32.

43. C. A. Bayly, *Empire and Information: Intelligence Gathering and Social Communication in India, 1780–1870* (Cambridge: Cambridge University Press, 1996), 158–159. The British in India and Central Eurasia would in many ways continue this interest in writing on the subject of horses. See, for example, William Moorcroft, *Travels in the Himalayan Provinces of Hindustan and the Panjab*, 2 vols. (London: J. Murray, 1841). In the 1830s, for instance, the officers of the Madras Cavalry Regiments wrote now archived reports on the Turkmen and Khurasan horses in their ranks, including descriptions of individual horses.

44. Manuscripts of these *farasnamas* are scattered in archival collections from Los Angeles to London to Tehran. See, for instance, British Library, India Office, Add. 14057, Folios 3–60, 'Abd Allah ibn Safi, "Tarjuma-yi Salotar"; British Library, India Office, Add. 16854, Folios 3–74, 'Abd Allah Khan Firuz Jang, "Tarjuma-yi Salotar-i Asban"; British Library, India Office, Add. 16854, Folios 75–121, Zayn al-'Abidin ibn Abu'l Hasan Karbala'i Husayni Hashimi, "Farasnama"; British Library, India Office, Add. 7716, Folio 47, Nizam al-Din Ahmad, "Mizmar-i Danish"; British Library, India Office, Add. 23562, Folios 49–67, Muhammad 'Ali Hazin, "Farasnama"; British Library, India Office, Or. 1762, 28, Folios 533–536, Mirza Buchchu Bayg Salotar, "Farasnama"; Bodleian Library, Oxford University, Fraser Collection, Ms. 173, Muhammad bin Muhammad, "Aspnama"; Bodleian Library, Oxford University, Or. 590, 'Abd Allah Khan Firuz Jang, "Farasnama Hindi"; University of California, Los Angeles, Young Research Library, Minassian Collection, Box 166, Ms. 1394, Muhammad Husayn ibn Muhammad Salih al-Husayni Khatunabadi, "Muhasin al-Hisan"; University of California, Los Angeles, Young Research Library, Minassian Collection, Box 166, Ms. 1395, Husayni Isfahani, "Farasnama-yi Arastu"; University of California, Los Angeles, Young Research

Library, Minassian Collection, Box 166, Ms. 1396, Anonymous, "Farasnama." I thank Ghazzal Dabiri of Columbia University for bringing the manuscripts in the Minassian Collection to my notice.

45. Sa'adat Yar Khan, *The Faras-Nama-e Rangin or The Book of the Horse*, trans. D. C. Phillott (London: Bernard Quaritch, 1911), viii. This printing is based on a nineteenth-century Urdu manuscript. For other published examples of the genre of the *farasnama*, see 'Ali Sultani Gird Faramarzi, ed., *Du Farasnama-yi Manthur va Manzum* (Tehran: University of Tehran, 1987); 'Abd al-Husayn Mahdavi, ed., *Ganjina-yi Baharistan: Majmu'a-yi 6 Farasnama* (Tehran: Majlis-i Shuray-i Islami, 2008); Ibn Sayyid Abu'l Husain "Hashimi," *The Faras-Nama of Hashimi*, ed. D. C. Phillott (Calcutta: Asiatic Society, 1910); Abdallah Khan Firoze Jung, *A Treatise on Horses, Entitled Saloter, or, A Complete System of Indian Farriery*, trans. Joseph Earles (Calcutta: George Gordon, 1788).

46. Abdallah Khan Firoze Jung, *A Treatise on Horses*, viii.

47. Asadallah Khan Khvansari, "Farasnama," in *Ganjina-yi Baharistan*, 442 and 506.

48. University of California, Los Angeles, Young Research Library, Minassian Collection, Box 166, Ms. 1396, Anonymous, "Farasnama-yi Tuhfa-yi Padishahan va Buzurgan-i Sahiban-i Asban," 1805–06. This fact is also noted, nearly verbatim, in other horse books. See, for instance, Anonymous, "Farasnama-yi Manthur," in *Du Farasnama-yi Manthur va Manzum*, ed. 'Ali Sultani Gird Faramarzi (Tehran: University of Tehran, 1987), 70–71; Muhammad bin Muhammad, "Farasnama," in *Ganjina-yi Baharistan: Majmu'a-yi 6 Farasnama*, ed. 'Abd al-Husayn Mahdavi (Tehran: Majlis-i Shuray-i Islami, 2008), 114.

49. Muhammad, "Farasnama," 75.

50. Moorcroft, *Travels in the Himalayan Provinces*, 1: xl–xli.

51. I thank Mohamad Tavakoli-Targhi for bringing to my attention the Oxford manuscript of Mir Izzat Ullah's travel book, on which this chapter is based. Mir Izzat Ullah, *Ahval-i Safar-i Bukhara*, Ms. Or. 745, Sir William and Gore Ouseley Collection, Bodleian Library, Oxford University, ff. 236. Additional manuscripts of Mir Izzat Ullah's voyage may also be found at the India Office Library in London, Ms. 2009 and Ms. 2769. Different manuscript versions may additionally be found at the Bibliothèque Nationale in Paris, under the title *Masir-i Bukhara*, Suppl. Persan 1346 and Suppl. Persan 1283. For an English translation of the manuscript, see Sayyid Izzat-Allah, *Travels in Central Asia by Meer Izzut-Oollah in the Years 1812–1813*, trans. Captain Henderson (Calcutta: Foreign Department Press, 1872).

52. On "homeless texts," see Mohamad Tavakoli-Targhi, *Refashioning Iran: Orientalism, Occidentalism, and Historiography* (New York: Palgrave Macmillan, 2001). For an analysis of Mir Izzat Ullah's travels and the journal he kept, as well as a summary of his itinerary and an explanation of the value of his narrative as a historical source on Central Eurasian trade, see Maria Szuppe, "En quête de chevaux turkmènes: le journal de voyage de Mîr 'Izzatullâh de Delhi à Boukhara en 1812–1813," *Cahiers d'Asie centrale* 1–2 (1996): 91–111.

53. Mir Izzat Ullah traveled during parts of his journey with an Afghan named Hafiz Muhammad Fazil Khan, who apparently traveled in advance to Balkh and

Bukhara and left an account of the stages of the road that was incorporated into Mir Izzat Ullah's journal. For the itinerary of this traveler, see Muhammad Fazil Khan, *Tarikh-i Manazili Bukhara*, trans. Iqtidar Husain Siddiqui (Srinagar: Centre of Central Asian Studies for the University of Kashmir, 1981).

54. In Persian: *Har chi khud dida u shinida bi qayd-i qalam avurda*. Mir Izzat Ullah, *Ahval-i Safar-i Bukhara*, ff. 2.

55. For other examples of surveys of Central Eurasia and its horse trade written by "native explorers" like Mir Izzat Ullah, see Mohan Lal, *Travels in the Panjab, Afghanistan, Turkistan, to Balk, Bokhara, and Herat* (London: W. H. Allen, 1846); Khwaja Ahmud Shah Nukshbundee Syud, "Narrative of the Travels of Khwaja Ahmud Shah Nukshbundee Syud," *Journal of the Asiatic Society of Bengal* 25 (1856): 344–358; T. G. Montgomerie, "Report on the Trans-Himalayan Explorations, in connexion with the Great Trigonometrical Survey of India, during 1856–7: Route-Survey made by Pundit———," *Proceedings of the Royal Geographical Society of London* 12 (1867–1868): 146–175; idem., "Report of the Mirza's Exploration of the Route from Caubul to Kashgar," *Proceedings of the Royal Geographical Society of London* 15 (1870–1871): 181–204; idem., "Report of 'The Mirza's' Exploration from Caubul to Kashgar," *Journal of the Royal Geographical Society of London* 41 (1871): 132–193; idem., "A Havildar's Journey Through Chitral to Faizabad in 1870," *Proceedings of the Royal Geographical Society of London* 16 (1871–1872): 253–261; idem., "A Havildar's Journey Through Chitral to Faizabad in 1870," *Journal of the Royal Geographical Society of London* 42 (1872): 180–201; idem., "Journey to Shigatz, in Tibet, and Return by Dingri-Maidan into Nepaul, in 1871, by the Native Explorer No. 9," *Journal of the Royal Geographical Society of London* 45 (1875): 330–349; G. S. W. Hayward, "Route from Jellalabad to Yarkand Through Chitral, Badakhshan, and Pamir Steppe, Given by Mahomed Amin of Yarkand," *Proceedings of the Royal Geographical Society of London* 13 (1868–1869): 122–130; Ibrahim Khan, "Route of Ibrahim Khan from Kashmir through Yassin to Yarkand in 1870," *Proceedings of the Royal Geographical Society of London* 15 (1870–1871): 387–392; H. Yule, Munphool Pundit, and Faiz Bukhsh, "Papers Connected with the Upper Oxus Regions," *Journal of the Royal Geographical Society of London* 42 (1872): 438–481; H. Trotter, "Account of the Pundit's Journey in Great Tibet from Leh in Ladakh to Lhasa, and of His Return to India via Assam," *Proceedings of the Royal Geographical Society of London* 21 (1876–1877): 325–350.

56. For his description of Bukhara, see Mir Izzat Ullah, *Ahval-i Safar-i Bukhara*, ff. 117–54. For the map of the city, see: ibid., ff. 124.

57. Ibid., ff. 119–120.

58. Ibid., ff. 124–125.

59. Ibid., ff. 117–125.

60. He, for instance, cautioned that the journey between Bukhara and Balkh was through a waterless desert and risked encounters with Turkmen in the habit of raiding (*qarat kardan*) caravans. Ibid., ff. 162.

61. Ibid., ff. 118–119; Sayyid Izzat-Allah, *Travels in Central Asia*, 58.

62. Mir Izzat Ullah, *Ahval-i Safar-i Bukhara*, ff. 134.

63. Ibid., ff. 125.

64. Ibid., ff. 162–163.

65. Ibid., ff. 168–169. The trade in these highly coveted Turkmen horses extended through the Qara Qum and Qizil Qum deserts and toward Iran. We know from nineteenth-century geographical chronicles, including *Mir'at al-Buldan*, compiled by I'timad al-Saltana, that the Turkmen horse trade reached the Iranian province of Khurasan. He writes, "The horses of the Turkmen, which are from the Turkmen tribes of Tekke at Marv, are taken to Mazandaran, Tehran, and other lands in Iran, and are sold and traded there." I'timad al-Saltana, *Mir'at al-Buldan*, 1: 352.

66. Mir Izzat Ullah fell sick in Qunduz in northern Afghanistan and later died on the road between Kabul and Peshawar while headed back to India. Moorcroft, *Travels in the Himalayan Provinces*, 1: lxviii; Szuppe, "En quête de chevaux turkmènes," 93.

67. On "regimes of circulation," see Claude Markovits, Jacques Pouchepadass, and Sanjay Subrahmanyam, eds., *Society and Circulation: Mobile People and Itinerant Cultures in South Asia, 1750–1950* (Delhi: Permanent Black, 2003).

7

Enclosing Nature in North Africa: National Parks and the Politics of Environmental History

Diana K. Davis

As the French conquered North Africa, they fabricated a tale of environmental change that held the local North African populations, especially nomads, responsible for ruining what many Europeans believed had been a lush, fertile, and forested environment in the classical past—before the "Arab invasions" of the eleventh century. For the last two to three thousand years, however, the North African environment has been arid to semi-arid with highly irregular rainfall; most of the region is more suited to pastoralism than to settled farming.[1] While far from accurate, this French colonial environmental history served, beginning in 1830s Algeria, to undermine the lifeways of the indigenous populations as it justified—in the name of environmental protection—the expropriation of their land and property, the transfer of tribal forests to the French state, and the sedentarization of nomads.

Some of the most enduring symbols of this transformation are the multiple national parks and nature reserves created by the French in Algeria, Tunisia, and Morocco. Foreshadowed in the nineteenth century by hunting reserves, official national parks began in the Maghreb with the 1921 law in Algeria instituting their formation. The national park created two years later in Algeria was the first on the African continent. Developed ostensibly to protect nature and provide areas for scientific study, this park and those that followed were primarily built to generate tourism revenue. They further served as bases from which to monitor and control "problematic populations."

This chapter explores the history of these national parks and the complex, frequently negative effects they had and continue to have on local populations and the environment.

Nature Reserves and National Parks in (North) Africa

The first "nature reserves" on the African continent were game reserves created in the late nineteenth century initially to preserve wildlife for hunting by the colonizing European powers.[2] Many of these game reserves were transformed into national parks in the 1920s and later, although they still functioned, in effect, as game preserves because hunting was allowed just outside their permeable borders. Parc National Albert, now Virunga National Park (most commonly, though incorrectly, cited as the first national park in Africa), was established in 1925 in the Belgian Congo by Prince Albert from what had been a gorilla reserve.[3] Similarly, Kruger National Park in South Africa was created in 1926 from what had been the Sabie Game Reserve since 1892.[4]

Governments banned hunting to varying degrees in these sub-Saharan African game reserves to protect breeding females and their young in order to ensure a supply of animals for hunting in the future. In the process, the hunting rights of indigenous populations were severely curtailed or eliminated and those of Europeans were regulated with licenses, permits, and other measures. Most scholars agree that for sub-Saharan Africa, hunting and the preservation of game were the largest motivating factors in the creation of nature reserves and national parks in the early twentieth century. Authorities took much of the land that was protected, first as game reserves and later as national parks, in sub-Saharan Africa from local African hunting grounds and multiuse land long used by indigenous Africans, effectively criminalizing their traditional livelihood activities.[5] Under colonial rule, whites were hunters but blacks were poachers.

In North Africa, however, the French had established several national parks before 1925, and for different reasons that have gone largely unrecognized in the literature. The motivation in French North Africa, though, was more tightly allied to forestry and concern for the preservation of trees and other plants. French interest in environmental conservation paralleled international interest in the same, as evidenced by a flurry of conferences on protecting nature that took place around the turn of the century.

Such conferences included the ornithological meeting in Paris in 1895 (its recommendations regarding migrating birds were included in the 1919 Treaty of Versailles), the International Congresses of Zoology in 1901 and 1910, and the First International Congress for the Protection of Nature held in Paris in 1923.[6] In 1913, a coalition of seventeen European states led by the energetic Swiss zoologist Paul Sarasin formed the Consultative Commission for the International Protection of Nature headquartered in Basel, Switzerland.[7] By 1948, the influential IUPN (International Union for the Protection of Nature) would be created.[8]

1913 was a particularly important year for the development of national parks in French North Africa because the International Forestry Congress held in Paris that year issued sweeping recommendations concerning national parks. The Congress, in its discussions of the role of the forest in the development of tourism, declared that every country should create new national parks and extend any existing ones. This applied equally to colonial possessions.[9] By 1921, the French government passed the law establishing national parks in Algeria. Following the 1912 recommendations of the Society for Natural History of North Africa, and under the guidance of the Algerian forest service, the first national park in Africa, Les Cèdres de Téniet el-Haâd, was created in Algeria on August 3, 1923.[10] This was followed quickly by Dar-el-Oued Taza National Park on August 23, 1923. In all, five national parks were established in Algeria before Parc National Albert was formed in the Belgian Congo in 1925.[11] National parks L'Ouarensis and Djebel Gouraya were established in 1924; L'Akfadou, Chrea, Le Djurdjura, and Les Planteurs in 1925 (of the parks founded in 1925, only L'Akfadou was established before Parc National Albert).[12] During the decade from 1921–1931, more than 27,000 hectares of national forest land had been designated as national parks in Algeria.[13] The authorities created all of these first thirteen national parks in Algeria on forest land.

Morocco followed the Algerian example. In 1934, the protectorate government promulgated the law to create national parks. The first of these, Toubkal, was established in 1942 and encompassed 36,000 hectares.[14] Tunisia had a similar decree in place by 1936, the year it created the state park Bou-Hedma.[15] According to the 1913 Congress, the flora and fauna within these parks were to be "allowed to evolve freely without human intervention," and "strict surveillance and very severe sanctions" were to be levied for the defense and the protection of the parks.[16] These recommendations were incorporated into the laws governing national parks in Algeria, Tunisia, and Morocco, several of them still regulating use today.

In Algeria, by virtue of the law of February 17, 1921, national parks were to be created in forests already under the Algerian forest code.[17] They were therefore to be managed and policed by the Algerian forest service. These parks were hence created in areas where the forest code and its multiple restrictions on grazing, gathering food and forest products, and using fire were already in place. Such restrictions on indigenous Algerians' use of forest resources had

Vue du chalet du Rond-Point

FIGURE 7.1. "Vue du chalet du Rond Point (Parc National Les Cèdres de Téniet-el-Haâd)." Despite strict restrictions preventing Algerians from keeping any livestock in the forests of the national parks, French forest guards were allowed to keep some as evidenced by the text accompanying this photo. "What a marvelous sight! What an enchanting view! At the end of a little dale, completely surrounded by cedars, we find a deliciously fresh prairie, of a tender green, where the grass grows thickly and vigorously. Some cows, belonging to the forest guard, graze this grass. It is truly a landscape from Switzerland or Savoie." Source: *Cahiers du centenaire de l'Algérie*, vol. 7 of M. le Général de Bonneval, *L'Algérie touristique*, ed. Comité National Métropolitain du Centenaire de l'Algérie, 12 vols. (Orléans: Imprimerie A. Pigelet et Cie., 1930), 43.

been in place in Algeria since at least the middle of the nineteenth century, motivated primarily by tenuous French conceptions of nature and environmental change in the Maghreb.

The French Environmental Imaginary of North Africa

In order to fully appreciate how and why the French established national parks in North Africa, it is first necessary to understand the prevailing conceptions about the Maghreb's environment at that time. The state of the environment and understanding of how it had changed were intimately tied to beliefs about local uses and "misuses" of the land by the North Africans that necessitated "protection." Thus within a decade of their conquest of Algiers in 1830, the French began to criticize the traditional land-use practices of the local Algerians. They claimed that nomads overgrazed and that Algerian farmers wantonly burned fields and forests alike, ruining the environment. Pre-conquest French conceptions of North African nature, conceptions of a lush, fertile, forested environment—the veritable granary of Rome—therefore changed within a short time after occupation to a story of a landscape ruined by the "natives."[18] The local Algerians were blamed for desertifying the environment with overgrazing and for deforesting the entire region with the reckless, even malicious, setting of forest fires. Such environmental destruction, the French story claimed, had been occurring since the Arab nomad "invasions" of the eleventh century, the so-called Hilalian invasion. This claim was based on a combination of observation of the landscape, unfamiliarity with traditional Algerian land-use practices, and a highly selective reading of certain medieval Arab texts like the works of

Région des Hauts Plateaux Constantinois.
Forêt en régression après incendie et pacage.

FIGURE 7.2. "Region of the Hauts Plateaux of Constantine. Forest in regression after fire and grazing." Source: Henri Marc, *Notes sur les forêts de l'Algérie, Collection du centenaire de l'Algérie, 1830–1930* (Paris: Librairie Larose, 1930), planche 14.

Ibn Khaldun.[19] By the middle of the nineteenth century, the colonial story of environmental ruin by the "natives" had been well developed in Algeria, and it soon became the predominant environmental history of the region. It included the entire Maghreb and was thus applied to Tunisia in 1881 and Morocco in 1912 as the French occupied and pacified these territories.

What the French never understood during the colonial period was that traditional North African land-use practices were largely ecologically appropriate and thus "sustainable" for the population levels of the early to mid-nineteenth century. Given the ecology of the region, in which a majority of plants were highly adapted to drought, fire, and grazing, regenerating pastures and clearing fields of weeds and pests with fire was well suited to the environment.[20] Such methods of "tending the land with fire" were complemented by an extensive grazing system that utilized large expanses of apparently unoccupied land since only small, localized areas of land had grass and other grazing plants available at any given time in this arid and semi-arid region of unpredictable rainfall. Herders' survival depended on mobility and large expanses of land. Most of this seemingly empty land was actually communally owned and managed by largely unwritten rules about who could graze where and when; these rules were enforced by various powerful members of the tribes in a variety of ways.

Perhaps even more important, the French idea that the North African landscape had been more lush, fertile, and forested before the "Arab invasions" of the eleventh century is highly questionable and based on a partial reading of classical Greek and Roman texts.[21] This supposed granary of Rome actually produced more grain, not less, during the colonial period than the classical period.[22] Contemporary paleo-ecological evidence, moreover, shows that French colonial claims that the Maghreb was between 50 percent and 85 percent deforested compared to the classical period are not true. Although there has been a reduction in some plant and tree species over the last two thousand years, other species have increased. In fact, there have been few significant changes and no overall trends over the last ten thousand years.[23] Indeed, the most significant and well-documented deforestation occurred during the colonial period due to Franco-European economic activities.

The French administration used the colonial environmental history of North African ruin by the "natives" to justify a host of legislation and policies regulating and restricting the use of natural resources and the actions of the local Maghrebi populations throughout the colonial period. Even before it was proclaimed in the authoritative *Exploration Scientifique de l'Algérie* of 1847 that "this land, once the object of intense cultivation, was neither deforested nor depopulated as today ... it was the abundant granary of Rome," the decree of

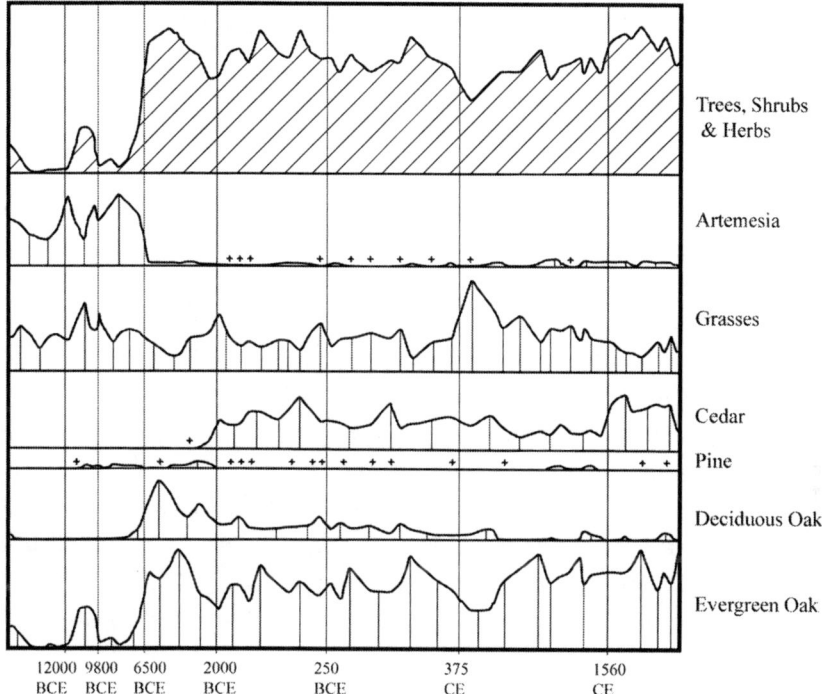

FIGURE 7.3. Diagram of pollen core data from Lake Tigalmamine in the Middle Atlas Mountains, Morocco. This diagram illustrates changing levels of plant pollen over the last fourteen thousand years. Based on: H. F. Lamb, U. Eichner, and V. R. Switsur, "An 18,000-Year Record of Vegetation, Lake-level and Climate Change from Tigalmamine, Middle Atlas, Morocco," *Journal of Biogeography* 1 (1989): 65–74. The pollen core is shown on p. 71. Diagram by Diana K. Davis.

1838 outlawed burning vegetation for any reason.[24] The forests, having been declared the property of the state the year of conquest (1830), were placed under the French Forest Code and later managed by the Algerian forest service established in 1838. A few small areas of forest were deemed communal forests in 1851, which gave the local populations limited use rights.[25] A number of laws and policies followed—like the 1874 and 1885 laws further restricting forest use and amplifying punishments for infractions—culminating in the 1903 Algerian forest code.[26] Much more restrictive and punitive than the French Forest Code, it banned nearly all grazing in the forests, all burning or gathering of firewood, and most other traditional uses of the forest like hunting or gathering food and medicinal plants.

Outside forested areas, the colonial administration enacted other laws and policies such as the 1844 and 1846 laws that facilitated the confiscation of large areas of communally managed grazing land defined as "waste land" by the state and thus available for colonization efforts.[27] Similarly, the Senatus Consultus (1863) and the Warnier Laws (1873 and 1887) in effect destroyed indigenous, communal land tenure regimes by instituting individual private property rights and a booming real estate market.[28] These changes and the imposition of multiple taxes pauperized a majority of the local pastoral and non-pastoral populations but facilitated French social and economic control. The protectorate administrations enacted very similar laws and policies in Tunisia and Morocco with equally devastating effects on the indigenous populations.

Whereas illegal grazing was recognized as the principal problem in Morocco, fire was considered the most significant challenge for Algerian forests, which were important economic resources for the French. This was because large forest fires occurred with alarming regularity in Algeria throughout the colonial period, often burning well over one hundred thousand hectares in a given year and destroying the opportunity for profits from timber or cork products in burned areas for years. The 1874 forest fire law formalized collective punishment for setting forest fires and required that Algerians either help fight fires or be fined and/or imprisoned, even if they did not set the fire.[29] The 1874 law later became an integral part of the 1903 Algerian forest code. This code formed the basis for the new colonial forest codes instituted in Tunisia (1915) and Morocco (1917).

National Parks, Profit, and Social Control

By the time the French were promoting and planning national parks in the Maghreb, then, the forests were already cleared of most human occupation and nearly all traditional uses by the local populations. Indigenous uses of other natural resources such as grazing lands had also been severely curtailed or criminalized, justified by the same French colonial environmental story that had been mobilized so effectively in instituting the colonial forestry laws.

The stated goal of the 1921 Algerian law establishing national parks, following the 1913 International Forestry Congress guidelines, was to "remove all plants and animals from all human intervention not related to conservation and protection."[30] The law specified that national parks were to be established on forest land already under the 1903 forest code and that they could be formed on state, private, or communal land.[31] Although the Algerian forest code restricted or criminalized nearly all traditional uses of the forest, some small areas had been designated "communal forests." In these areas, some traditional uses such as limited grazing were allowed under certain circumstances. The national parks law of 1921 outlawed "all use (exploitation)" of nature in national parks. These restrictions could be lifted in some cases, but only by the governor general of Algeria. In most areas, grazing was completely outlawed. In others, the locals had retained some usage rights and grazing was controlled by the "selling back" of grazing rights to traditional users on a limited basis.³² The national parks law also decreed hunting and the destruction of "pests" illegal, except when the forest service deemed it necessary to control "harmful" species.³³ Certain parks or sections of parks, however, were to be defined as *"parcs renforcés"* (reinforced parks), due to special scientific or touristic value and were placed strictly off-limits to all usage with no exceptions.³⁴ Of the original thirteen parks covering 27,600 hectares created by 1931, only one, Chrea National Park, was established on communal land. The rest were implemented on state forest land; two, Les Planteurs and Saint-Ferdinand, were created in planted stands of trees that were "completely artificial" forests.³⁵

Although scientific interest and the conservation of nature were invoked to justify the creation of these parks, a primary concern—reflecting a major point of the 1913 Forestry Congress—was the "legitimate desire to create through tourism new sources of wealth."³⁶ The development of the national parks required new work in the forests, including the construction of houses

2. — Maison forestière de Tala Kitan

Forêt domaniale et Parc National d'Akfadou
(Grande Kabylie)

FIGURE 7.4. A House for Forest Guards in Akfadou National Park. Source: Henri Marc, *Notes sur les forêts de l'Algérie, Collection du centenaire de l'Algérie, 1830–1930* (Paris: Librairie Larose, 1930), planche 9.

for more forestry personnel, who in turn were necessary to increase the surveil-lance needed to impose fines for infractions such as grazing and to better fight forest fires.[37] Telephone lines were hung, roads built, and refuges established. Such changes were also designed to increase "safety and security" in these areas to further encourage tourism.

The administration created nearly half of the early Algerian national parks in the forests of Constantine Province, which experienced more destructive fires during the late nineteenth and early twentieth centuries than any other province. Within the province of Constantine, the city of Bône (now Annaba) was ringed with three areas that suffered the worst and most frequent fires, including the forests of La Calle (now El Kala), Beni-Salah, and Edough.[38] This area around Annaba supported some of the largest numbers and highest den-sities of livestock, particularly cattle, during the colonial period in Algeria and still does today.[39] Algerians would sometimes set forest fires to stimulate pas-ture growth; this was an ecologically sound method integral to their traditional agro-pastoral economy. Setting fires was also one way rural Algerians may have protested the erosion of their way of life and the confiscation of their land, par-ticularly their forests, by the colonial government.[40] The government deemed many of these fires arson—although many were not—and meted out steep fines and collective punishments on those they held responsible.[41]

In the Edough and Beni-Salah Mountains, rates of forest fires and ille-gal grazing and their attendant punishments were very high.[42] By 1930 the French had established a two-thousand-hectare national park in the Edough Mountains, which included the summer retreat of Bugeaud.[43] Another park, La Mahouna, was created in the region of Guelma (not far from Beni-Salah and also considered a problem fire area), which entailed 1,055 hectares.[44] French naturalists called for the creation of a scientific reserve, if not a national park, in a third zone of intense fire activity: La Calle.[45] In the province of Alger, just over 8,000 hectares of parks were created, including the 1,500-hectare park Les Cèdres de Téniet-el-Haâd and the 1,030-hectare park l'Ouarensis (located stra-tegically in another problem fire area, the Ouarensis mountains).[46] Téniet-el-Haâd also became notorious in the eyes of the forest administration for the large amount of forest grazing that took place there. Ain-N'Sour, a compara-tively small park of only 279 hectares, was established near another problem fire area in the region of Cherchell. Two other national parks were planned in areas that were problematic in terms of forest grazing.[47] These were to become, in the post-colonial period, Belezma National Park, near Batna, and Tlemcen National Park, in Oran Province.

These newly established national parks justified further restrictions of tra-ditional forest uses, including grazing and setting fires to create pasture and clear fields, and stricter enforcement of punishments for transgressions. They also justified more intense monitoring of local populations near problem fire or grazing areas. Though they may have protected nature and generated revenue, the parks also necessitated and heightened surveillance of "problem populations" and further restricted traditional modes of existence for these local populations—both interests of the French colonial regime in North Africa. The placement of so many of the earliest parks in areas problematic for the French colonial administration strongly suggests that they were intended to help maintain social control.

French colonial authorities had far fewer difficulties with forest fires and forest grazing in Tunisia than in Algeria or Morocco, although they considered grazing a significant problem.[48] The first Tunisian national park was created in 1936 in the acacia forest of Bled Talha, in the mountains of Bou Hedma, an area considered especially degraded by many centuries of grazing.[49] This was also the location of the caravanserai of Bou Hedma, built in 1892, attesting to its importance in the pastoral economy.[50] Many believed that the Bled Talha forest was the last refuge of tropical elephants in North Africa and that it had been much larger during the classical period; therefore, its presumed destruction by the "Arabs" received blistering condemnation during the colonial period, especially by naturalists.[51] The establishment of the park in this area was likely motivated in large part by the desire to control grazing, and therefore the movements, of the nomads and other pastoralists in the region.[52]

The overriding French environmental preoccupation in Morocco was not forest fires but transhumance—the movement of livestock in mountains—and illegal grazing by the multitude of highly mobile tribal groups raising ruminants. The predominant French view of the Moroccan landscape, especially in areas categorized as forest land, was that it was severely degraded and deforested by overgrazing, especially by small stock.[53] The protectorate administration passed the law establishing national parks in 1934, approximately six months after pacification was officially achieved in Morocco. As in Algeria, the existing forest code in Morocco (dating to 1917) was applied in national parks, and thus most traditional uses of these areas were forbidden. The rationale for the creation of national parks in Morocco, which included the protection of nature for scientific and tourist purposes, went a step further than in Algeria and also noted reasons of "social utility."[54]

The first park created was Toubkal in the High Atlas Mountains in 1942. This came late in the French colonial venture in North Africa, and the administration established the park in an area long coveted by winter sports enthusiasts and other tourists. It was also located in the middle of the "bled siba," the "dissident" territory historically outside the reach of the Moroccan sultan and, until

1934, also outside the control of the French colonial authorities. The inspector general of forestry for southern Morocco developed this park and most of the forestry laws were automatically applied. The stated purposes for this park and others created later were, as in Algeria, to provide a refuge for the scientific study of flora and fauna and, importantly, to develop tourism in order to generate profit for the province.[55]

Although harvesting forest products and any kind of burning was forbidden in Toubkal, limited grazing, according to strictly monitored and controlled modified traditional transhumant practices, was allowed at the inception of the park. However, the restrictions placed by the authorities on the timing of grazing in the pastures, the limited number of tribes that were allowed to graze, the prohibition on building corrals, and a strict and inflexible limit on the kinds and numbers of animals allowed severely handicapped traditional practices.[56] In effect, the newly created park excluded thousands of local pastoralists from their traditional summer pastures within its thirty-six thousand hectares.[57] These restrictions reflected the colonial government's official policy in Morocco, from the 1930s until the 1950s, of trying to completely suppress transhumance.[58]

The next several protected areas created in Morocco were all placed in zones, especially in the Middle Atlas region, that were perceived as highly problematic with respect to transhumance and forest grazing. These included Aguelmam Affenourir, established as a hunting reserve in 1943, and Tazekka National Park, created in 1950.[59] The justifications for protecting the land from overgrazing and restoring pasture and forest health were invoked in part so the government could pursue its goals of destocking and reducing the mobility of the transhumant tribes in order to try to control the local populations. The French regime was also very concerned with the rise of nationalism in Morocco during this period; the Middle Atlas region was especially troublesome from a colonial perspective since it harbored several tribal groups thought to be sympathetic to independence and perhaps revolution.[60] As in Algeria, and to a certain extent Tunisia, the protection of nature and the generation of tourism revenue went hand in hand with the social control of populations long considered problematic for, or hostile to, French colonial authorities.

Post-Colonial Nature in the Maghreb

Since winning independence, the Maghreb countries have retained and expanded many of the colonial national parks and created new ones. These post-colonial governments have also maintained many of the original colonial goals of park creation, such as revenue generation and control of difficult populations. Moreover, they have retained the underlying French colonial vision of nature and environmental change. Morocco and Algeria stand out for their active engagement with the protection of nature along these lines since independence, especially since the late 1970s. Algeria has retained, as of 2006, five of its original thirteen colonial national parks (Téniet El-Had, Djurdjura, Chréa, Taza, and Gouraya) and created five new national parks since 1983. Originally conceived and advocated during the colonial period, national parks were not established at El Kala until 1983 nor at Belezma until 1984.[61] The most recent national park is Tlemcen (1993), and the largest are Tassili and Hoggar (1987), which together cover over five hundred thousand square kilometers of protected area in the Sahara. This brings the total area protected in ten national parks to nearly a million square kilometers or approximately 39 percent of the surface area of Algeria—a comparatively large amount of land.[62] The Algerian government created all of these new parks in areas considered problematic in terms of "destructive" grazing and/or control of pastoralists.

In the case of El Kala, for example, claims of overgrazing and land degradation—which both allegedly harm habitat and wildlife in the park—justify the government's continuing efforts to decrease pastoralism in the area despite convincing evidence that grazing is not causing land degradation.[63] Contemporary ecological research has demonstrated that "the environmental benefits of the El Kala livestock system lie in its complementary use of resources essential to wildlife survival."[64] The government, though, relies on its own studies of the ecology and economy of the region that are "of dubious validity ... [and] may play into the hands of other interests seeking to expropriate the underlying resource[s]."[65] More than a decade after this warning, grazing is still being suppressed and livelihoods are still being needlessly damaged in El Kala and other national parks in Algeria by officials who claim that traditional management practices are archaic and destructive.[66]

Similar claims are made in Morocco, where the two original national parks from the French colonial period—Toubkal and Tazekka—still exist and eight new parks have been established. New national parks were added in the 1990s: Sous Massa (1991) and Iriqi (1994). A flurry of park creation followed in the 2000s, with four parks established in 2004 alone: Al Hoceima, Talassemtane, Ifrane, and Haut Atlas Oriental, followed by Khenifiss in 2006 and Khénifra in 2008.[67] The Moroccan government established three-quarters of these new parks—including Iriqi, Talassemtane, Ifrane, Haut Atlas Oriental, Khénifra, and Khenifiss—in areas long perceived as problematic for overgrazing or pastoral incursions. Khenifiss National Park is located between Tan Tan and Tarfaya, at the border with the occupied Western Sahara. It is in a coastal pastoral area that has obvious geopolitical significance to the Moroccan government, as does the planned national park deep within the Western Sahara at Dakhla.[68] The Moroccan state has long been preoccupied with the control of nomads in this area. The creation of these new parks, then, must be considered not only in light of the Moroccan government's stated interest in environmental protection but also in the context of its political interests in retaining the Western Sahara and controlling nomads who are seen as a threat to national security.

The colonial national parks law in Morocco was abrogated in 2008 with a new law regulating protected areas, including national parks. The primary goals of nature protection in Morocco now include "sustainable development," protecting biodiversity, and conserving natural and cultural heritage—in addition to the earlier goals of tourism and preventing ecological degradation.[69] The new law includes significantly elevated fines and penalties for violations.[70] Eighty percent of national parks in Morocco are located in seasonal or drought pasture areas important to different pastoral groups, and the prohibition on even passing through these areas has impoverished their way of life. While the enforcement of rules prohibiting grazing and gathering firewood, for example, is said to be far from perfect, fines are frequently levied and are often very high.

FIGURE 7.5. Tazekka National Park, Morocco. Photo by Randolph Self. Reproduced with permission.

In the case of Tazekka National Park, for example, estimates of the percentage of the local population fined for illegally grazing or gathering firewood were as high as 34 percent per year for the period 1980–1992.[71] These fines ranged from 200 to 1,500 Moroccan dirhams.[72] The average minimum wage for a day's work in Morocco in 1990 was only 27 dirhams.[73] These fines, therefore, represented between one week's and two months' work at the average minimum wage. The 2008 law has dramatically raised fines for illegal grazing (or even having animals in proscribed spaces) to between "1,200 to 10,000 dirhams and imprisonment of 15 days to 3 months."[74] Although the average minimum wage has doubled to 55 dirhams per day since 2009, these fines are still ruinously high.[75] In addition to prison time, the new fines would take from three weeks' to three years' total wages at the 2009 rates. Particularly in poor, rural areas, such fines over time could necessitate people selling their herds and could also drive poor families into deep poverty, forcing them to sell their land and migrate to cities.[76]

Elsewhere in Africa and the global south, the effects of national park creation on local peoples have been similar, and in many places they have been worse. Wholesale evictions of indigenous populations from newly instituted national parks, and the subsequent pauperization and immiseration of these local peoples during the colonial and post-colonial period, have been documented around the world.[77] Nor is the developed world immune to such state-led appropriation of land and resources in the name of environmental protection, as evidenced by the histories of national park creation in Yellowstone, the Grand Canyon, and the Adirondacks in the United States.[78] In fact, the parallels between the reasons for and the effects of the creation of national parks in the United States and in the Maghreb in the early twentieth century are uncanny. Many policies put in place in national parks globally before the Second World War remain in place today, although sometimes in modified form. Underlying many of these policies are certain Western visions of the natural world that are increasingly being revealed to be social constructions largely divorced from ecological, social, and historical realities.

As is clear from this examination of the development of national parks in the Maghreb, the French colonial environmental imaginary of a formerly fertile landscape degraded by anarchic and destructive "traditional" grazing and farming practices has been retained to a large degree by elites and government officials in the post-colonial period. Despite a growing body of scientific evidence that points to the ecological appropriateness of many herding practices and other "traditional" livelihood tools in the Maghreb, official claims of environmental degradation persist in large part because they are politically, socially, and economically useful.[79] While such environmental stories facilitate the control and sedentarization of nomads long deemed dangerous by the state, they also help to generate revenue in the form of funding from international environmental organizations like the United Nations Environment Programme, the World Bank, and international non-governmental organizations (NGOs) interested in preserving biodiversity, for example. Few scholars of the MENA have analyzed nature preservation from a social or historical perspective, despite warnings like the one issued by a pastoralism expert in a keynote speech to the United Nations Food and Agriculture Organization (UNFAO) not long ago asserting that "the marginal lands that were previously the province of pastoralists are increasingly coming into focus as reserves of biodiversity."[80] The MENA is home to more pastoralists than almost anywhere else in the world. To enclose more of their productive arid grazing lands in the name of biodiversity or environmental protection will lead almost certainly to further social injustice and likely to greater ecological harm.

Notes

1. Except in the narrow coastal zones. See note 20.

2. See, for example, William M. Adams, *Green Development: Environment and Sustainability in a Developing World*, 3rd edn. (London: Routledge, 2009); William Beinart and Lotte Hughes, *Environment and Empire*, in *The Oxford History of the British Empire Companion Series*, ed. Wm. Roger Louis (Oxford: Oxford University Press, 2007); Mark Cioc, *The Game of Conservation: International Treaties to Protect the World's Migratory Animals* (Athens: Ohio University Press, 2009); Roderick P. Neumann, "Dukes, Earls, and Ersatz Edens: Aristocratic Nature Preservationists in Colonial Africa," *Environment and Planning D: Society and Space* 14 (1996): 79–98.

3. The first national park in Africa, Les Cèdres de Téniet el-Haâd, was actually created in Algeria in 1923. This fact is nearly always overlooked in histories of national parks, perhaps because it was not originally a nineteenth-century game preserve.

4. Adams, *Green Development*, 29–33; Cioc, *The Game*, 14–57.

5. Edward I. Steinhart, *Black Poachers, White Hunters: A Social History of Hunting in Colonial Kenya* (Athens: Ohio University Press, 2006); Roderick P. Neumann, *Imposing Wilderness: Struggles over Livelihood and Nature Preservation in Africa* (Berkeley: University of California Press, 1998); Jane Carruthers, *The Kruger National Park: A Social and Political History* (Pietermaritzburg: University of Natal Press, 1995); Neumann, "Dukes, Earls, and Ersatz Edens"; Cioc, *The Game*.

6. Jean Baer, "Aperçu Historique de la Protection de la Nature," *Biological Conservation* 1 (1968): 7–12. For a history of international conferences specifically related to hunting and its regulation, see Cioc, *The Game*.

7. Baer, "Aperçu," 9.

8. Jean-Paul Harroy, "L'Union Internationale pour la Conservation de la Nature et de ses Ressources: Origine et Constitution," *Biological Conservation* 1 (1969): 106–110. The IUPN changed its name to the IUCN (International Union for the Conservation of Nature) in 1956. Adams, *Green Development*, 33–35.

9. Henri Marc, *Notes sur les forêts de l'Algérie, Collection du centenaire de l'Algérie, 1830–1930* (Paris: Librairie Larose, 1930), 528–535. The former director of the Algerian Forest Service, Henri Marc, makes it clear in this volume that the 1913 Forest Congress was very influential in the shaping of the national park system in Algeria.

10. P. de Peyerimhoff, "Les 'Parcs Nationaux' d'Algérie," in *Contribution à l'étude des réserves naturelles et des parcs nationaux*, ed. Société de Biogéographie (Paris: Paul Lechevalier, 1937), 127–138. For a detailed discussion of the 1912 recommendations of the Society for Natural History of North Africa under the guidance of the botanist and physician René Maire, see Service du Tourisme Gouvernement Général de l'Algérie, *Rapports et Études de la Commission du Tourisme* (Alger: Imprimerie Algérienne, 1920), 43–54.

11. For information on the establishment of Parc National Albert, see E. Cartier, "A National Park in the Belgian Congo," *Science* 16 (1925): 623–624.

12. de Peyerimhoff, "Les 'Parcs,'" 134–135.

13. Ibid.

14. Théophile Delaye, "Le Parc National du Toubkal," *Revue de Géographie Marocaine* 28 (1944): 3–14. For an example of the early reasoning involved in the creation of national parks in Morocco, see Georges Carle and Jean Gattefossé, "Réserves naturelles et parc chérifiens," *Revue Scientifique Illustrée* 17 (1933): 622–628. The French governed Morocco and Tunisia as protectorates rather than colonies.

15. Bernard Bousquet, *Guide Des Parcs Nationaux d'Afrique: Afrique du Nord, Afrique de l'Ouest* (Paris: Delachaux et Niestlé, 1992), 301.

16. Marc, *Notes sur les forêts*, 528.

17. Ibid., 530.

18. For details of the French colonial environmental history of the Maghreb, its development, and its various uses, see Diana K. Davis, *Resurrecting the Granary of Rome: Environmental History and French Colonial Expansion in North Africa* (Athens: Ohio University Press, 2007).

19. For more details, see: ibid., 54–57.

20. For details on the regional ecology, see ibid., 177–186. For information on traditional land use practices, see ibid., 27–32; André Nouschi, *Enquête sur le niveau de vie des populations rurales constantinois de la conquête jusqu'en 1919* (Paris: Presses Universitaires de France, 1961).

21. Davis, *Resurrecting the Granary*, 16–23.

22. Ibid., 4–5.

23. For details of the paleo-ecological and other evidence on past environments, see ibid., 8–12.

24. For the quote, see J.-A.-N. Périer, *Exploration scientifique de l'Algérie: Sciences médicales: de l'hygiène en Algérie*, 2 vols. (Paris: Imprimerie Royale, 1847), 1: 29. For the 1838 decree, see: Marc, *Notes sur les forêts*, 208.

25. Davis, *Resurrecting the Granary*, 33–34. Communal forests caused such problems for the French in Algeria that they did not recognize them in Morocco, where all forest land was decreed state property.

26. For details, see ibid., 81–83 and 120–123.

27. Charles-André Julien, *Histoire de l'Algérie contemporaine: La conquete et les débuts de la colonisation (1827–1871)* (Paris: Presses Universitaires de France, 1964), 240–241.

28. For details, see: Davis, *Resurrecting the Granary*, 86–87 and 97–100.

29. For more information, see ibid., 80–82.

30. Marc, *Notes sur les forêts*, 530–531. This is from article 4 of the 1921 law. See also, de Peyerimhoff, "Les 'Parcs,'" 132–133.

31. Establishment on private or communal land required formal agreement of the proprietors. Marc, *Notes sur les forêts*, 530.

32. Ibid., 531.

33. de Peyerimhoff, "Les 'Parcs,'" 133.

34. Marc, *Notes sur les forêts*, 531.

35. de Peyerimhoff, "Les 'Parcs,'" 134–136.

36. Marc, *Notes sur les forêts*, 529.

37. Ibid., 531.

38. Paul Boudy, *Économie forestière nord-africaine: Milieu physique et milieu humain*, vol. 1 (Paris: Éditions Larose, 1948), 647–648; David Prochaska, "Fire on the Mountain: Resisting Colonialism in Algeria," in *Banditry, Rebellion and Social Protest in Africa*, ed. Donald Crummey (London: James Currey, 1986), 229–252.

39. Katherine M. Homewood, *Livestock Economy and Ecology in El Kala, Algeria: Evaluating Ecological and Economic Costs and Benefits in Pastoralist Systems*, Pastoral Development Network Paper no. 35 (London: Overseas Development Institute, 1993). See also Samir Ouelmouhoub, "Gestion multi-usage et conservation du patrimoine forestier: Cas des subéraies du Parc National d'El Kala (Algérie)" (Master of Science, CIHEAM-IAMM, 2005).

40. Prochaska, "Fire on the Mountain."

41. Collective punishments were also enforced for forest fires in Morocco, but not in Tunisia. Boudy, *Économie forestière*, vol. 1. Some of these fires were human generated but some were naturally occurring.

42. Fines for alleged arson were destructively high for rural Algerians. Beginning in 1877, collective punishment was levied for fires thought to be arson. These punishments often entailed the confiscation of tribal lands, which in at least one case amounted to more than half of the tribe's arable land. Stiff penalties were also meted out for grazing infractions. Approximately 50 percent of these fines were for illegal pasturing in forests, and the fines were often higher than the value of the livestock involved. Prochaska, "Fire on the Mountain."

43. Marc, *Notes sur les forêts*.

44. Ibid.

45. de Peyerimhoff, "Les 'Parcs,'" 131. El Kala did not become a national park until the postcolonial period.

46. Bousquet, *Guide Des Parcs*.

47. Marc, *Notes sur les forêts*, 532; Boudy, *Économie forestière*, 1: 603–606.

48. Paul Boudy, *Économie forestière nord-africaine: Description forestière de l'Algérie et de la Tunisie*, vol. 4 (Paris: Editions Larose, 1955), 465–466.

49. Ibid., 4: 457. See also Bousquet, *Guide Des Parcs*, 301.

50. See the World Conservation Monitoring Center's protected areas data sheet for Bou Hedma: http://sea.unep-wcmc.org/sites/pa/0402v.htm. (Accessed May 1, 2010, and since removed. Hard copy in the possession of the author.) A caravanserai is an inn or other place to rest along a travel route, frequently for caravans carrying goods for sale from one place to another.

51. For example, see Louis Lavauden, "La Tunisie et les Reserves Naturelles," in *Contribution à l'Étude des Réserves Naturelles et des Parcs Nationaux*, ed. Société de Biogéographie (Paris: Paul Lechevalier, 1937), 139–150, especially 146–147. The acacia is one of the arid zone trees best adapted to drought and grazing. It reproduces only under relatively rare conditions of high temperature and high rainfall, and thus its regeneration is not even in time or space. Therefore it is highly unlikely that overgrazing "destroyed" the Bled Talha "forest." As in other parts of the Maghreb, deforestation estimates by French colonial authorities were exaggerated. Davis, *Resurrecting the Granary*.

52. It is also interesting to note that Tunisia had the smallest number of parks created during the colonial period in the Maghreb. This has been attributed, at least in part, to the fact that European colonists strongly opposed the sites of several suggested parks. Raphael Larèrre, Bernadette Lizet, and Martine Berlan-Dargué, *Histoire des parcs nationaux: Comment prendre soin de la nature?* (Paris: Éditions Quai, 2009), 53–54.

53. Boudy, *Économie forestière*, 1: 609–617.

54. "Morocco," a Country Sheet Produced by the World Conservation Monitoring Center: http://ims.wcmc.org.uk/ipieca/countrysht/Morocco.html. (Accessed July 2000 and since removed. Hard copy in the possession of the author.)

55. Delaye, "Le Parc National."

56. Ibid.

57. Dawn Chatty, "Enclosures and Exclusions: Conserving Wildlife in Pastoral Areas of the Middle East," *Anthropology Today* 14 (1998): 2–7. See also Said Boujrouf, Mireille Bruston, Philippe Duhamel et al., "Les conditions de la mise en tourisme de la haute montagne et ses effets sur le territoire," *Revue de Géographie Alpine* 86 (1998): 67–82.

58. Commandant Ruet, *La Transhumance dans le moyen Atlas et la haute Moulouya* (CHEAM unpublished report, 1952).

59. For Aguelmam Affenourir, see the World Conservation Monitoring Centre's protected areas data sheet: http://sea.unep-wcmc.org/sites/pa/1425v.htm. (Accessed April 29, 2010, and since removed. Hard copy in the possession of the author.) For Tazekka, see Bousquet, *Guide Des Parcs*.

60. Ruet, *La Transhumance*. Highly mobile groups are difficult to monitor for insurrectionist activity, and so it is not surprising that Commandant Ruet and others in the military wanted as many of them sedentarized as possible.

61. Parc national de Théniet El Had and Direction Générale des Forêts, *Atlas des parcs nationaux algériens* (Cité administrative de Théniet El Had, Algérie: Parc national de Théniet El Had, 2006).

62. Bousquet, *Guide Des Parcs*, 73. This compares with about 2 percent of national territory in Morocco and Tunisia. See ibid., 203 and 281. The country statistics are available on the website of the Convention on Biological Diversity (CBD): http://www.cbd.int/countries (accessed May 27, 2011). For a wider comparison, consider that 6 percent of national territory is protected in national parks in South Africa, as is 8 percent in Kenya. Note that statistics vary with sources, collection methods, and dates of aggregation.

63. Homewood, *Livestock Economy and Ecology*. See also Boujrouf, et al., "Les conditions."

64. Homewood, *Livestock Economy and Ecology*, 14. See also: Avi Perevolotsky and No'am Seligman, "Role of Grazing in Mediterranean Rangeland Ecosystems," *BioScience* 48 (1998): 1007–1017.

65. Homewood, *Livestock Economy and Ecology*, 1–2.

66. See, for example, Mohammed Sahli, "Protection de la nature et développement: Cas du Parc national du Belezma," *New Medit. A Mediterranean Journal of Economics, Agriculture and Environment* 3 (2004): 38–43.

67. For more information, see the Centre d'Echange d'Information sur la Biodiversité du Maroc's website: http://ma.chm-cbd.net/manag_cons/esp_prot (accessed May 2, 2010).

68. See the following: http://www.worldheritagesite.org/sites/t1183.html (accessed May 19, 2012). It is interesting that although this park was still in the planning stages as of May 5, 2010, it had already appeared on Google Maps.

69. See the text of law No. 07–22 on protected areas at: http://ma.chm-cbd.net/manag_cons/esp_prot/stat_nat/projet-de-loi-n-07–22-relative-au-aires (accessed May 2, 2010).

70. Ibid., Chapter 5.

71. This number is based on my calculations of statistics provided in the following: Randolph Self, "Community-based Conservation and Sustainable Development in Tazekka National Park, Morocco" (Master's Thesis, Western Washington University, 1997).

72. Ibid.

73. E. Maciejewski et al., "Morocco: Selected Issues" (Washington, DC: International Monetary fund, 1997). The current exchange rate is 1 U.S. dollar to 8.5 Moroccan dirhams. This has not fluctuated much over the last decade, staying roughly between 7.5 and 9 dirhams to the dollar.

74. Law No. 07–22 on protected areas, Chapter 5.

75. For the new minimum wage numbers, see Siham Ali, "La Revalorisation du SMIG au Maroc ne satisfait pas les salariés," *Maghrebia*, July 8, 2009. Available at: http://www.magharebia.com/cocoon/awi/xhtml1/fr/features/awi/features/2009/07/08/feature-02 (accessed May 6, 2010).

76. Of course, there are and have been many other drivers of outmigration from rural to urban areas in Morocco and around the Mediterranean for a long time. See the detailed discussion of some of these factors in J. R. McNeill, *The Mountains of the Mediterranean World: An Environmental History* (Cambridge: Cambridge University Press, 1992).

77. Brian King, "Conservation Geographies in Sub-Saharan Africa: The Politics of National Parks, Community Conservation and Peace Parks," *Geography Compass* 4 (2009): 14–27; Charles Zerner, ed., *People, Plants and Justice: The Politics of Nature Conservation* (New York: Columbia University Press, 2000); Adams, *Green Development*; Neumann, *Imposing Wilderness*.

78. Karl Jacoby, *Crimes against Nature: Squatters, Poachers, Thieves, and the Hidden History of American Conservation* (Berkeley: University of California Press, 2001).

79. For a fuller discussion, see Davis, *Resurrecting the Granary*.

80. Roger Blench, *"You Can't Go Home Again." Extensive Pastoral Livestock Systems: Issues and Options for the Future* (London: Overseas Development Institute, 2000), 55. Work on nature preservation from a social historical perspective in the Middle East includes Chatty, "Enclosures and Exclusions"; Diana K. Davis, "Environmentalism as Social Control? An Exploration of the Transformation of Pastoral Nomadic Societies in French Colonial North Africa," *The Arab World Geographer* 3 (2000): 182–198. For analyses in other regions of the world, see Adams, *Green Development*; Paul Robbins, *Political Ecology* (Oxford: Blackwell Publishing, 2004); Nancy Peluso, "Coercing Conservation?: The Politics of State Resource Control," *Global Environmental Change* 3 (1993): 199–217; King, "Conservation Geographies in Sub-Saharan Africa"; Charles Zerner, *People, Plants and Justice.*

8

Building the Past: Rockscapes and the Aswan High Dam in Egypt

Nancy Y. Reynolds

The High Dam on the Nile River at Aswan in Upper (or southern) Egypt, built between 1960 and 1971, represented a highpoint of nationalist and developmental planning in the early postcolonial period. Egyptian president Jamal 'Abd al-Nasir depicted the hydroelectric project as a deliberate break with the past—a bold technological initiative to bring Egypt irrevocably out of colonial underdevelopment and into industrial modernity. Intended to greatly expand agriculture, ensure the security of water and food supplies, and jumpstart heavy industrialization, damming several years' worth of Nile floodwaters above Aswan offered the Egyptian state hydrological control of the river within its own national boundaries. As this chapter demonstrates, postcolonial public officials and popular supporters of the High Dam portrayed it as an instrument to restore Egypt "to its youth"—to create a national rebirth enabled by Egyptian sovereignty over the country's natural resources. The dam's building materials, its massive scale, and the concurrent international archaeological salvage campaign caused many Egyptians to cast in Pharaonic symbolism the dam's promise to restore the nation's youth. Many Egyptian writers have stressed that the dam's construction utilized seventeen times the amount of material used to build the Great Pyramid at Giza, and 'Abd al-Nasir himself claimed, "In antiquity, we built pyramids for the dead. Now we will build new pyramids for the living."[1] Like the pyramids, the High Dam is constructed primarily from local granite available at Aswan; nevertheless, the dam represented a new rupture in national space, severing areas north of Aswan from the long stretches of Nubia that linked Egypt to Sudan but would become submerged under the new reservoir.

The colossal environmental manipulation demanded by the High Dam prompted a rewriting of geological time that paralleled and supported the achievement of "national rebirth" and development in this new geography of the nation. Novel frameworks for understanding the relationships among various geomorphic formations of the Nile Valley emerged in the decades around the High Dam's construction, and these discoveries created a new geological chronology of Egyptian development that ultimately transformed the High Dam into a new geological seam of the nation. Geology has not often been the focus of studies of dams or modern hydrological planning, which tend to focus instead on riparian transformation.[2] The sheer massiveness of the High Dam, however, deeply reordered the structure and use both of the surrounding hills, physically stripping the rockscape around Aswan, and of the area to the south that would become flooded by the new reservoir. The movement of rock in the various campaigns around the High Dam, including the crushing of the granite hills around Aswan to create the fill and anchor for the dam; the transfer and preservation of ancient temples and statuary from the area to the south; and the general disregard for the mud-brick houses, shrines, graves, and archaeological sites deeper in Nubia or in current use by the local residents forged a link between the contradictory official narratives of a rupture with the past and the restoration of time. These narratives were conjoined by an interlocking characterization of rocks as elements of wilderness yet also building blocks of civilization, both ancient (archaeological/Pharaonic) and modern (the dam itself). The resulting geology of national development ultimately fueled popular recruitment campaigns for dam construction; helped shape the agenda for national geological research; sharpened Nubian claims about the injustice of the dam; and created expectations of the dam's ability to restore time, agency, and sovereignty to Egypt.

Control of the Nile

Efforts to control the waters of the Nile have a long history in Egypt, as agriculture in the nearly rainless Nile Valley depends on river water for irrigation. The Nile has two primary sources in the equatorial lakes of the highlands of Africa—the White Nile, which contributes a constant but modest stream of water, and the Blue Nile-Atbara river system, which drains the Ethiopian highlands and creates the monsoon-fed annual flood that until 1964 reached Egypt in late summer. Carrying extra water and silt from the volcanic rocks of the Ethiopian mountains, the flood replenished and flushed Egyptian soil, making Egyptian fields among the most fertile in the world. The flood also brought unpredictability to Egyptian agriculture. Heavy rains caused devastating floods that destroyed buildings, livestock, irrigation works, crops, and people. Low floods resulted in crop failures and famine.

Although various irrigation schemes have existed for millennia, the series of small, local barrages and canals for irrigation in Egypt expanded significantly in the nineteenth century as a result of huge increases in the cultivation of cotton and other cash crops grown primarily for export. The first or low dam at Aswan was completed in 1902 during the formal British occupation (1882–1924) and widely influenced landholding arrangements and the switch from basin to perennial irrigation.[3] Agricultural demand for water continued to escalate, and the British raised the dam in 1912 and again in 1933–1934. The low dam moderated the flow of the Nile and retained some water for summer cropping, but it could not protect the country from extended drought nor hold back a large flood and its silt. It could only capture about one-fifth of the annual flood.[4] British efforts at more efficient colony-wide basin control, such as those related to the Century Storage Scheme project of dams that would impound water in a variety of nation–states under the direct or indirect control of Britain (Uganda, Sudan, Egypt, and—to a lesser extent—Ethiopia), intensified Egyptian nationalist feelings in the 1930s and 1940s.[5]

Officially embraced just after the Free Officers' coup in 1952, the High Dam was the first Egyptian effort to entirely stem the annual flood and convert the river into "an enormous irrigation ditch."[6] Building the High Dam shaped domestic, regional, and international politics for the coming decades; propelled Egypt to the center of the Cold War; and ignited the Suez Crisis in 1956. Although Soviet support, in the form of large-scale loans and technical expertise, was critical to building the dam, 'Abd al-Nasir ultimately relied on European machinery and Egyptian contractors, in addition to United Nations monetary and technical assistance for the project's archaeological and cultural aspects. One of the biggest challenges involved the relocation of people and monuments from Nubia, an area that would be submerged under the reservoir behind the new dam.[7]

Although 'Abd al-Nasir's state muted much domestic criticism, sharp debate ensued about the technical plans and ecological impact of the dam.[8] Fifty years later, opinion on the environmental effects of the dam remains split. Many experts are wary of the long-term consequences, such as slow desiccation of the Mediterranean due to the loss of fresh water, erosion of the coastline, shrinking of the delta, elimination of the sardine population off the northern coast, and the long-term environmental and financial costs of artificial fertilizers and pesticides. Others argue that many of the short-term effects have been mitigated. For example, improvements in village drinking water and health education have actually reversed earlier high rates of diseases from parasites like *Schistosomiasis*. As one scientist recently explained:

> It now appears that none of the environmental disasters predicted before the dam was built has occurred. And although some negative consequences have been encountered, most people would probably feel now, as indeed most experts did before dam construction, that the benefits have outweighed the costs. The major objectives have been met: flood control, conversion from basin to perennial agriculture, and energy production. The major negatives of water saturation and salt buildups in fields should be solved by better drainage and more efficient application of irrigation water.[9]

The language of the balance sheet so often invoked in assessments of the dam obscures the social and natural changes that evolved in tandem from the new scale of Nile irrigation works. In a recent study of new technologies of power and expertise, Timothy Mitchell masterfully captures the complexity of these sorts of interactions in the years leading up to the High Dam, examining the "connections between a war, an epidemic, and a famine [that] depended upon connections between rivers, dams, fertilizers, food webs, and...several additional links and interactions."[10] Mitchell's integrated analysis provides a good example of how socio-environmental history can help to counter tendencies "to embrace 'the stance of engineers and managers contemplating a problem to be solved.'"[11] The social impact of environmental change on the daily lives of Egyptians has in fact been contradictory, at times sharpening existing social injustices and yet also creating new classes, social movements, and sociocultural forms.

Examination of Aswan High Dam rockscapes points methodologically to the implausibility of a sharp dichotomy between wilderness and human intervention in the landscape.[12] Many social science accounts of the High Dam exhibit a Nasirist and technological triumphalism that emphasizes human ability to overcome nature.[13] Conversely, environmental historians and scientists have tended to stress a certain transcendence of nature over human activity.[14] Human manipulation of rock at Aswan has been continuously practiced and celebrated since ancient times. Highlighting people's relationships—especially those of indigenous populations—to nature through labor and work has helped historians document how the invocation of a stark distinction between natural wilderness and human landscapes functioned as a tool of colonial power to discredit local, non-European land-use practices.[15] Conceptions of time in the relationship of human agency to nature, specifically geology, clearly reflect the political realities of colonial and postcolonial regimes and—as Ted Steinberg puts it—"the way power operates through and across landscapes."[16]

The Nature of the Dam

An embankment dam composed of granite boulders, gravel, sand, and clay, the High Dam is notable for how fully it blends into the landscape of Aswan. Commonly described as "a mountain of a dam,"[17] it has large wings that merge into the hillsides on each bank. One can be already on top of the dam without realizing it.[18] The dam is one of the largest ever built and measures 111 meters above the riverbed. It is nearly a kilometer wide at its base, which consists of a clay core covered by rock fill and is supported by walls of impacted sand. A grout curtain that reaches over two hundred meters down to the granite beneath the riverbed secures the dam into the sediment.[19] Cofferdams, the temporary structures designed to create a dry work environment during construction on the upstream and downstream sides of the main dam, were eventually incorporated into the dam itself rather than demolished at the completion of construction, further adding to the massive scale and gradual slope of the dam. Lake Nasir, the reservoir behind the dam that reached its maximum storage in 1975, is five hundred kilometers long and an average of ten kilometers wide.[20] That such a massive dam could blend into the local landscape indicates the enormous scale of transformation at Aswan.

The specific material morphology of Egyptian stone and the physical properties of the rockscape surrounding Aswan shaped the structure of the dam and the ways in which the cultural, political, and environmental narratives about it could unfold. Travelers and geologists have long been fascinated by the complex rock formations that suddenly appear in the Nile Valley at Aswan and continue south through the various cataracts into Sudan. W. F. Hume, who directed the Geological Survey of Egypt in the early twentieth century, regarded this as "the duplicate character" of the riverbed south of Aswan in which a "complex geological structure gives rise not only to hard bands (granite) in soft rocks (shists, etc.), traversing the river at right or acute angles, but also to soft veins in hard rock, trending parallel to the river's direction.... Between these regions of complex igneous and metamorphic constituents are long stretches of Nubian sandstone country; the duplicate character of the Nile (cataracts and smooth stretches) being thus directly due to the varying geological structure."[21] The local granite, which formed the

rocky gorges and islands that block the smooth flow of the river at the cataracts south of Aswan, was particularly suited for use in monumental construction projects.[22]

Although completely different in scale, structure, and appearance, both
the low and high dams at Aswan were constructed primarily from local granite.
The first or low dam at Aswan, built from 1898–1902, was a masonry dam
faced with granite and filled with granite rubble.[23] Sand from the surrounding
area was used to make mortar for the dam, although engineers preferred to use
the coarse granitic sand from an old riverbed in the south near Shellal because
the dry disintegration of granite resulted in irregularly sized and angular par-
ticles that could tightly compact in mortar, more so than the blown sand of
the desert that had weathered into rounder and more regularly sized grains.[24]
Transportation of granitic sand required the building of a railroad line from
Shellal to Aswan through the granite hills, further altering the landscape. The
low dam, which did not hold back the silt of the river and increased the river's
water action and erosive forces, accelerated the weathering of riverbed rocks.[25]
Another significant impact of the low dam was the reintroduction of granite
quarrying at Aswan. The Aswan granite quarries are among the earliest in
the historical record, being worked from about five thousand years ago until
Ptolemaic times to provide ornamental stone for temples and other buildings.
Largely abandoned from the first to the twentieth centuries of the Common
Era, since limestone from Lower Egypt was easier and cheaper to mine, work
on the low dam kindled new interest in the economic value of Aswan granite
as a construction material.[26]

Building the High Dam, located four miles to the south of the old dam,
required an even more massive reordering of the local rockscape. The 42.7
million cubic meters of construction materials,[27] most of which were granite
and sand taken from the surrounding area, had to be transported from the
hillsides to the riverbed in complex stages. Excavation of the diversion channel, one of the first stages of dam building, removed twelve million cubic
yards of solid granite from the east riverbank. "The building plan," according
to Tom Little, a British journalist chronicling the dam's construction in the
early 1960s, "was based on using the excavated granite as rock-filling on the
dam and its impregnation with sand which was available on the west bank
roughly opposite to Khor Kundi and in the desert at Shellal some miles to
the north. The 12 million cubic yards of rock excavated from the channel
had therefore to be stored at prepared sites and then again transported and
dumped accurately, in position and in quantity, in the river. The two coffer
dams alone required no less than 6.5 million cubic yards and the upstream
coffer required nearly 5 million cubic yards of sand filling."[28] Transportation

routes to move the rock and sand also had to be constructed through the granite hills; "this brought the total amount of rock to be excavated to nearly 14 million cubic yards."[29] The cofferdam construction used almost all the rock excavated from the diversion channel by 1964, "so three quarries [were] started in the vicinity of the dam to provide another 30 million cubic yards of rock for the rock-fill over the sand."[30]

Although the nearby hills provided strong granite, the riverbed unexpectedly held deep levels of unstable, loose sediment rather than accessible granite bedrock. A buried valley of the Nile lay under the water at Aswan, hollowed out by a change in the gradient of the river caused by the drying up of the Mediterranean Sea in the late Miocene period "that gave the river great erosive power," according to the geologist Bonnie Sampsell. "Within a few hundred thousand years, the enormous river—called the Eonile—carved a vast canyon into the layers of soft sedimentary rock. This canyon stretched from the shore of the declining Mediterranean southward as far as Aswan and perhaps as far as Wadi Halfa [186 miles away at the Sudanese border]. As the river cut deep into the rock layers, other erosion processes widened the valley."[31] The Nile Canyon then became a branch of the Mediterranean in the early Pliocene period until the Nile grew strong enough in the late Pliocene era to push back the sea, leaving behind cliffs on both sides of the river valley—the sides of the ancient canyon. This deep gorge made necessary the massive scale of the dam and the long grout curtain that anchored it to the valley floor. Discovery of the deep gorge during dam construction helped geologists posit the theory of Miocene Mediterranean desiccation.

To Restore the Nation's Youth

The massive geographic, geological, and hydrological changes wrought by the Aswan High Dam necessitated a remapping of Egyptian space through new narratives of time. The Egyptian state itself worked hard to mobilize ordinary Egyptians behind the dam project, and many responded enthusiastically. In the winter and spring of 1960, Egypt's popular singer 'Abd al-Halim Hafiz, who was patronized by the 'Abd al-Nasir regime and sang its nationalist anthems, performed a new hit song called "The High Dam [al-Sadd al-'Ali]."[32] Recounting the state's story of how the new regime redeemed the nation from colonialism and locating the revolution in a populist nationalist narrative that began at Dinshaway in 1906 and crested with the nationalization of the Suez Canal, the song extolled the challenge to colonialism created by new state-initiated, monumental technological projects.[33] As symbol and catalyst for economic development and national renewal, the dam could restore time and land in Egypt, the lyrics claimed.

> We said we will build, and indeed we built, the High Dam
> O Colonialism, we built it with our hands, the High Dam....
> With our money, with our workers' hands...
> That's the word; indeed we built it....
> The good, beloved earth returned to the hands of its owners.
> We were reunited with the might [al-'izz] in it and the treasures lost in
>   its soil
> We said, "We must build its future as we return its youth."[34]

The song echoed throughout the streets of Cairo in 1960, playing in taxi cabs, coffee shops, record stores, and on transistor radios in outdoor markets.[35] Downtown cinemas all screened the same short documentary, *The Story of the High Dam!*, before their various feature performances.[36]

Tom Little, the British journalist who spent time on the dam site, remarked that in 1961 "truck-load after truck-load of labourers arrived [at the dam site to work on the project], chanting the new song, 'Bineneyn el-sadd el-aali'—'We are building the High Dam.'"[37] In addition, "hundreds of university youths volunteered to work at the dam during their 1963 vacations for a wage of £E 15 a month and were praised by the engineers for their enthusiasm. Their efforts were probably more valuable to public morale than in actual work at the site, but many of them returned to Aswan when they finished school in order to take training courses."[38] In 1963, a local committee in Alexandria organized a caravan that traveled from Alexandria to Aswan to raise money for the archaeological campaign to save the monuments of Nubia. The caravan mounted soccer matches and theatrical performances in the capitals of the governorates, using five railroad cars between Alexandria and Aswan to move around the country. Public lectures were also held in Cairo about the techniques and problems involved in the archaeological campaign.[39] Although some dam workers went on strike in 1962, by 1963 there were no fewer than thirty thousand workers on the project. According to Little, "Above all there was the feeling of achievement. It reached down to the most ignorant workers, who felt that *they* were the dam builders, and among the more educated men was reflected in a sense of mission."[40]

State-sponsored discourse at times suggested that Egyptian mastery of the rocky landscape would enable the country to grow and modernize, that the destruction of rock would create new life and youth for Egyptian peasants. Newspaper coverage of the inauguration of work on the diversion channel in 1960 made clear this connection. An *al-Ahram* profile by Ahmad Bahjat featured the story of a peasant man named Dandarawi 'Abd al-Mun'im Hasan who worked on the hills above the dam, breaking up the granite cliffs with dynamite. Directly linking the construction work "atop one of the mountains of Aswan" to the high price and relative unavailability of farm land in the Aswan area, the article noted that the need for water to till agricultural land propelled peasants to "come from the village and respond to the calls of the mountain to work in dynamite." The destructive crushing of granite promised to restore peasant livelihood.

Everyday [Dandarawi ʿAbd al-Munʿim Hasan] drills into the mountain and inserts fingers of dynamite while his heart beats fast. He
spreads out the fuse as he turns away. He reflects sometimes on the
ceiling of the mountain above him, the millions of rocks above this
ceiling. What if they were spread out on top of his chest suddenly?
Fear colors his life. Sometimes, when he spreads the fuse he risks
touching fifteen other fuses. The siren of danger rings out after the
ignition of the first fuse.

The siren dislodges him from behind like a hand pushing him to
the side. One minute and a quarter in the summer and one minute
and three quarters in the winter, then the dynamite explodes. He
goes out a long while, since he doesn't have a watch, and then he
lies down prostrate with his face down. He trembles with life while
he clings to the mountain that he has demolished, and he trembles
with life while he collects 45 piasters a day, and he lives. His frame
of mind changes. Thick shadows hide his past that had laid claim to
him like statues of the Pharaohs, seated on the ground. Everything in
his life changes.[41]

The rock-workers struggled in this account, then, to impregnate the rock with dynamite and emerge from the womblike granite mountain. The images of birthing emphasized expectations that the dam would create a new life and youth for Egyptian peasants, allowing them to overcome the servitude of Pharaonic times. Human work on the environment, in this narrative, was lib-erating and gave agency to peasants as well as the nation more broadly.

State-patronized discourses also depicted rocks as a life-giving force to the Nubians. On a Ministry of Culture tour for artists to "record" Nubia before it was completely flooded by the reservoir, the painter Tahiya Halim portrayed the alignment of food and stone in the dam project; her painting titled "Bread from Rock" eventually won a State Promotion Award. "When I lived with the Nubians," Halim later explained, "I saw how poor they are. They work all day long toiling under the sunlight to earn their simple livelihood, yet they are satisfied. When I began painting [the] 'Bread from Rock' portrait, I imagined them getting their bread from rocks."[42] In a similar way, her "High Dam Rejoices" painting depicted 'Abd al-Nasir on a boat in the Nile marking the inauguration of the dam by offering wheat to a Nubian delegation.[43]

Local geologists were also recruited to the High Dam project in various ways that reinforced a linking of the building of the High Dam with the restoration of national sovereignty and renewal. The expected benefits of the High Dam initiated an extensive search by locally trained geologists for domestic mineral and fossil deposits; new mandates to restructure the academy and local companies from European leadership to the hands of Egyptians resulted in professional opportunities; and the direction of several scientific debates about Egyptian geology reinforced the geographic separation of Nubia and Sudan from Egyptian settled life. Rushdi Said, one of the most important postcolonial Egyptian geologists, attributed the trajectory of geological research in Egypt to the unjust conditions of work in the old-regime's mining industry and science academies. Educated at Fu'ad (now Cairo) University in the 1940s, Said recounted in his memoirs his surprise at the dominance of Europeans on the sciences faculty.[44] His most nationalistic feelings, however, grew out of his work in the phosphate mines on the Red Sea coast (at Qusayr) after 1941. Noting that the mining sector was under the control of foreigners, especially Italian companies at Qusayr, he pointed to the oppressive conditions for unskilled Egyptian laborers in the mines, including working in deep and harsh underground conditions for long hours and little pay, and the unfair profits earned by the company. Said was also struck by the paradox created by the partial mechanization of the industry: the area was blasted with dynamite and then cleared of the debris by workers using palm-leaf baskets.[45] By contrast, Said remembered his geological explorations around the preparations for the High Dam as one of the "happiest spiritually and productive scientifically" times of his life.[46]

The novel conditions created by the 1952 revolution, Said argued, provided the context for a "new life" for Egyptian geological studies. This period, he wrote later,

> was one of great expansion. Nineteen sixty two saw the publication of Said's treatise on *The Geology of Egypt* (1962) in which an attempt was made to find some order in the large amount of information that had accumulated over the previous years by fitting it into a conceptual framework. Between 1954 and 1976, the Geological Survey of Egypt conducted an aggressive program of exploration for economic mineral deposits which were sought for the fulfillment of the five-year

industrialization plan of the country....The number of scientists engaged was unprecedented, and new methods were introduced including geophysical and geochemical surveying as well as drilling and mining techniques.[47]

At the height of building the High Dam in 1963, Said directed a large-scale geological mission, sponsored by the Egyptian Ministry of Industry, to search for lead in the Egyptian oases of Kurkur and Dungul. Despite being outfitted with eight cars and one hundred workers, Said approached UNESCO officials of the archaeological salvage campaign to borrow instruments for carbon dating the geological layers he was excavating.[48] This encounter points to the informal ways scientists attempted to redirect international resources into national economic development.

Egyptian geological research in the 1950s and 1960s also sharpened the geological segmentation of the Nile Basin below the dam in Egypt from the riverbed above it in Nubia and Sudan. This rested on two changes in understanding the morphology of the rocks lining the riverbed. The first entailed a more nuanced classification of the rocks that had been previously grouped under the term "Nubian sandstone."[49] Research and debate about the differentiation of Nubian sandstone, which lacked rich fossil deposits that could help geologists easily date the various formations, continued from at least the 1940s to the 1970s. In his 1944 master's thesis, Rushdi Said, for example, examined the Nubian sandstone at Aswan.[50] In 1976, teams of geologists subdivided the previously monolithic Nubian sandstone beds for the first time.[51] Against mostly European geologists who advocated for the complete abandonment of the term "Nubian sandstone," local Egyptian geologists argued that the name should be retained to describe the beds in southern Egypt and northern Sudan, thereby highlighting the geological specificity of the region.[52] Ultimately linked to the search for oil deposits in Egypt, this debate was complicated by the fact that the original sample of sandstone came from the scarp of Abu Simbel in the 1830s, an area that by the late 1960s was no longer accessible to geologists because it lay under the new Lake Nasir.[53]

A second geological revision focused on the history of various parts of the Nile and the dating of its riverbeds. Whereas earlier geologists described the Nile as passing "in succession from the older to the younger [geological] strata," due to the "general inclination of the dip of the strata being northward, and nearly twenty times as great as the slope of the valley,"[54] Egyptian scholars in the 1950s and 1960s began to find geomorphic evidence that the current Nile Valley had joined radically different areas at a rather late date. This new history reversed the chronological flow of the river: the area north of Aswan was now considered very old, whereas the Nubian areas, even though their exposed rocks were older, were in fact geologically newer.[55] Both geological revisions facilitated the hydrological severing of Egypt at Aswan and corresponded to the submersion of Nubia under the new reservoir. Rewriting the geological chronology of the Nile basin undergirded the new infrastructural projects, population movements, and environmental manipulation of the 1960s.

Moving Rocks in Nubia

Faced with the impending total submersion of Egyptian Nubia and most of Sudanese Nubia, the Egyptian and Sudanese governments initiated a number of campaigns of preservation, documentation, and resettlement. UNESCO's "International Campaign to Save the Monuments of Nubia" (1959–1980) included more than forty archaeological missions from Africa, America, Asia, and Europe, and large financial contributions from more than fifty UNESCO member states. Although the most elaborate temple removals were the final ones at Abu Simbel and the Island of Philae, three other structures were actually moved closer to the High Dam to take advantage of higher ground (the temples of Bayt al-Wali, Kalabsha, and the Qertassi Kiosk), six other temples and tombs were relocated in Egypt, and four smaller temples were donated to countries participating in the salvage. Four temples were relocated to the garden of the national antiquities museum in Khartoum (Aksha, Buhen, Semna East, and Semna West). Extensive documentation work was also carried out at a large number of sites containing prehistoric, Pharaonic, Ptolemaic, Nubian, and Christian ruins.

Significantly, the UNESCO campaign rested on a claim of the international right to the area's ancient heritage. While this claim accorded ancient Egypt a foundational role in Western civilization, it nevertheless replicated a Eurocentric and teleological narrative that dismissed modern Egypt as culturally insignificant and challenged its national sovereignty. The official UNESCO appeal, launched on March 8, 1960, by Dr. Vittorino Veronese, the director-general of UNESCO, argued, "These monuments, whose loss may be tragically near, do not belong solely to the countries who hold them in trust. The whole world has the right to see them endure. They are part of a common heritage."[56] These rhetorical claims justified economic interests: nations that mounted archaeological missions were entitled to half of the artifacts found on their sites. Moreover, the Egyptian monuments received much more support than those in Sudan; the granite and sandstone monuments around Aswan were more likely to survive relocation, whereas many of the Sudanese sites were constructed from less durable mud brick. Many more Sudanese sites were also prehistoric and more modest in terms of their museum display value.[57] In fact, the prehistoric salvage campaigns were in many ways the most contentious and divisive among archaeologists, antiquities administrators, and UNESCO officials.[58]

The most prominent of all the archaeological operations occurred at Abu Simbel from 1963–1968, costing nearly $40 million.[59] In this huge rock-moving project, temples that had been originally carved into the sandstone cliffs during the reign of Ramses II (1290–1224 BCE) were cut out in blocks and reconstructed on a site sixty-five meters above the original location. Over one thousand blocks of stone were sawed from the cliffs and transported; each block weighed between twenty and thirty tons and measured roughly eighty centimeters thick, three meters high, and five meters long. The sandstone itself was fairly fragile, so many of the blocks needed substantial reinforcements in the process. In addition, some 150,000 cubic meters of rock above the temples were mechanically removed. A cofferdam, composed of 380,000 cubic meters of sand and rock, was constructed to protect the site from the rising reservoir. The rockface cutting was very precise, executed by experts from Italian marble quarries working with specially designed handsaws. In addition, support facilities were built at the site, including a town to house two thousand workers and their families.[60]

Egyptian state press coverage framed Abu Simbel in terms of the restoration and new life of worked rock. The newspaper al-Musawwar ran a report on April 16, 1965, about the removal of the 250,000 tons of upper rock material above Abu Simbel, what it called "the mountain on top of the two temples." "From a special plane we were able to document the new life in Abu Simbel.... Mechanical tools have started working together with 1,500 labourers and engineers on the most dangerous operation... by which rock and sand weighing as much as the three largest ships in the world shall be cut into pieces." The article interviewed the assistant head engineer for the campaign about the technicalities of stone transportation; the language of delicacy and precision stood in stark contrast to the movement of (unhewn) rock by blasting at Aswan near the dam: "They did not start on the cutting of these excavated rocks until after having built supports inside the temples to save the temples from collapsing.... Before the cutting up of the stones of the temple itself, they shall be injected with a special cement liquid so that no harm befalls them during the operation. The blocks... shall be transported by means of a large derrick crane of two tractors to transport them to the new site atop the hill. They shall be numbered and given horizontal and vertical markings to ensure them correct re-erection."[61] A careful and deliberate project of restoration and rescue of the human-manipulated rockscape, Abu Simbel allowed UNESCO participants to leave an imprint on the stone as they preserved its man-made originality.

In the same years, the Egyptian government initiated a series of cultural surveys of contemporary Nubia, as it prepared for the transfer of about fifty thousand Egyptian Nubians to Kom Ombo, a region to the north of Aswan, to live on land that would be reclaimed as a result of the new dam.[62] An ethnic group that lived mostly along the Nile between the first and third cataracts, Nubians had a distinct identity, despite speaking several different dialects.[63] Although Nubians in Egypt farmed the land between the desert and the Nile, the heightening of the old Aswan dam had already forced many Nubian men to migrate to urban areas for work and many Nubian villages to be relocated. The area just south of Aswan contained a number of saints' shrines and local cemeteries, in addition to many spacious houses, mature palm groves, and small but rich agricultural fields. Despite 'Abd al-Nasir's visit to Nubia in January 1960 to quell fears about the relocation plans and demonstrate "official and national concern for the Nubians,"[64] the move in both Egypt and Sudan was traumatic for a variety of reasons. Widespread despair was reported at leaving the graves of family members as well as the shrines of important local saints.[65] Other Nubians expressed appreciation for the increased social services that relocation would bring, and some Nubian doormen in Cairo in 1960 wore "under their white turbans, skullcaps decorated with the new High Dam design, a pattern in yellow and green and orange thread created by a Nubian housewife."[66]

Nubians also had a contradictory experience with the Aswan rockscape. Many Nubians worked on the construction of the High Dam.[67] In the many Nubian petitions to the Egyptian state about compensation or the various injustices of relocation, Nubian writers displayed a keen knowledge of comparable hydraulic projects from around the world, drawing parallels between their own relocations and those required by, for example, North American and European dam projects.[68] The nature of granite weathering also made possible less formal relationships between Nubians and the rockscape. Despite its strength and suitability for monumental construction projects, granite's morphology made it usable in more intimate ways not dependent on the massive mobilization of labor and technology by a well-organized central state. Granite tends to weather into discrete boulders that can be easily worked. Formed as a solid mass from molten magma deep underground and cooled under great pressure, granite rises closer to the surface over time through tectonic movements and erosion. As a result, geologist Sampsell explained, the "external pressure on the mass is reduced and the granite begins to fracture along horizontal and vertical planes at right angles....This produces joints that define nearly perfectly rectangular blocks...[that] weather along the joints into more rounded shapes. Then when the rocks are exposed by erosion of the overburden [by the river or other processes], the blocks are revealed ready for use without the difficulties attendant to quarrying them."[69] The granite boulders—almost like natural canvases—thus existed somewhere between the wild and the worked as elements of the landscape. The British writer Edward Lane noted in the 1820s that an island at Aswan, "like the other islands in its vicinity, is mainly composed of heaps of granite rocks, which appear as if they had been thrown together by some convulsion of nature. Upon one of these natural piles of granite rocks, at the southern end of the island, are many hieroglyphic inscriptions and other sculptures, rudely and slightly cut; but in a good style. They are very ancient."[70]

Yusuf Chahine's 1968 film al-Nass wa'l-Nil (The People and the Nile), an Egyptian-Soviet coproduction, echoed Nubian protest over changes to the local rockscape and was banned in Egypt.[71] The early part of the film follows a young Nubian man about to leave his village to work as an unskilled laborer on the High Dam in 1961 or 1962. Although excited for a new urban life in Aswan, he must say good-bye to his fiancé. In his farewell, he carves a Pharaonic-style relief of them and their names onto a loose granite boulder at the edge of the river. "There," he tells her, "Our names will live longer than the pyramids." The cinematography also focused on how easily Nubians move over the granite rocks by the river. Later in the film, another Nubian villager rides a donkey to the Abu Simbel worksite, where he sees cranes lifting the face of Ramses up the mountain amid neat piles of numbered stones. He calls out in anger that he will not be relocated to New Nubia until "they move the entire village. They can only do what they did at Abu Simbel, from the basement to the attic, the men and the animals, stone by stone. None of these stones [at Abu Simbel] are worth our village."[72] The film depicted Nubians as original and informal workers in stonescape. It used the natural canvases of granite boulders and carved sandstone blocks to record a comparison between state-initiated monumental projects and the needs of everyday people.

In a similar way, Nubian novelist Idris 'Ali offers a critique of the official lack of concern for Nubians that he contrasts with the elaborate and careful program of architectural rescue. In Dongola: A Novel of Nubia, the main character confronts local officials near his relocated village at Silsila Mountain, an important ancient quarry near Kom Ombo. He declares: "Give us back our old homeland....Give us land for planting....Build us factories....Pay us compensation equal to what we lost....Aren't we worth just as much as temples and statues?" At another point in the story, he "cursed them all...all the ministers of irrigation who had ever lived and those still to come, and anyone who had placed a single stone in the first and second dams. He cursed the river that had surrendered to the dam, and he cursed the whole world, which had helped to save the temples, while leaving the people to their fate."[73]

The politics and practices around actual rock quarrying near Aswan also point to the complex relationship of Nubians with the local rockscape. Rocks have been continuously quarried around Aswan for the past sixteen thousand years. Although the majority of quarrying sites on the West Bank produced grinding stones, especially from silicified sandstone,[74] sandstone and granite quarries were also worked over thousands of years, and some later converted to tombs and religious buildings, such as the monastery of St. Simeon. In the twentieth century, mining granite for Aswan dam construction reopened industrial mining activities near Aswan, especially by international firms. British geologists in the early twentieth century enthusiastically supported new granite mining because of the wide range of uses of the stone—from the ornamentation of important buildings in Cairo to road paving and street curbs—and because of the industry's proximity to the river.[75] Egyptian geologists, however, were less interested in stone mining than the extraction of other minerals. When Said's 1962 *Geology of Egypt* first appeared, an American reviewer for *Science* commended it as "an excellent summary of a large amount of information, generously documented," although he noted, "[A] final section discusses the economic mineral deposits, with the disappointing exceptions of the famed building stones and the precious ground water that is so often considered a mineral resource."[76] Attuned to Pharaonic history, many Europeans and Americans saw Egypt through the lens of built architectural stonescapes, whereas Egyptians tended to harness worked stonescapes as sources of natural resources to build a new, modern, postcolonial civilization. Nevertheless, the output of Egyptian quarries grew substantially in the 1950s and 1960s.[77] Much new mining was organized industrially and damaged the local environment, as many of the modern quarries for sandstone and granite in Aswan used blasting techniques and deep, open pits to dislodge the rocks.[78] According to archaeologists of ancient Egyptian geology, today at Silsila on the West Bank lies a quarry for sandstone, "a remarkable characteristic of which is the fact that waste material is dumped into the Nile...a practice that was generally avoided in Pharaonic times so as to keep the river navigable as well as to prevent any threat to loading conditions on the riverbank."[79]

Modern industrial quarrying techniques contrast sharply with local artisanal uses of the rock, most of which are carried out by Nubian workers. The granite quarry of Gebel Ibrahim Pasha, owned by the Egyptian state today, uses wedge-splitting techniques similar to those of ancient times, whereas international companies in the area rely primarily on drilling and blasting to remove the granite.[80] Quarry archaeologists have recently documented the long-term connection quarrying has to local cultures, especially those of the Nubians. The Aswan quarries, Elizabeth Bloxam has argued, show traces of rock inscription and graffiti that date from the late Palaeolithic to the contemporary era, and thus "represen[t] a 'lived' landscape" of continuous use for the past fifteen thousand years. "These traces that people left," she continues, "may be linked, in antiquity, to the production of grinding stones but also suggest the consistent human intervention in this landscape, with a strong Nubian influence....A distinct Nubian identity [including Nubian housing] remains on the West Bank, even to the point where small artisan quarrying is specifically linked to local use of the stone for building their houses. There is a social embeddedness in this landscape ...given the evidence of a Nubian presence to at least 6,000 years ago."[81] Despite the sense of injustice Nubian villagers felt over the stones of ancient monuments receiving more careful relocation, in actual practice, their own relationships with the Aswan rockscape have also been complex and contradictory.

Conclusion

Rocks have figured specifically into several of the environmental consequences of the building of the High Dam. The downstream temples, many so carefully removed from the banks of the Nile that would be flooded, are increasingly threatened with deterioration due to salting of their foundations by the overuse of irrigation waters in nearby fields since construction of the dam.[82] Similar salt deterioration is expected the length of the Nile Valley south of Fayyum. A second consequence of the High Dam has been the destruction of the raw material of the domestic mud-brick industry. A locally available, durable, and low-impact building material, mud-brick has long been in common use throughout Egypt. Nile mud, the silt that was deposited on the banks of the Nile, is essentially "minute chemically-unaltered fragments of igneous rock" that, because of their irregular sizing, interlock tightly on drying.[83] Four-thousand-year-old mud-brick walls were still standing at Semna in 1964.[84] The total loss of flood-borne silt after completion of the High Dam necessitated a search for new raw materials by the owners of the nation's seven thousand brick kilns, which produced more than a billion bricks in the early 1980s. Kiln operators first excavated farmland by removing the soil down to the groundwater table, destroying the very commodity—cultivable land—that the dam was designed to expand. When the state prohibited the practice in 1984, after the destruction of nearly one thousand square kilometers, brickmakers turned to raw materials from the desert.[85] A third geological consequence of the High Dam has been a possible increase in regional seismicity due to pressure from the large volume of water held in Lake Nasir. During the dam's construction, a UNESCO-sponsored international team of seismologists predicted that the dam itself was not at risk but that the lake might cause tremors.[86] A strong earthquake did occur at Lake Nasir in 1981; although attributed to tectonic activity, it may have been triggered by the lake.[87] The lake itself has also altered the climate of Aswan, bringing rain and clouds to the previously dry climate.[88] Archaeologists expect this to cause erosion to sandstone monuments as well as perhaps wider erosion of banks and silting of the river.[89] Dam construction has also increased development and population density at Aswan, leading to extensive new urbanization and transportation works. Rockscape erosion and mining thus altered the region in complex ways.

The way that rocks figure as both elements of nature and building blocks of civilization allows historians to articulate the immensity of geological time with the more minute histories of the construction of the High Dam and the politics of the relocation of people and rock monuments from Nubia. A new geological narrative firmly partitioned Nubia from Egypt north of Aswan and supported the idea that older, northern Egypt be home to Egypt's archaeological heritage. The unwieldy attempt to restore time, youth, and sovereignty in Egypt through modern, massive, technological projects such as the High Dam calls for a reassessment of 'Abd al-Nasir's slogan that it would be a "pyramid for the living." Rather than simply evidence of the megalomania of rulers like the Pharaohs and 'Abd al-Nasir, or empty words of state propaganda, "a pyramid for the living" does capture the state's effort to restore time to the nation through geological manipulation. The emphasis on restoration of the human-manipulated landscape in different ways by a variety of groups offers a challenge to frameworks that dichotomize wilderness and humanscapes and downplay human interaction with nature through work and labor.

Debate about the consequences of muting human agency in environmental history is echoed in riparian contests in other parts of the Middle East. Israeli and Palestinian narratives about the environment and territorial sovereignty exhibit a deep disjuncture.[90] Turkey's control of the upstream area of Iraq's major rivers, and the politics of using national development schemes such as dams as an opportunity to relocate ethnic groups that the state has been trying to control for the past century—such as the Kurds—present clear comparisons to the High Dam case.[91] The framing of these debates as (good) wild nature versus (destructive) human culture tends to silence disenfranchised groups such as the Kurds, the Nubians, or the "marsh-dwellers" of Iraq, whose complex relationships with the local riparian landscapes are reduced to a narrative of victimization. To break out of these dualistic modes of thinking, environmental historians must attend to the complex and contradictory relationships among local land use, rockscape practices, and geological theories.

Acknowledgment

The author gratefully acknowledges the following people for helpful comments on various drafts of this chapter: Alan Mikhail, Khaled Fahmy, Will Hanley, Jean Allman, Andrea Friedman, Tim Parsons, Jennifer Collins, and two anonymous reviewers for Oxford University Press. This chapter greatly benefited from comments during presentations at Florida State University and Washington University in St. Louis. Research for the study was undertaken during a junior faculty sabbatical from Washington University in St. Louis.

Notes

1. Hussein M. Fahim, *Dams, People, and Development: The Aswan High Dam Case* (New York: Pergamon Press, 1981), 14. See also J. R. McNeill, *Something New under the Sun: An Environmental History of the Twentieth-Century World* (New York: W. W. Norton, 2000), 168–169; Sandra Postel, *Pillar of Sand: Can the Irrigation Miracle Last?* (New York: W. W. Norton, 1999), 54. Structurally, the dam actually resembles a pyramid when viewed from its north–south cross-section. Bonnie M. Sampsell, *A Traveler's Guide to the Geology of Egypt* (New York: American University in Cairo Press, 2003), 63.

2. See, for example, David Blackbourn, *The Conquest of Nature: Water, Landscape, and the Making of Modern Germany* (New York: W. W. Norton, 2006); Matthew D. Evenden, *Fish versus Power: An Environmental History of the Fraser River* (Cambridge: Cambridge University Press, 2004); Donald Worster, *Rivers of Empire: Water, Aridity, and the Growth of the American West* (New York: Pantheon, 1985).

3. E. R. J. Owen, *Cotton and the Egyptian Economy, 1820–1914* (Oxford: Clarendon Press, 1969); Helen Anne B. Rivlin, *The Agricultural Policy of Muḥammad 'Alī in Egypt* (Cambridge: Harvard University Press, 1961).

4. McNeill, *Something New under the Sun*, 168.

5. Emil Ludwig, *The Nile: The Life-Story of a River*, trans. Mary H. Lindsay (New York: Viking Press, 1937), 295–296.

6. John Waterbury, *Hydropolitics of the Nile Valley* (Syracuse, NY: Syracuse University Press, 1979), 12.

7. Ibid.; Wm. Roger Louis and Roger Owen, eds., *Suez 1956: The Crisis and Its Consequences* (Oxford: Clarendon Press, 1989); Elizabeth Bishop, "Talking Shop: Egyptian Engineers and Soviet Specialists at the Aswan High Dam" (Ph.D. diss., University of Chicago, 1997); Tom Little, *High Dam at Aswan: The Subjugation of the Nile* (New York: John Day, 1965).

8. Ahmad Shokr, "Watering a Revolution: The Aswan High Dam and the Politics of Expertise in Mid-Century Egypt" (M.A. thesis, New York University, May 2008); Waterbury, *Hydropolitics*, 117; Little, *High Dam*, 53–56.

9. Sampsell, *Guide to the Geology of Egypt*, 46. For a less optimistic view, see McNeill, *Something New under the Sun*.

10. Timothy Mitchell, "Can the Mosquito Speak?" in *Rule of Experts: Egypt, Techno-Politics, Modernity* (Berkeley: University of California Press, 2002), 27.

11. Ted Steinberg, "Down to Earth: Nature, Agency, and Power in History," *American Historical Review* 107 (2002), 804, citing Elizabeth Blackmar, "Contemplating the Force of Nature," *Radical Historians Newsletter* 70 (1994): 4.

12. William Cronon, "A Place for Stories: Nature, History, and Narrative," *Journal of American History* 78 (1992), 1349. On agency in environmental history, see also Richard C. Hoffman, Nancy Langston, James C. McCann, Peter C. Perdue, and Lise Sedrez, "*AHR* Conversation: Environmental Historians and Environmental Crisis," *American Historical Review* 113 (2008): 1431–1465; Mitchell, "Can the Mosquito Speak?" 29–31; Donald Worster, "Doing Environmental History," in *The Ends of the Earth: Perspectives on Modern Environmental History*, ed. Donald Worster (New York: Cambridge University Press, 1988), 293; Carolyn Merchant, "Shades of Darkness: Race and Environmental History," *Environmental History* 8 (2003): 380–394; Richard White, *Land Use, Environment, and Social Change: The Shaping of Island County, Washington* (Seattle: University of Washington Press, 1980).

13. Shokr has argued that many of these histories have "bound together the history of the river and the history of the nation-state." Shokr, "Watering a Revolution," 3. Political-diplomatic accounts have dominated the literature on the Aswan High Dam. Among the most comprehensive is Waterbury, *Hydropolitics of the Nile Valley*. Tom Little's *High Dam at Aswan* provides a rich, firsthand narrative of the early years of dam building. More specific studies examine the politics of expertise in engineering, statistics, economics, and public health that emerged as a result of the dam. See Shokr, "Watering a Revolution"; Bishop, "Talking Shop"; Clement Henry Moore, *Images of Development: Egyptian Engineers in Search of Industry*, 2nd edn. (Cairo: American University in Cairo Press, 1994); Mitchell, "Can the Mosquito Speak?"; Yusuf A. Shibl, *The Aswan High Dam* (Beirut: The Arab Institute for Research and Publishing, 1971). Although most writings about the dam mention the scheme's "drawbacks," the High Dam also regularly features in broader environmental surveys and in environmental science. McNeill, *Something New under the Sun*; Gilbert F. White, "The Environmental Effects of the High Dam at Aswan," *Environment* 30 (1988): 5–11 and 34–40; Steven Soloman, *Water: The Epic Struggle for Wealth, Power, and Civilization* (New York: HarperCollins, 2010).

14. McNeill, *Something New under the Sun*, 49.

15. Paul Sutter, "What Can U.S. Environmental Historians Learn from Non-U.S. Environmental Historiography?" *Environmental History* 8 (2003), 9 and 11. Alfred Crosby has famously used the term "ecological imperialism" to discuss European migrations and their environmental consequences in the Americas and the Pacific Ocean. Alfred W. Crosby, *Ecological Imperialism: The Biological Expansion of Europe, 900–1900* (Cambridge: Cambridge University Press, 2004). Good discussions of the literature on European colonialism and environmentalism can be found in the following: Caroline Ford, "Reforestation, Landscape Conservation, and the Anxieties of Empire in French Colonial

Algeria," *American Historical Review* 113 (2008): 341–362; Diana K. Davis, *Resurrecting the Granary of Rome: Environmental History and French Colonial Expansion in North Africa* (Athens: Ohio University Press, 2007). For postcolonial critiques from South Asia, see Ramachandra Guha, *The Unquiet Woods: Ecological Change and Peasant Resistance in the Himalaya* (Berkeley: University of California Press, 1990).

16. Steinberg, "Down to Earth," 803.

17. Little, *High Dam*, 5; Peggy Parks, *The Aswan High Dam* (New York: Blackbirch Press, 2004), 5; Bishop, "Talking Shop," 12–13.

18. Bishop, "Talking Shop," 12.

19. Waterbury, *Hydropolitics*, 111.

20. Ibid., 111–112. The southern part of the reservoir lies in Sudan and is called Lake Nubia.

21. W. Willcocks and J. I. Craig, *Egyptian Irrigation*, 2 vols. (New York: Spon and Chamberlain, 1913), 1: 3.

22. Ibid., 1: 4.

23. John Ball, *A Description of the First or Aswan Cataract of the Nile* (Cairo: National Printing Department for the Ministry of Finance Survey Department, 1907), 46.

24. Ibid., 57 and 64.

25. Ibid., 110–111.

26. Ibid., 74.

27. Waterbury, *Hydropolitics*, 111.

28. Little, *High Dam*, 86.

29. Ibid., 87. See also ibid., 89 and 121.

30. Ibid., 222.

31. Sampsell, *Guide to the Geology of Egypt*, 29–30.

32. Martin Stokes, "Listening to Abd al-Halim Hafiz," in *Global Soundtracks: Worlds of Film Music*, ed. Mark Slobin (Middletown, CT: Wesleyan University Press, 2008), 309–333.

33. The Suez battle anthem from the 1957 movie *Port Sa'id* expressed a similar conception of the restoration of time by state nationalization of the Suez Canal: "O Egypt, by nationalization let us renew the ages." Joel Gordon, *Revolutionary Melodrama: Popular Film and Civic Identity in Nasser's Egypt* (Chicago: Middle East Documentation Center, 2002), 76. See also ibid., 95.

34. These partial Arabic lyrics are from the following: http://www.arabicmusictranslation.com/2007/11/abdel-halim-hafez-high-dam.html (accessed January 13, 2010).

35. Elizabeth Warnock Fernea and Robert A. Fernea, with Aleya Rouchdy, *Nubian Ethnographies* (Prospect Heights, IL: Waveland Press, 1991), 7–8.

36. Ibid., 8.

37. Little, *High Dam*, 102.

38. Ibid., 118.

39. Copy of article "La caravane pour le sauvetage des vestiges de Nubie," *Journal d'Egypte* 18 (1963), in CLT/CH/ARB 292, UNESCO Paris.

40. Little, *High Dam*, 117–118.

41. Ahmad Bahjat, "Wajhun fi'l-Ziham," *al-Ahram*, January 8, 1960, 4. Founded in 1875, *al-Ahram*, by 1960, had become an instrument of the government. On the lethal effects of granite dust for stonecutters, see Mari Tomasi and Roaldus Richmond, *Men Against Granite* (Shelburne, VT: The New England Press, 2004), 3.

42. "Tahia Halim…Nubian Love," online Egyptian magazine from the State Information Service, 60–63. http://www.sis.gov.eg/VR/magazine%20english%20 43/16.pdf (accessed May 24, 2011). See also Liliane Karnouk, *Contemporary Egyptian Art* (Cairo: American University in Cairo Press, 1995), 15.

43. On the Ministry of Culture cruise, see Hussein M. Fahim, *Egyptian Nubians: Resettlement and Years of Coping* (Salt Lake City: University of Utah Press, 1983), 34; Karnouk, *Contemporary Egyptian Art*; Jessica Winegar, *Creative Reckonings: The Politics of Art and Culture in Contemporary Egypt* (Stanford, CA: Stanford University Press, 2006), 177.

44. Rushdi Sa'id, *Rihlat 'Umr: Tharawat Misr bayna 'Abd al-Nasir wa'l-Sadat* (Cairo: Dar al-Hilal, 2000), 44.

45. Ibid., 58–62.

46. Ibid., 74. He has a similarly nationalist periodization of the history of geological research in Egypt. Rushdi Said, ed., *The Geology of Egypt* (Rotterdam: A.A. Balkema, 1990), 3–5. See also Rosemarie Klemm and Dietrich D. Klemm, *Stones and Quarries in Ancient Egypt* (London: British Museum Press, 2008), 1–2.

47. Said, ed., *Geology of Egypt*, 3–5. See also Derek Hopwood, *Egypt, Politics and Society, 1945–1981* (London: Allen and Unwin, 1982), 123–125.

48. Christophe to Service des Monuments de la Nubie, Le Caire, 15 May 1963, "a/s Survey préhistorique de la Nubie: Oasis de Kurkur et de Dungul," AN/LE/ Mémo 615, CLT/CH/ARB/292, UNESCO Paris.

49. Klemm and Klemm, *Stones and Quarries*, 212.

50. Completed for Fu'ad I University and advised by Professor N. M. Shukri, the thesis was entitled "A Contribution to the Geology of the Nubian Sandstone." Sa'id, *Rihlat 'Umr*, 44 and 64; Klemm and Klemm, *Stones and Quarries*, 348. Shukri himself published a number of papers on the characteristics of Nubian sandstone, several coauthored by Said, in the middle 1940s and early 1950s. See, for example, N. M. Shukri, "Geology of the Nubian Sandstone," *Nature* 156 (July 28, 1945): 116. See also Klemm and Klemm, *Stones and Quarries*, 348.

51. E. Klitzsch and Heinz Schandelmeier, "South Western Desert," in *The Geology of Egypt*, ed. Rushdi Said (Rotterdam: A. A. Balkema, 1990), 249.

52. On the characterization of the sandstone, see Rushdi Said, *The Geology of Egypt* (New York: Elsevier, 1962), 129–131. Extensive debate on the subject occurred in the *American Association of Petroleum Geologists Bulletin [AAPGB]*. See specifically the following volumes: 52/4 (April 1968), 53/1 (January 1969), 54/3 (March 1970), 55/6 (June 1971). The debate about sandstone was related to the larger search for oil in Egypt. Egyptian limestone and sandstone was often overlaid with shale deposits, which can be oil-bearing, and sandstone often holds deposits of oil that has migrated from other rocks. The Nubian sandstone unit in Libya, for example, was found to be oil-producing. A. J. Whiteman, "Nubian Group: Origin and Status," *AAPGB*

54/3 (March 1970): 522–526. For a defense of the retention of the term "Nubian sandstone," see Bahay Issawi, "Nubia Sandstone: A Discussion," *AAPGB* 55/6 (June 1971): 885–887.

53. Issawi, "Nubia Sandstone," 886.

54. Willcocks and Craig, *Egyptian Irrigation*, 1: 3–4.

55. M. S. Abu al-Izz, *Landforms of Egypt*, trans. Y. A. Fayid (Cairo: American University in Cairo Press, 1971), 77–78. In a footnote on p. 96, the author notes that he wrote the text in 1966.

56. *A Common Trust: The Preservation of the Ancient Monuments of Nubia* (Paris: UNESCO, 1960).

57. No permanent UNESCO committee was established for Sudan. See Torgny Säve-Söderbergh, "International Salvage Archeology: Some Organizational and Technical Aspects of the Nubian Campaign," *Annales Academiae Regiae Scientiarum Upsaliensis (1971–72)* (Stockholm: Almquist and Wiksell, 1971–1972), 116–140, enclosed in CLT/CH/ARB/384, UNESCO Paris.

58. Reed to Christophe, Aswan, 2 March 1963, CLT/CH/ARB/292, UNESCO Paris. See also: Christophe to Service, Cairo 23 February 1963, CLT/CH/ARB/292, UNESCO Paris; Sa'id, *Rihlat 'Umr*, 74; Fred Wendorf, *Desert Days: My Life as a Field Archaeologist* (Dallas: Southern Methodist University Press, 2008), 122–146.

59. Torgny Säve-Söderbergh, *Victoire en Nubie: La Campagne internationale de sauvegarde d'Abou Simbel, de Philae and d'autres trésors culturels* (Paris: UNESCO, 1992), 112 and 121.

60. Ibid., 122–129. See also Christiane Desroches-Noblecourt and Georg Gerster, *The World Saves Abu Simbel* (Vienna: Verlag, 1968); UNESCO report dated Paris, 12 July 1965, "Summary of the Progress of Work on Saving the Abu Simbel Temples, March, April, May 1965," in CLT/CH/ARB/294, UNESCO Paris.

61. Cited and translated in letter from Cairo Liaison Office to Service for the Monuments of Nubia, UNESCO-Paris, 11 April 1965, "Subject: UAR Press Clippings," CLT/CH/ARB/292, UNESCO Paris.

62. "The Ethnological Survey of Egyptian Nubia," coordinated by the Social Research Center of the American University in Cairo (AUC) from 1960–1964, was directed by Robert A. Fernea, then an anthropologist at the AUC. The Ford Foundation provided the major funding for the project, and it was supported both by the Egyptian Ministry of Social Affairs, which was responsible for the Nubian resettlement project, and by UNESCO. Fernea's main study coming out of the project was the following: Robert A. Fernea, *Nubians in Egypt: Peaceful People* (Austin: University of Texas Press, 1973).

63. On the development of an ethnic identity for Nubians, see Robert A. Fernea and Aleya Rouchdy, "Contemporary Egyptian Nubians," in *Nubian Ethnographies*, ed. Elizabeth Warnock Fernea and Robert A. Fernea, with Aleya Rouchdy (Prospect Heights, IL: Waveland Press, 1991), 183–202. On the construction of Nubian identities in Kenya, see Douglas H. Johnson, "Tribe or Nationality? The Sudanese Diaspora and the Kenyan Nubis," *Journal of Eastern African Studies* 3 (2009): 112–131.

64. Fahim, *Egyptian Nubians*, 36.

65. Ibid., 43; Little, *High Dam*, 136. On Nubian saint worship, see Fernea and Fernea, with Rouchdy, *Nubian Ethnographies*.

66. Fernea and Fernea, with Rouchdy, *Nubian Ethnographies*, 8. On the Nubian experience of resettlement, see Fahim, *Egyptian Nubians*.

67. Little, *High Dam*, 88.

68. "Inundated Nubians" and attached petitions with translations, 1933; The National Archives of the United Kingdom, Foreign Office 141/699/3.

69. Sampsell, *Guide to the Geology of Egypt*, 59–60.

70. Edward William Lane, *Description of Egypt: Notes and Views in Egypt and Nubia, Made during the Years 1825, 26, 27, and 28*, ed. and intro. Jason Thompson (Cairo: American University in Cairo Press, 2000), 431–432.

71. On Nubians in Egyptian history and film, see Viola Shafik, *Popular Egyptian Cinema: Gender, Class, and Nation* (New York: American University in Cairo Press, 2006), 64–78.

72. Yusuf Chahine, *al-Nass wa'l-Nil (The People and the Nile)*, 1968. A coproduction with Soviet filmmakers, the screenplay was written by Hasan Fuad and Nicolai Virkroski, and it starred Suad Husni, Salah Zoulficar, and 'Izzat al-Ayali. Ibrahim Fawal, *Youssef Chahine* (London: British Film Institute, 2001), 20 and 230; Magda Wassef, ed., *Egypte, 100 ans de cinéma* (Paris: Institut du Monde Arabe, 1995), 302.

73. Idris 'Ali, *Dongola: A Novel of Nubia*, trans. Peter C. Theroux (Fayetteville: University of Arkansas Press, 1998), 20 and 34. See also ibid., 36. 'Ali also plays with the theme of Nubian drowning in his 2005 novel, *Tahta Khatt al-Faqr*, recently translated into English as the following: Idris 'Ali, *Poor*, trans. Elliott Colla (Cairo: American University in Cairo Press, 2007).

74. Tom Heldal and Per Storemyr, "The Quarries at the Aswan West Bank," in "Quarryscapes Report: Characterisation of Complex Quarry Landscapes, an example from the West Bank Quarries, Aswan," no. 4., ed. Elizabeth Bloxam, Tom Heldal, and Per Storemyr (Trondheim: QuarryScapes Project, 2007), 129. Available at: www.quarryscapes.no/publications.php (accessed June 1, 2011).

75. Ball, *First or Aswan Cataract of the Nile*, 75.

76. Louis C. Conant, "Review: Geology of North Africa," *Science*, New Series, 140 (April 5, 1963): 41.

77. United Arab Republic, *The Year Book, 1965* (Cairo: Information Department Press, 1965), 116–117.

78. Klemm and Klemm, *Stones and Quarries*, 236.

79. Ibid., 181.

80. Ibid., 241.

81. Elizabeth Bloxam, "QuarryScapes Report: The Assessment of Significance of Ancient Quarry Landscapes—Problems and Possible Solutions. The Case of the Aswan West Bank," no. 5 (Trondheim: QuarryScapes Project, 2007), 20–21. Available at: www.quarryscapes.no/text/publications/QS_del5_report.pdf (accessed June 1, 2011).

82. Beverly Karplus Hartline, "Irrigation Threatens Egyptian Temples," *Science* 209 (August 15, 1980): 796; G. Burns, T. C. Billard, and K. M. Matsui, "Salinity Threat to Upper Egypt," *Nature* 344 (March 1, 1990): 25.

83. Ball, *First or Aswan Cataract of the Nile,* 61.

84. Ibid., 60.

85. White, "Environmental Effects of the High Dam," 11. See also Rushdi Said, *The River Nile: Geology, Hydrology, and Utilization* (New York: Pergamon Press, 1993), 248.

86. White, "Environmental Effects of the High Dam," 10.

87. Edward Goldsmith and Nicholas Hildyard, *The Social and Environmental Effects of Large Dams* (San Francisco: Sierra Club Books, 1984), 116.

88. Hopwood, *Egypt, Politics and Society,* 128–129. See also Klemm and Klemm, *Stones and Quarries,* 7.

89. Per Storemyr, "Outline of the Geography and Environmental History of the West Bank at Aswan," in "Quarryscapes Report: Characterisation of Complex Quarry Landscapes, an example from the West Bank Quarries, Aswan," no. 4., ed. Elizabeth Bloxam, Tom Heldal, and Per Storemyr (Trondheim: QuarryScapes Project, 2007), 15–16. See also: Ball, *First or Aswan Cataract of the Nile,* 45.

90. Samer Alatout, "Narratives of Power: Territory, Population, and Environmental Narratives in Palestine and Israel," in *Palestinian and Israeli Environmental Narratives,* ed. Stuart Schoenfeld (Toronto: The York Centre for International and Security Studies, 2005), 300; idem., "Towards a Bio-Territorial Conception of Power: Territory, Population, and Environmental Narratives in Palestine and Israel," *Political Geography* 25 (2006): 601–621.

91. Hilal Elver, "Turkey's Rivers of Dispute," *Middle East Report* 254 (2010): 14–18. This issue of *Middle East Report* presents an excellent overview of a number of water conflicts in the Middle East, including in Iraq, Yemen, and Saudi Arabia. On the politics of the Euphrates and Tigris Rivers, see also Hilal Elver, *Peaceful Uses of International Rivers: The Case of the Euphrates and Tigris Rivers Dispute* (Ardsley, NY: Transnational, 2002); John F. Kolars and William A. Mitchell, *The Euphrates River and the Southeast Anatolia Development Project* (Carbondale: Southern Illinois University Press, 1991). Other sources on more general Middle East water disputes include the following: J. A. Allan, *The Middle East Water Question: Hydropolitics and the Global Economy* (London: I. B. Tauris, 2002); Hussein A. Amery and Aaron T. Wolf, eds., *Water in the Middle East: A Geography of Peace* (Austin: University of Texas Press, 2000); Thomas Naff and Ruth C. Matson, eds., *Water in the Middle East: Conflict or Cooperation?* (Boulder, CO: Westview Press, 1984); Jeffrey K. Sosland, *Cooperating Rivals: The Riparian Politics of the Jordan River Basin* (Albany: State University of New York Press, 2007); Jan Selby, *Water, Power and Politics in the Middle East: The Other Israeli-Palestinian Conflict* (London: I. B. Tauris, 2003); Miriam R. Lowi, *Water and Power: The Politics of a Scarce Resource in the Jordan River Basin* (Cambridge: Cambridge University Press, 1993); Hillel Shuval and Hassan Dweik, eds., *Water Resources in the Middle East: Israeli-Palestinian Water Issues, from Conflict to Cooperation* (Berlin: Springer, 2007).

9

The Rise and Decline of Environmentalism in Lebanon

Karim Makdisi

> A land of golden beaches and stunning mountain landscapes.
> Landscapes that change with the seasons, but are always bathed
> in the warm sun. The daylight rising over Lebanon brings end-
> less opportunities of fun, beach, nature and outdoor activities.
> You'll never know what to choose. Whatever it will be, an unfor-
> gettable time awaits you under the shiny Lebanese sun.[1]

The environment occupies an important place in the construction not
only of Lebanon's natural heritage but also its national and cultural
identity and its political economy. Nestled in a narrow 240-kilometer
strip of land averaging only 500 meters in width along the eastern
Mediterranean, the Lebanese mountains ascend majestically to
reach, at their peak, over 3,000 meters. Extensive forests of fir, pine,
juniper and, most notably, the famous Lebanese cedar once covered
three-quarters of this surface area. A steep descent then leads to the
fertile green carpetlike Beqa'a Valley, which has served as an important agricultural zone and farming area since the days of the Roman
Empire and today contains 40 percent of Lebanon's arable land. The
Beqa'a is sandwiched on its eastern side by another high mountain
range forming the border with Syria. Overall, Lebanon's small total
surface area of 10,500 square kilometers comprises notable climatic and ecological diversity within five distinct geo-morphological
regions: the coastal zone reaching 250 meters above sea level (representing 13 percent of its territory); the Mount Lebanon mountain

range from the Akkar plains in the north to Jabel Amel in the south (47 percent); the Beqa'a Valley that feeds Lebanon's two main rivers, the Orontes (Assi) River in the north and the Litani River to the south (14 percent); the anti-Lebanon mountain range that includes Mount Hermon to the south (19 percent); and the elevated plateau of South Lebanon (7 percent).[2]

Despite the symbolic and economic importance of Lebanon's ecological diversity, the country's modern environmental history is one of rapid decline—a decline that has greatly accelerated during the post-civil war reconstruction period that began in the early 1990s. The most recent official *State of the Environment* report (2010) makes it clear that Lebanon's land resources have been seriously threatened by high population density in urban areas (with the area of greater Beirut accounting for 50 percent of the country's residents), an explosion of unplanned urban expansion and building construction (the total surface area promised to construction permits, for example, nearly doubled from 2007 to 2008), and the rampant building of road networks that have carved up previously inaccessible mountain areas to accommodate vast commercial residential projects.[3] A high urbanization rate and environmental degradation have also resulted in dangerous levels of stress to Lebanon's water resources. Greater Beirut, for instance, now only receives three hours of drinking water daily during the summer months. Rivers, springs, and groundwater resources are also severely impacted by the unregulated discharge of raw sewage and other waste and effluents that are dumped by households, farmers, and industries—largely without treatment. Indeed, there are over fifty major sewage outfalls along Lebanon's coastline that together absorb roughly 65 percent of Lebanon's total waste load.[4] Moreover, thermal power plants and heavy industries along the coast result in coastal pollution and contribute, along with the transport sector, to the rapid deterioration of air quality in Lebanon. Forest areas today cover no more than an estimated 7–13 percent of Lebanon's territory, and massive deforestation has not only resulted in a loss of timber resources utilized for centuries by local communities and their foreign overlords but has also led to wildlife extinction, biodiversity loss, and dramatic soil erosion.[5]

Over the past decade, there has been an impressive increase in the available data, knowledge, and public awareness of Lebanon's chronically ill environment and shrinking natural resources. Global funding has enabled basic national data stocktaking exercises in such areas as biodiversity loss, land degradation, desertification, the chemical trade, and—more recently—climate change.[6] Meanwhile, the number of registered environmental NGOs has hugely increased in the postwar years, reaching roughly three hundred organizations. Moreover, since 2001, sixteen academic and research institutions have also been established to specifically address various environmental issues.[7] The principle national daily newspapers, such as *al-Safir*, *al-Akhbar*, *al-Nahar*, and *The Daily Star*, regularly feature environmental news. Lebanon also hosts the main pan-Arab environmental magazine, *al-Bi'ah wa al-Tanmiyya*. In line with this greater public knowledge and awareness of the environment, official discourse has also changed as key national environmental issues such as quarrying, water resource and solid waste management, and energy policies have become highly politicized. Indeed, Lebanon now boasts two recently established green parties.

This increased interest in environmental issues, however, has translated into a mostly technical and shallow policy literature that has remained, with some notable exceptions, depoliticized and dehistoricized. The literature dealing with conceptual aspects of environmentalism or the history of the environmental movement in Lebanon thus remains sparse, though this state of affairs seems to be changing.[8] For writer and activist Habib Ma'alouf, for instance, Lebanon's current environmental travails are intimately related to the palpable loss of a philosophical environmental framework that Lebanese society, he claims, traditionally depended on.[9] Elsewhere, I have put forth a concept of "flows" to connect global environmental processes such as the trade in hazardous waste with the response of transnational civil society.[10] Paul Kingston articulates a more rigorous framework showing that an analysis based on patron–client links rather than state–society relations is more productive in understanding Lebanon's postwar environmental politics.[11] Some scholars are now also researching and theorizing the formulation, advocacy, and implementation of environmental policy in Lebanon.[12]

In general, however, there remains a clear bifurcation between the increasingly dynamic body of work by scholars assessing modern Lebanon's political and socioeconomic history and that of scholars exploring environmental issues within larger social, political, and historical contexts.[13] While the former type of scholarship has tended not to address environmental problems directly, the latter literature has largely been dominated by studies carried out by the United Nations and other international agencies, as well as by NGO reports and workshop proceedings.[14]

This chapter addresses the gap between these two bodies of work by exploring the evolution of environmentalism in Lebanon in the period since its independence in 1943. It contends that two main strands of environmentalism, challenging various aspects of the state and state policy, emerged prior to Lebanon's long civil war in response to inherent contradictions within the Lebanese state itself and the nature of its post-independence socioeconomic and political development. The first strand was a pioneer environmentalism rooted in liberal civil society that began to coalesce in the 1960s and represented elite concerns over the disappearance of nature; this brand of environmentalism ultimately took a more technical rather than political approach to environmental problem-solving. The second kind of environmentalism was a broad social movement largely embedded within the disenfranchised southern Lebanese and Beqa'a Valley Shi'a community that advocated for the more equitable redistribution of natural resources and state services in Lebanon. There was also a third—perhaps less-pronounced, but no less important—strand of environmentalism that emerged in Lebanon during its civil war to confront emergency situations and that served in an ad hoc and temporary manner to address the inability of the Lebanese state to protect particular communities in times of war or perceived local environmental catastrophe. This chapter thus explores how Lebanese society, operating within the contexts of sectarianism and (neo)liberalism, has responded to environmental threats, and how apparently distinct strands of environmentalism briefly coalesced into a coherent national environmental movement in the immediate post-civil war period before declining and ultimately fragmenting.

The Emergence and Consolidation of Sectarianism: From the Ottoman Nineteenth Century to the French Mandate

Concerns about the environment in modern Lebanon, including the manifestation of various strands of environmentalism, have of course been mediated by the state. In order to understand the nature of this relationship and environmentalism more broadly, therefore, the Lebanese state must first be historicized within the context of the sectarian and clientalist systems that emerged in the nineteenth century against the backdrop of Ottoman reform and European colonial intervention.[15] While the struggle among various groups and communities seeking patronage goes back to the Ottoman period (if not earlier), sectarianism soon became the dominant mode through which such patronage worked and became structurally embedded in twentieth-century Lebanon's liberal political economy, state institutions, and social relations.[16]

Prior to the emergence of sectarianism, leading feudal families dominated society, a phenomenon the historian Albert Hourani terms the "politics of the notables."[17] These notables controlled the self-representation of their respective communities and formed an "interdependent trans-sectarian elite" that was separated, collectively, from the *ahali*, or common people, by an "almost impenetrable barrier, which was reinforced by customs of clothing, language, title, land holdings and marriage alliances."[18] However, beginning in the 1820s, sustained peasant insurrections against their feudal landlords in Mount Lebanon, along with increasing capitalist penetration, ultimately led to an agreement among European powers and the Ottomans to administratively divide Mount Lebanon, for the very first time, along expressly sectarian lines of Maronite Christian and Druze.[19] Such sectarian administrative division did nothing to respond to peasants' socioeconomic demands and clearly ignored the social reality on the ground: nearly all districts contained mixed populations; thus, the root problem was manifestly not sectarian in nature.[20] This new form of political and administrative sectarian division set many of the parameters for modern Lebanon.[21] Lebanese state administration was thus, even before formal independence, captured by a sectarian elite—one which was initially dominated by Maronites but whose makeup gradually changed over time. This elite has effectively used systems of patronage to maintain both national divisions—important for preventing social movements (environmental or otherwise) from challenging the system—and the subservience of its "clients": the Lebanese citizenry.[22] Henceforth, social movements led by the *ahali*—and later by middle-class civil society—would be deflected and fractured by the sectarian elite via the construction of sectarian agitation.

As sectarianism became increasingly consolidated institutionally and socially in Ottoman Mount Lebanon beginning in the second half of the nineteenth century, the appointments of governors and district heads, the allocation of seats in the newly created administrative council, and the recruitment of the gendarmerie were all divided along explicitly sectarian lines. French colonial influence continued to markedly favor Maronite Christians in administrative and other appointments, which in turn helped advance the emerging Maronite bourgeois and its patronage networks in the era just before formal independence.[23]

The division of Ottoman Arab territories following the Conference of San Remo led to the creation of "Greater Lebanon" in 1920 under a formal French mandate.[24] France—which had occupied Lebanon in 1918—combined the previously separate and largely autonomous region of Mount Lebanon with the main coastal provinces of Beirut, Sidon, and Tripoli to help foster sustained economic viability for the Maronite political elite who would continue to benefit from French protection.[25] The resulting union, consummated against the wishes of a significant number of the country's population,[26] was part and parcel of a growing gap in access to resources among the Lebanese and led to serious disagreements over national identity. As the political scientist Sami Ofeish notes:

> Greater Lebanon engulfed two areas unequal in their level of capital-ist development and their access to services and resources: the more advanced area of Beirut and Mount Lebanon, constituting the center, and the less advanced areas of northern, eastern, and southern Lebanon, constituting the peripheries...The different concentrations of sectarian communities in the center versus the peripheries also meant that Christians, predominately in the center, had better access to resources while Muslims, predominately of the peripheries, had less. This access also varied with class differences, with the upper classes of various religious affiliations in both regions having much better access to resources.[27]

From the outset, then, the territorial integration of peripheral areas was ham-pered by Lebanon's heterogeneous confessional structure.[28]

Furthermore, even as the French initiated key infrastructural developments in Lebanon during their mandate, these projects only accelerated the imbalances among sectors of production and among people. Following the pattern of other Arab countries, Lebanon undertook a series of modernization programs during the mandate period that strengthened the central authority's grip on key resources while subsuming perceived "softer" environmental concerns under the rubric of public health.[29] These latter concerns centered on water quality, pesticide residue in food, and waste disposal for rapidly urbanizing population centers. There was little national legislation on matters like nature conservation or biodiversity, with the exception of species considered important for agricultural purposes or for hunting, a favorite pastime of the national elite. Indeed, unlike the management of natural resources such as water and fossil fuels, which under both Ottoman rule and the subsequent European mandates remained highly centralized because of their obvious economic and political significance, the management of public health and environmental issues in the twentieth century were considered politically insignificant and were thus left to be regulated by municipalities and local communities that could respond more swiftly to local needs.[30]

The first official Lebanese legislation dealing with the environment was passed in 1925. It defined water as being within the "public domain" and declared territorial waters and the Mediterranean shore to be the property of the state. The state thus asserted its authority by ensuring that this important natural resource could only be used for public, not commerial, purposes, with some limited exceptions like the distribution of sanctioned concessions and acquired rights to water.[31] Further laws passed during the mandate period empowered the newly established Ministry of Health to oversee all matters related to water and public health regulations; the Ministry of Agriculture to oversee irrigation matters; and the Water Department within the Ministry of Public Works to carry out geological surveys, set standards for potable water, and build the necessary water infrastructure. With regard to water distribution and use, traditional water customs based on earlier Ottoman and Islamic codes were used, in particular *haqq al-shirb* (the right for cattle to drink and land to be irrigated) and *haqq al-shifa'* (the right to drink water and to carry such water to one's family in small receptacles).[32]

From Independence to Civil War

The exclusion of the periphery in the consolidation of an eager Maronite and hesitant Sunni elite became enshrined on the eve of independence in 1943 in an informal agreement known as the "National Pact." The National Pact, though only an informal understanding rather than a formal legal document, effectively negated the 1926 constitution's more politically and socially liberal elements calling for individual citizenship rights and responsibilities. Instead, it institutionalized Lebanon's sectarian political system with a particular, and very rigid, hierarchical ranking that clearly favored Christians. Henceforth, Maronites alone were entitled to the presidency (which carried exclusive executive authority), the Sunnis were allotted the prime ministership, Shi'a leaders accepted the speakership of the parliament, and the Greek Orthodox took the vice-presidency of the Council of Ministers and the vice-presidency of the Chamber of Deputies. A series of constitutional laws that had handed overwhelming power to the Maronite president were also included in the amended constitution of 1943. Representation in the parliament was set at a six to five ratio of Christians to Muslims.

Sectarianism in postcolonial Lebanon also entrenched the economic position and interests of the elite, enabling an economic system that—unlike the rest of the twentieth-century Arab world—could be described as neoliberal from the start. French mandate policy had led to a "rising discrepancy between a large repressive apparatus and a fairly rudimentary civil administration," which meant that administrative services were cut to a bare minimum.[33] Most medical, educational, social, and any existing environmental services were left in the hands of communal organizations. This process continued after formal independence and soon accelerated as market mechanisms were entrenched in the daily activities of the state. The social components of Lebanon's public administration were ceded to private entities and corporations.[34] Indeed, from these early days an "implicit economic social contract" was constructed among the elite across the main sects that dominated the state.[35] This informal agreement ensured little or no income or profit from taxes and favored heavy public sector investment in the construction of extensive infrastructure, including roads, trading routes, airports, ports, and communication networks. All of this came at the expense of promoting competing commodity-producing sectors such as agriculture.

During this early post-independence period, the Lebanese state largely ignored environmental issues, focusing instead on a national development strategy aimed at supporting the rapidly growing export, construction, and service-oriented coastal economy centered around Beirut. In 1948, for instance, the state licensed the private Lebanese Oil Company to explore oil and gas reserves. In the early 1950s, the Maronite Christian-dominated state began to enact limited decentralization plans, passing legislation authorizing local authorities to manage commons areas—largely woodlands—even as the Ministry of Agriculture worked with these same local authorities in limited reforestation plans.

Most significantly, the Litani River Authority (LRA) was created in 1954 following a decade of work and lobbying by Ibrahim 'Abd el-'Al, a Lebanese engineer and early pioneer on behalf of the rational management of Lebanon's water supply. Henceforth, the LRA would manage the development of irrigation systems, potable water supplies, and hydroelectricity projects along Lebanon's most significant river, which runs the entire length of Lebanon before heading west to the Mediterranean. The LRA would also regulate the distribution and usage of the river's water. A series of regional water authorities was also established during this period, starting in 1951 with the Beirut Water Authority that replaced expiring private concessions distributed by French authorities during the mandate period.

The Lebanese state, controlled by a cross-sectarian elite with financial and economic interests focused on Beirut and other major coastal cities, undertook a series of significant water development projects to produce the energy needed to fuel these urban centers and thereby serve elite interests. Although the Water Department within the Ministry of Public Works was linked from the start to electricity generation, a separate Ministry of Hydraulic and Electric Resources was established in 1966; by then, about 50 percent of the country's power came from hydroelectric dams, a percentage that was expected to grow in the ensuing decade.[36] At the same time, plans to modernize irrigation infrastructure, so important to the agricultural livelihoods of the majority of the population in Lebanon's poorer southern and northern peripheries, did not materialize. This partially fueled the political awakening of the Shi'a in South Lebanon.

These policies of the 1950s and 1960s resulted in an economic boom for the relevant patrons and clients in the services and banking sectors, but it also led to the further marginalization of regions, sects, and citizens outside of these sectors.[37] It was this social and political disenfranchisement and the resulting social and political mobilization of the 1960s that led to the emergence of two major forms of environmentalism in Lebanon. The first was rooted in early elite concerns for nature that gained momentum as part of the social movements of the 1960s and that would eventually be led by Lebanon's globalizing elite liberal civil society. The other was an environmentalism of the poor that originated in the social and sectarian periphery of the postcolonial state.

Elite Concerns for Nature and Pioneer Environmentalism

Concern for nature among those in elite circles in Lebanon has a long history and even predates the country's formal independence. The cedar tree, in particular, had significant symbolic meaning for the country's founding fathers and for their French patrons. The tree's prominence derived partly from its antiquity: cedars are mentioned in the Bible over seventy times, and nationalist narratives self-consciously connected modern Lebanon to ancient (and, significantly for the Christian elite, pre-Islamic) cultures such as the Phoenician. Early nationalist histories, not to mention contemporary tour guides, regularly tell of how the Phoenicians used cedar wood to construct their renowned commercial and military fleets. The ancient Egyptians, among them Ramses II and Cleopatra, used Lebanon's cedar wood to build their ships and tombs, while Gilgamesh, hero of the ancient Sumerian epic tale, undertook one of his great adventures to challenge the guardian of Lebanon's cedar forests. Thus Lebanon's oldest cedars, located in the northern mountains of Bsharre and measuring up to 40 meters (131 feet) in height and 14 meters (46 feet) in diameter, have served as grand images useful to a proudly nationalist narrative of an ancient "Lebanese" society.

During the mandate period, as part of its "complete tutelage" over Lebanon, France designed a national flag using the French tricolor with the Lebanese cedar in the middle.[38] After independence, Lebanon endowed the cedar on the flag—now given a second red stripe to replace the blue one—with specific nationalist character by officially fixing the tree as a symbol of "immortality and tolerance."[39] Indeed, Emile Edde, who was to become Lebanon's president in 1936, was a founding member and the first head of the Lebanese Friends of the Tree Association that was created in the early 1930s.[40] Each December, both the president and prime minister would attend Tree Day (*Eid al-Shajara*) ceremonies as a symbolic gesture to celebrate the importance of nature.[41] The construction of Lebanon's early national identity, in other words, was firmly built on its most visible natural asset. Kamal Jumblatt, the leader of the Progressive Socialist Party and scion of the Druze community in Lebanon, has also been identified among these early elite environmental "pioneers" for his role in passing hunting laws in 1952, as has Hussain Kaed Bey for establishing the Lebanese League for Bird Preservation.[42]

International meetings such as the seminal UNESCO Conference on the Protection of Nature in the Middle East, held in Beirut in 1954, added further scientific impetus to elite concerns for the region's natural habitat that was already clearly threatened with increased pollution and rapid urbanization from various modernization programs.[43] Still, such early initiatives for the environment remained limited in scope and membership and did not gain momentum until the late 1960s. It was only then that a wide spectrum of civil society movements, including those concerned with environmental issues, emerged in Lebanon.

The main inspiration and impetus for these pioneer environmentalists—mostly professional, middle class, and Western educated (or, at the very least, deeply influenced by international norms)—was the success of the environmental movement in Europe and North America during the 1960s that culminated in the 1972 United Nations Conference on the Human Environment in Stockholm. Lebanon was one of fourteen Arab countries to participate in the Stockholm Conference, which was the first global meeting devoted solely to human-induced environmental problems. After the conference, Lebanon gave its National Council for Scientific Research (NCSR) the general charge of following up on environmental issues and the specific task of combating pollution. This official attention, coupled with the increasing availability of donor funds, gave further impetus to a number of civil society organizations that began to form around questions of environmental health and pollution.[44]

Perhaps most importantly, the impact of this global environmental awareness created demand for and interest in environmentalism among a younger generation of educated and liberal elites at institutions like the American University of Beirut (AUB) who were to come of age in the post-civil war period. Ricardus Haber, a noted botanist at AUB, was considered a leading pioneer environmentalist in Lebanon and was arguably the "father" of its nascent environmental movement.[45] His earliest and most formative encounters with nature came during childhood walks in the forest wilderness near his hometown of Bhmandun in the lush region of the Shouf Mountains and from his long experience with the Boy Scouts. Haber was specifically influenced by his reading of Boy Scout founder Robert Baden-Powell's works that emphasized traditional, and largely apolitical, scouting virtues based on the exploration of the outdoors and frontierism.

By 1970, Haber had become a Scouts commissioner, "guiding his troops to the fields to show them the unbound life in nature" and giving them the experience of living "in harmony with their surrounding" wilderness.[46] After entering university as a biology student, Haber set up one of the first explicitly environmental NGOs in Lebanon in 1972: the Friends of Nature (FON). By the middle of the 1990s, this association had gained over three thousand members and an organizational structure representing various scientific, engineering, and humanities disciplines.[47] The most notable early activities of the FON were nature walks and camping expeditions led by Haber, as well as educational lectures on the need to preserve the environment. Inspired by his reading of works by the U.S. naturalist and preservationist John Muir, Haber sought above all else to preserve Lebanon's wilderness. Indeed, his main passion and achievements were extensively recording and photographing Lebanon's natural bounty—then still largely inaccessible and unknown to most Lebanese—and connecting these natural assets to Lebanon's cultural heritage.

Haber's form of pre-civil war pioneer environmentalism was thus focused on protecting and preserving the Lebanese wilderness from encroaching development, rapid urbanization, and population growth, as well as increased grazing in commons areas. His activism was, however, not overtly political in form or substance. Haber's main policy achievement was his lobbying in the late 1980s of Sulaiman Frangieh—Lebanon's former president and then scion of the Frangieh family and patronage network in the Zghora region of northern Lebanon—for the establishment of a nature reserve in one of Lebanon's most diverse and beautiful forests, Horsh Ehden. Similar lobbying campaigns by Haber and other notables from the northern port city of Tripoli also resulted in the protection of the ecologically important Palm Island. These lobbying efforts by Haber and his colleagues thus directly led to the establishment of Lebanon's first two legally protected nature reserves in 1992. Again, rather than explicit political confrontation, Haber's strategies for these important conservation battles were public awareness campaigns and cooperation with the relevant notables in both regions. For many postwar environmentalists within liberal civil society circles, Haber served as the "father" of the environmental movement.

Environmentalism of the Poor as a Crisis of Participation: Shi'a Mobilization and the Anti-Establishment Politics of Musa al-Sadr

Contrary to the Lebanese environmental movement's standard understanding of itself, the roots of modern Lebanese environmentalism, in its wider sense, are most accurately traced to the social and political mobilization of the Shi'a community in southern Lebanon and the Beqa'a Valley that began in the 1950s and 1960s. Excluded politically from the initial Maronite-Sunni Pact of 1943, the Shi'a community—whose members comprised only 3.5 percent of Beirut's residents but 70–85 percent of rural southern Lebanon's population—was thereafter neglected economically and socially by the state.[48] As a "community-class," the Shi'a thus constituted the most disadvantaged segment of Lebanese society in the post-independence period.[49] Lebanon's much-vaunted "economic miracle" of the 1950s and 1960s was built largely around tourism, commerce, finance, banking, education, and other services in the tertiary sector centered mostly in and around Beirut. However, agriculture was by far the main source of livelihood for the Shi'a in southern Lebanon and the Beqa'a Valley; until the late 1950s, the sector employed around 90 percent of the available workforce in the two regions.[50] Under the regime of President Fuad Shihab in the late 1950s, modernization and national development programs sought, for the first time, to raise rural infrastructure standards and establish schools and hospitals. But efforts to liberalize Lebanon's agricultural economy—including, for example, the replacement of traditional practices such as tobacco farming with the cultivation of cash crops for export—produced a mass migration of rural Shi'a to Beirut's suburbs, where they settled in what came to be known as "misery belts."[51] Lebanon's midcentury boom thus exposed huge—and rising—political, economic, social, and cultural disparities between modernizing urban Beirut (along with adjacent Christian-dominated Mount Lebanon and, to a lesser extent, other major coastal cities) and stagnating rural Shi'a communities in the south and Beqa'a Valley.

Moreover, in historian Albert Hourani's words, the transformation of Lebanon from "an agrarian republic into an extended city state" in this period was accompanied by a large demographic increase in the country's Shi'a population and, crucially, the gradual disintegration of rural society.[52] Of course, the wider Lebanese system of clientalism and patronage extended into the Shi'a community as well. Indeed, the domination of a small band of families representing the Shi'a in Lebanese politics was so complete that from independence in 1943 to the outbreak of civil war in 1975, only four different individuals occupied the highest post in government allotted to the Shi'a, that of speaker of parliament.[53]

Echoing similar movements globally, the Shi'a thus came to perceive environmental problems as a "crisis of participation"; as an excluded and poorer community, they sought more equitable distribution of environmental resources and more meaningful participation in societal decision making.[54] Expressions of Shi'a discontent with and alienation from the Lebanese political system were initially channeled into various nationalist and leftist movements that proliferated in the 1950s and 1960s, especially the Nasserist and Palestinian revolutionary struggles. However, these radical movements ultimately failed to break the system or to secure better communal rights within Lebanon's stubbornly sectarian political structure. Indeed, initial strong popular support for the

Palestinian liberation struggle among the Shiʻa soured as Israel's increasingly violent strikes against Palestinian and Lebanese leftist forces took a huge toll on southern Lebanon's residents and infrastructure in the late 1960s. To gain more participation in the political system, the challenge for a Shiʻa leader was to combine the "mobilizational value of the Shiʻa cultural heritage" in Islamic terms with the logic of Lebanese sectarianism.[55] This challenge was taken up by Imam Musa al-Sadr.

It is perhaps ironic and surprising that the beginning of meaningful environmentalism in Lebanon is to be found in the person of an Iranian-born imam who became the father of Lebanon's Shiʻa political mobilization. Neither al-Sadr himself nor his followers would have considered themselves environmentalists, and at no point did the Imam, or his political heirs, take on environmental causes in any meaningful sense. Still, the literature on environmentalism by scholars like Ramachandra Guha makes clear that al-Sadr's form of "environmentalism of the poor" clearly echoed that of other environmentalisms around the globe in combining concerns (direct or implicit) for the environment with more visible concerns for social and economic justice.[56] In this sense, opposition by impoverished local communities to interventions that deprive them of their economic and natural livelihoods and resources is as much an environmental movement as a social one—the two are inseparable.

Al-Sadr's main political message was that deprivation and exclusion were not to be fatalistically accepted but rather could and should be challenged. He observed that "whenever the poor involve themselves in a social revolution it is a confirmation that injustice is not predestined."[57] He tapped into the dramatic story of Hussain, the Prophet Muhammad's grandson and, according to the Shiʻa, heir to the Prophet's leadership. Hussain's passion and martyrdom, reenacted annually in the rituals of ʻAshura, tell the story of his battle and death and helped to make Imam al-Sadr's political message clear—just as Hussain challenged injustice in his time, so too must the Shiʻa do in theirs.

The Shiʻa community remained largely rural in the first three decades after independence, and thus its exclusion from the boom of Beirut could be exploited as a sectarian issue rather than being interpreted solely as a consequence of the urban-rural divide. There was, to be sure, plenty of non-Shiʻa poverty in Lebanon as well. Indeed, in 1960 al-Sadr joined forces on issues of environmental justice with another religious figure, Gregoire Haddad, a parish priest who would become the bishop of Lebanon's Greek Catholic community.[58] Haddad founded the *Mouvement Social Libanais*, an avowedly secular volunteer organization with the agenda of bettering the social conditions of the deprived across Lebanon. Haddad's message was at its heart a moral one influenced by traditions of social justice found among many liberation theologians in Latin America. The essential motivating principle behind his work was an absolute commitment to humanistic values and the power of learning, without any social, sectarian, class, or gender segregation of any kind. However, because his message was universal—he explicitly distanced the role of the Lebanese church from affairs of the state to such an extent that he was temporarily defrocked in 1975—he could not attract the kind of communal support so integral to al-Sadr's message.

Both al-Sadr and Haddad thus organized themselves around an environmentalism of the poor. However, whereas al-Sadr used the demographic realities and logic of Lebanon's sectarian system to claim communal rights from within this system, Haddad was effectively calling for a more radical secularization of the state that would empower citizens across all social divides not just along sectarian lines. Haddad's more radical form of environmentalism thus required longer-term education and knowledge to counter more than a century's worth of sectarianism in the hopes of creating a national social movement representing all the *ahali* of Lebanon. This difference between Haddad and al-Sadr ultimately sealed the fate of another movement the two founded—*Harakat al-Mahrumeen* (Movement of the Deprived)—a non-confessional organization formed on the eve of the Lebanese civil war that aimed to achieve social justice for all the people of Lebanon's peripheries. Despite its initial large popular base, this movement nevertheless eventually morphed into the more explicitly Shi'a political party and militia, Amal. While Amal has maintained an officially secular platform, it gradually became—especially after al-Sadr disappeared while on a trip to Libya in 1978—the personal platform for an ambitious lawyer from the newly empowered Shi'a middle class: Nabih Berri, a militia leader during the 1980s and speaker of parliament since 1992.

Emergency Environmentalism during the Civil War

The 1975 onset of Lebanon's long civil war resulted in massive environmental damage. Armed conflict during the fifteen-year war and multiple Israeli invasions, along with state neglect and ineffective management of the environment, led to forest burnings, land displacement, land degradation, intrusion of seawater into ground and spring water resources because of the large-scale drilling of boreholes all over the country, solid waste discharge into coastal waters and soil, coastal water pollution due to unmonitored and untreated garbage dumps, and air and noise pollution caused by the overuse of loud diesel-fueled generators.[59] Industrial effluents—mostly from major electricity, fuel, and cement factories, but also from small-scale projects—added greatly to coastal water pollution, and the unmonitored use of chemicals in industry and agriculture left underground water sources dangerously polluted. Agricultural lands were abandoned as male workers emigrated or joined militias. Poverty in rural areas forced many into real estate speculation or construction, turning large tracts of agricultural land into "forests of cement."[60] By the end of the war in 1990, Lebanon had suffered an estimated $25 billion in losses to its infrastructure. The United Nations estimated that up to 80 percent of the country's water resources had been polluted.[61]

The outbreak of civil war also largely halted the progress of Lebanon's environmental movement. The machinery of government, however, continued during the frequent lulls in fighting. In 1981, in the midst of the war, Lebanon managed to appoint, for the first time in its history, a state minister for the environment. His mandate included trying to prevent all forms of pollution, deforestation, forest fires, and the use of dangerous pesticides. He was charged as well with the modernization of solid waste disposal methods, the protection of biodiversity, and the organization and control of massive urbanization so as to bring it into "equilibrium" with the environment.[62] These lofty ideals were not backed up by financial, technical, or administrative support—the minister was "lent" a single employee from the Ministry of Economy and Trade—and the fledgling ministry collapsed under the weight of the 1982 Israeli invasion and the subsequent chaos of the second half of the 1980s.[63]

For the most part, then, the state had little effective control over its territory and environmental resources. The focus for many thus became what can be called emergency environmentalism, as local communities—sometimes aided by communal or social organizations such as Haddad's *Mouvement Social*—banded together to salvage basic environmental services that the state could no longer provide or to rescue themselves from unfolding environmental disasters such as a forest fire or sewage spill that the state was ill prepared to handle. A particularly infamous case of such emergency environmentalism occurred in 1988 when it was discovered that an Italian company had capitalized on an informal agreement between the Italian mafia and Lebanese militias to export nearly sixteen thousand barrels and twenty large containers of toxic waste from Italy to Beirut.[64] After passing through customs rather quickly, the barrels and containers were stored in secret locations east of Beirut. The contents of these barrels were then sold as raw materials to factories and used as fertilizers, pesticides, and paints. "Barrels were emptied, re-painted so as to hide their origin, and sold as to be reused. People used them out of ignorance to store food and drinking water. Waste barrels were burnt or emptied and their contents dumped into several areas...and into the Mediterranean Sea."[65] A large number of these barrels were emptied into streams in the Lebanese mountains that fed groundwater supplies. Some were sold directly to individual citizens for the price of US$5 per barrel. Mostly poor, these villagers and farmers sought out the barrels because of their unusual thickness, which was seen as especially useful for storing water for domestic or agricultural uses.[66] The Lebanese media exploded with harrowing tales of this "death cargo," as it came to be known locally. Lebanese authorities banned people from bathing in the sea when suspicious-looking barrels washed ashore. The public outcry was so strong that the Lebanese prime minister Salim al-Hoss called for an official investigation that ultimately resulted in the passage of a law banning such waste.

The Rise and Fall of Lebanon's National Environmental Movement in the Postwar Period of the 1990s

The 1989 Document Of National Understanding, popularly known as the Ta'if Accords, which officially ended the fifteen-year civil war, once again entrenched sectarianism in Lebanon's postwar social relations. It modified the 1943 pact's details so as to adjust to new demographic and economic realities that now favored Muslims over Christians. Parliamentary seats were divided equally among Christian and Muslim communities, as well as between the country's geographic regions,[67] and executive powers were largely transferred from the Maronite-held presidency to the Council of Ministers (dominated by a Sunni prime minister).[68] The basic assumptions that underlie the Ta'if Accords do not differ significantly from those of the 1943 pact in that they embrace—even enhance—the sectarian logic that necessitates the distribution of public offices among the various communities, effectively giving the more powerful majority the prerogative to veto public appointments.

With the end of the civil war, the public's urgent pleas to address the environmental catastrophes that had resulted from the conflict led to the appointment in December 1990 of the second state minister for the environment. Once again, lofty words by the first postwar government in the early 1990s were not matched by deeds. For example, when the minister took on his new post, there was no building or even dedicated offices in which he and his staff of four could work. To make matters worse, all official documentation relating to the environment prior to 1990 had "disappeared," leaving him with a "clean slate" and a budget of only 50 million pounds (about US$30,000).[69] With such limited financial and institutional support, efforts in this period concentrated mostly on much-needed general public environmental awareness and emergency response.

Rafiq Hariri's ascension to power in May of 1992 was the start of a new era in Lebanon. Ambitious plans to "return" Lebanon to its previous role as the financial center of the Middle East led to the adoption of unprecedented large-scale development and construction projects whose pace and aggressiveness took most Lebanese by surprise. Hariri's zeal for development and the atmosphere of political patronage that ensued came at enormous costs to the environment. Mountains were literally cut open to feed the concrete and construction industries, while the coast was ravaged first by the removal of fertile soil and later by land reclamation for building and development schemes. One environmentalist lamented in the middle of the 1990s that Lebanon's geography would literally have to be redrawn, sarcastically adding that Lebanese politicians made people wish they were being ruled by Ali Baba and the Forty Thieves.[70] Hariri's neoliberal blurring of the economic public and private spheres dipped into the management of the environment as well. His appointment of one of his business partners, Samir Muqbel, in 1992 as the new state minister for the environment brought repeated protests from the media and NGOs. Though Muqbel convened monthly meetings with environmental NGOs,[71] his reputation was tarnished when he was openly accused of protecting and profiting from another instance of the importation of hazardous waste from Italy to be buried in Lebanese soil.[72]

The ambitious and costly reconstruction plans begun in Lebanon under Hariri's government necessitated huge donor funding and investments; during the post-1992 Earth Summit era, this also meant required environmental assessments. In April 1993, Law No. 216 established Lebanon's first official Ministry of Environment. Muqbel remained in office and became the first environment minister. The law itself is curiously short and, significantly, failed to vest the Ministry of Environment with the requisite authority to enforce its rather ambitious scope.[73] Article 5 of the law established the administrative organization of the ministry. It consisted of one director-general—who was to oversee the ministry's administration and departments and to coordinate between the ministry and all other public and private institutions dealing with environmental protection—and three main departments—the Departments of Nature Preservation, Residential Environmental Protection, and Protection from Technological Effects and Natural Hazards. The Ministry of Environment's structure and organization has evolved over the nearly two decades since its creation, but its effectiveness has in general remained limited.

Conclusion

The environmentalism of the pre-civil war period did not lead to a self-consciously coordinated national movement across sectarian lines that sought to curb Lebanon's brand of development and its many deleterious impacts on the environment. Musa al-Sadr's pragmatism and desire to increase the political standing of the Shi'a within Lebanon's sectarian political system precluded any genuine partnership with progressive elements both within the Shi'a community—those who joined communist or leftist groups—and outside, such as Gregoire Haddad's *Mouvement Social*, which has left a vibrant legacy only within a relatively small circle. As for the early environmentalist efforts of pioneers like Ricardus Haber, the onset of the civil war in 1975 precluded the possibility of producing a deeper reaction within Lebanese civil society for environmental matters and stalled any growing national environmental consciousness among the Lebanese middle class. Therefore, a genuinely national environmental movement can only be said to have emerged after the end of the civil war in 1990.

The real impetus behind this Lebanese environmentalism was the Earth Summit in Rio de Janeiro in 1992. Following the peace agreement that ended the civil war, signed in Ta'if in 1989, a sharp growth in the environmental movement in terms of numbers and projects began in parallel with global movements inspired by the Rio Earth Summit. For the first time, networks of NGOs began to cluster around common themes in order to positively impact development policies and to create more-professional associations. This trajectory of the environmental movement in Lebanon in the 1990s again illustrates the fundamental connection between civil society initiatives and global trends and norms.

By the middle of the 1990s, however, this environmental movement had hit a wall in the form of Harirism (the dominance of political and financial figure Rafiq Hariri). The emerging national environmental movement of the early 1990s, which led to a short-lived environmental network unified in vision and strategy, thus fell apart under the weight of Hariri's reinvigorated state-led neoliberal imperatives, elite-controlled sectarianism, and donor influence. The result was internecine struggles within the environmental movement and the subsequent retreat by elite environmentalists into technical projects and specialization that fulfilled donor and state agendas. With the Shi'a sectarian elite now better integrated into Lebanon's political and economic system, al-Sadr's environmentalism of the poor was also largely extinguished as a movement that resonated within the logic of sectarianism. Today, there are more environmental NGOs and greater public awareness and knowledge about Lebanon's environment than ever before. The challenge for Lebanon's environmentalists, nevertheless, is to learn the lessons of the past and to find a way to unify the various political and communitarian strands embedded in the country's postcolonial history.

Notes

1. This quote is the welcome note on the website of the Republic of Lebanon Ministry of Tourism: http://www.lebanon-tourism.gov.lb/Default.aspx (accessed January 30, 2012).

2. Republic of Lebanon Ministry of Environment, United Nations Development Programme, and ECODIT, *State and Trends of the Lebanese Environment 2010* (Beirut: Ministry of Environment, 2011), 190–191.

3. Ibid., 187–190.

4. Ibid., 53–64.

5. Rania Masri, "Environmental Challenges in Lebanon," *Journal of Developing Societies* 13 (1997), 75.

6. See, for example, the following collection of National Capacity Self-Assessment documents of the Republic of Lebanon Ministry of the Environment: http://www.moe.gov.lb/Pages/ncsa-public.aspx (accessed January 16, 2012, and since removed).

7. Republic of Lebanon Ministry of Environment et al., *State and Trends of the Lebanese Environment*, 28.

8. In addition to the works cited in the next few notes, see also Karam Karam, "Revendiquer, Mobiliser, Participer: Les Associations Civiles dans le Liban de l'Apres-Guerre" (Ph.D. thesis, Aix Marseille III, 2004). Most of these studies are of the environment and environmental politics in the postwar period. There are also a number of scholars and scientists who have chronicled the state of Lebanon's environment and its management. See, for example, Salpie Djoundourian, "Environmental Movement in Lebanon," *Environment, Development and Sustainability* 11 (2009): 427–438.

9. Habib Ma'alouf, *'Al al-Haffa: Madkhal ila al-Falsafa al-Bi'aya* (Beirut: Markaz al-Thaqafi al-'Arabi, 2002).

10. Karim Makdisi, "Trapped Between Sovereignty and Globalization: Implementing International Environmental and Natural Resource Treaties in Developing Countries, The Case of Lebanon" (Ph.D. diss., Tufts University, 2001).

11. Paul Kingston, "Patron, Clients, and Civil Society in Lebanon: A Case Study of Environmental Politics in Post-War Lebanon," *Arab Studies Quarterly* 23 (2001): 55–72.

12. Paul Kingston, for example, is writing a book examining policy issues in postwar Lebanon, while the American University of Beirut's Issam Fares Institute for Public Policy and International Affairs has two research programs assessing various aspects of Lebanon's (and the wider Arab region's) public policy processes, including environmental policy.

13. The recent book by the noted Lebanese political and social historian Fawwaz Traboulsi illustrates this point well. Fawwaz Traboulsi, *A History of Modern Lebanon* (London: Pluto Press, 2007).

14. Lebanon serves as a hub for many United Nations and other international institutions, including the regional Economic and Social Commission for Western

Asia (ESCWA). It also hosts over one hundred environmental NGOs, a relatively large number in regional terms.

15. For a detailed analysis of the origins of sectarianism in nineteenth-century Lebanon, see Ussama Makdisi, *The Culture of Sectarianism: Community, History, and Violence in Nineteenth-Century Ottoman Lebanon* (Berkeley: University of California Press, 2000). For a discussion of patronage in Lebanon, see Samir Khalaf, "Changing Forms of Political Patronage in Lebanon," in *Patrons and Clients in Mediterranean Societies*, ed. Ernest Gellner and John Waterbury (London: Duckworth, 1977), 186.

16. See, for instance, Fadia Kiwan, "The Formation of Lebanese Civil Society," *The Beirut Review: A Journal on Lebanon and the Middle East* 6 (1993): 72–73.

17. Albert Hourani, "Ottoman Reform and the Politics of the Notables," in *The Modern Middle East: A Reader*, ed. Albert Hourani, Philip S. Khoury, and Mary C. Wilson (Berkeley: University of California Press, 1994), 89. Hourani defines "notables" as "those who can play a certain political role as intermediaries between the government and people, and—within certain limits—as leaders of the urban population."

18. Ussama Makdisi, "Reconstructing the Nation–State: The Modernity of Secularism in Lebanon," *Middle East Report* 200 (1996): 23–26 and 30.

19. The European intervention in Mount Lebanon took a decidedly religious dimension. Among the major players, France strongly backed the Maronites, Britain supported the Druze, and Russia defended the Greek Orthodox.

20. The historian Leila Fawaz writes that "the new administrative units created more problems than they solved because they did not correspond to the social realities of nineteenth century Lebanon." Leila Tarazi Fawaz, *An Occasion for War: Civil Conflict in Lebanon and Damascus in 1860* (Berkeley: University of California Press, 1994), 28.

21. Makdisi, "Reconstructing the Nation–State."

22. On this point see, for instance, A. Nizar Hamzeh, "Clientalism, Lebanon: Roots and Trends," *Middle Eastern Studies* 37 (2001): 167–178.

23. Samir Khalaf, *Persistence and Change in 19th Century Lebanon: A Sociological Essay* (Beirut: American University of Beirut, 1979), 95.

24. The secret 1916 Sykes-Picot Agreement dividing up the Levant between Britain and France, in addition to the 1917 Balfour Declaration paving the way for Zionist aspirations in Palestine, were central to Western hegemony in the Arab world. The division of Arab lands despite European pledges—those of the Husayn-McMahon Correspondence for instance—to grant independence in return for much-needed Arab resistance against the Ottoman-German Axis remains of crucial importance to understanding conflict in the Arab world today. The question of Palestine, Syrian attitudes toward Lebanon, and Iraqi claims on Kuwait all stem from this colonial history. The attachment of the interior Beqa'a plain to Lebanon rather than to Syria, and the even more rupturing division of the Galilee were quite traumatic for the residents of these areas. As Elizabeth Picard writes, "of all the Arab countries born of the dismantling of the Ottoman Empire, perhaps none had borders less natural than those of Greater Lebanon." Elizabeth Picard, *Lebanon, A Shattered Country: Myths and Realities of the Wars in Lebanon* (New York: Holmes & Meier, 1996), 23.

25. Ibid.

26. Ahmad Beydoun, "Lebanon's Sects and the Difficult Road to a Unifying Identity," *The Beirut Review: A Journal on Lebanon and the Middle East* 6 (1993): 16.

27. Sami Ofeish, "Lebanon's Second Republic: Secular Talk, Sectarian Application," *Arab Studies Quarterly* 21 (1999), 102.

28. Farid el-Khazen, "Lebanon's Communal Elite-Mass Politics: The Institutionalization of Disintegration," *The Beirut Review: A Journal on Lebanon and the Middle East* (1992): 61.

29. United Nations Economic and Social Commission for Western Asia, "Governance for Sustainable Development in the Arab Region: Institutions and Instruments for Moving Beyond an Environmental Management Culture," E/ESCWA/SDPD/2003/8 (October 23, 2003), 3.

30. Ibid., 12–13.

31. Hyam Mallat, "Water Laws in Lebanon," in *Water in the Middle East: Legal, Political, and Commercial Implications*, ed. J. A. Allan and Chibli Mallat, with Shai Wade and Jonathan Wild (London: I. B. Tauris, 1995), 151–176.

32. Ibid.; Karim Makdisi, "Towards a Human Rights Approach to Water in Lebanon: Implementation beyond 'Reform,'" *International Journal of Water Resources Development* 23 (2007), 374.

33. Picard, *Shattered Country*, 39.

34. Mona Fawaz, "Neoliberal Urbanity and the Right to the City: A View from Beirut's Periphery," *Development and Change* 40 (2009), 839.

35. H. A. Amery and A. A. Kubursi, "The Litani River Basin: The Politics and Economics of Water," *The Beirut Review: A Journal on Lebanon and the Middle East* 1 (1992): 95–107.

36. Nasser Nassrallah (current Lebanese member of parliament and former director of the Litani River Authority), interview with author, Beirut.

37. Ibid.

38. Kamal S. Salibi, *The Modern History of Lebanon* (London: Weidenfeld and Nicolson, 1965), 172.

39. "Anthem and Lebanese Flag," Presidency of the Republic of Lebanon Official Website: http://www.presidency.gov.lb/English/Pages/AnthemLyrics.aspx (accessed June 20, 2011).

40. Another founding member of the association was the notable Beiruti Badr Dimishquieh, a Sunni Muslim, who took over the group in 1936 and led it until his death in 1952.

41. Nadim Dimishquieh, *Mahatat fi Hayati al-Diblomassiyah: Dhikriatun fi al-Siyassa wa al-'Alaqat al Dawaliyya* (Beirut: Dar al-Nahar, 1995), 18.

42. Tala Hasbini, "Environmental NGOs as Mediators in the Policy-Making Process: Exploring the Establishment of Nature Reserves in Lebanon" (M.A. thesis, American University of Beirut, 2009), 76.

43. Malik Ghandour, "Tajribat al-Tajamu' al-Lubnani li-Himayat al-Bi'ah," Unpublished Paper, 1988.

44. Kamal Mudawar, *al-Intihar aw Hadm al-Bi'ah* (Beirut: Dar al-Abjadiyya, 1983), 18–19.

45. A recent website set up by Haber's friends and followers refers to him as the "father" of the environmental movement in Lebanon. http://www.ricardushaber.com (accessed November 15, 2010).

46. "In the Troops." http://www.ricardushaber.com/index.php?option=com_content&task=view&id=69&Itemid=42 (accessed November 15, 2010).

47. Naji Chamieh and Jihad Issa, "Policy Paper on the Environment in Lebanon" (Beirut: Lebanese Center for Policy Studies, 1996), 37.

48. Lara Deeb, *An Enchanted Modern: Gender and Public Piety in Shi'i Lebanon* (Princeton, NJ: Princeton University Press, 2006), 73.

49. Amal Saad-Ghorayeb, *Hizbu'llah: Politics and Religion* (London: Pluto Press, 2002), 7.

50. Majed Halawi, *A Lebanon Defied: Musa al-Sadr and the Shi'a Community* (Boulder, CO: Westview Press, 1992), 52.

51. Deeb, *Enchanted Modern*, 73.

52. Albert Hourani, "Lebanon from Feudalism to Modern State," *Middle Eastern Studies* 2 (1966), 263.

53. Farid el-Khazen, *The Breakdown of the State in Lebanon: 1967–1976* (Cambridge, MA: Harvard University Press, 2000), 42.

54. Robyn Eckersley, *Environmentalism and Political Theory: Toward an Ecocentric Approach* (Albany: State University of New York Press, 1992), 7–8.

55. Saad-Ghorayeb, *Hizbu'llah*, 7.

56. Ramachandra Guha, *Environmentalism: A Global History* (New York: Longman, 2000), 105.

57. Quoted in Richard Augustus Norton, *Hezbollah: A Short History* (Princeton, NJ: Princeton University Press, 2007), 20.

58. See Bishop Gregoire Haddad's website: http://www.gregoirehaddad.com (accessed January 15, 2012).

59. Republic of Lebanon Ministry of Environment, "Dirasat Haykaliya al-Istratijeyah al-Bi'iyah" (Beirut: No. 3230/B, November 1998), 1.

60. United Nations Environment Program (UNEP/Plan Bleu), "Lebanon: Environment and Sustainable Development Issues and Policies," 17.

61. UN Development Programme, *Sustainable Development Profile*.

62. Republic of Lebanon, Decree No. 3842 (March 6, 1981), Article 3, quoted in Hyam Mallat, *Le droit de l'urbanisme, de la construction, de l'environnement et de l'eau au Liban* (Paris: Delta, 1997), 142.

63. Nancy Khoury (Public and International Relations Affairs, Ministry of the Environment), interview with author, Antelias, Lebanon, February 19, 1999.

64. For a detailed account of this incident, see Makdisi, "Trapped between Sovereignty and Globalization," 212–240.

65. Fouad Hamdan, "Toxic Attack Against Lebanon, Case One: Toxic Waste from Italy, A Chronology," (Beirut: Greenpeace, 1997), 1.

66. Pierre Attallah, *al-Nifayat al-Sama fi Bilad al-Arz: al-Qissa al-Kamila* (Beirut: Dar al-Nahar, 1998), 40.

67. The constitutional amendments of 1990 added a clause to Article 24 concerning the composition of the Chamber of Deputies. In the original constitution, no numbers were specified. However, as already noted, the 1943 Pact led to a six to five ratio of Christians to Muslims. In 1990, the following clause was added to Article 24: "Until the Chamber of Deputies adopts an electoral law without confessional community apportionment, parliamentary seats will be allotted according to the following rules: A. Equally between Christians and Muslims; B. Proportionally between the communities of the two groups; and C. Proportionally between the regions."

68. Article 17 (as amended in 1990) reads: "Executive power is entrusted to the Council of Ministers which exercises it in accordance with the provisions of the present Constitution." The original text of Article 17 reads: "Executive power is entrusted to the President of the Republic who exercises it with the assistance of the Ministers, according to the conditions established by the present Constitution."

69. Hagop Joekhadarian (member of parliament), interview with author, Beirut, February 22, 1999.

70. Ara Alain Arzoumanian, "Activist: Top Leaders Don't Care about Environment," *Daily Star* (June 21, 2003).

71. Malik Ghandour, "Lebanese Environment Forum Publication," 1.

72. Najah Wakeem, *al-Ayadi al-Soud* (Beirut: Sharikat al-Matbu'at lil-Tawzi' wa al-Nashr, 1998). See also Attallah, *al-Nifayat al-Sama*.

73. Republic of Lebanon, Law No. 216 (April 2, 1993). The original text is in Arabic, and the English translation (a loose one intended to clarify the meaning) is my own. The original language set forth in this law is, at times, poor and confusing.

IO

State of Nature: The Politics of Water in the Making of Saudi Arabia

Toby C. Jones

Power and sovereignty have long been closely tied to energy and the environment in Saudi Arabia.[1] Home to the world's largest oil reserves, the Kingdom of Saudi Arabia has been the most important supplier of the prized natural resource since at least the 1970s. Its massive deposits of crude oil have helped fuel a global energy regime that is almost wholly reliant on the power of oil for transportation and industrial production. The kingdom has benefited handsomely from its abundance of petroleum, reaping untold billions of dollars in revenue and profit in the last half century. In addition to enjoying this financial windfall, Saudi Arabia has also benefited from the political protection offered by the United States, which has made protecting the security of Saudi Arabia a high priority.[2] Although the Americans have cloaked their military engagements in the Persian Gulf in the language of freedom and the war on terrorism, protecting the flow of Saudi Arabia's oil has been their preeminent concern. Oil has also secured the political fortunes of the kingdom's ruling family, the Al Saud, at home. The accumulation over the second half of the twentieth century of petrodollars tightly controlled by the small ruling elite enabled the Saudis to shore up their political authority and to build a political system entirely beholden to them and their wishes. Oil's singular importance in the twentieth century's global economy and Saudi Arabia's privileged place within this economy have brought great wealth and political fortune for the kingdom's rulers. Indeed, it is hard to overstate the primacy of oil in Saudi Arabia's modern history.

But while oil and the wealth it has generated have been hugely important, the history of the modern Saudi state and the consolidation of the power of the Saudi royal family in the twentieth century had more complex environmental foundations. Water, agriculture, and the broader pursuit of mastery over other non-petroleum natural resources all figured in important ways in the making of modern Saudi Arabia. In the first half of the twentieth century, in fact, it was the convergence of several environmental factors—most notably the pursuit of control over both oil and water—that most shaped the contemporary political order in the kingdom. Control over both would prove necessary to secure the fortunes of the Al Saud. Until the middle of the century, the kingdom was politically fragile, vulnerable to internal rivalries within the royal family, and hampered by the state's own limited reach. In part this had to do with the vast size of the country. Forged through conquest and violence in the first three decades of the twentieth century, the foundation for Saudi authority was limited and stretched thin. Aside from the backing of a community of religious scholars based in central Arabia, the Al Saud had no significant social base of support. Over the course of the century, the kingdom's rulers would strive to overcome the obstacles to their power, most importantly by building a modern state and by mastering the environment and the large number of subjects who depended on it. Often these goals went hand in hand. By the end of the century, the Al Saud had overcome considerable challenges to its power and was firmly in control of a strong, centralized state.

Given the scarcity of life-sustaining natural resources, particularly water, on the devastatingly arid Arabian Peninsula, the Saudi ambition to control them is hardly surprising. Both settled and nomadic communities have depended on and often struggled violently for access to water resources for their survival. No less important was the role of agriculture, occupying the energies of the vast majority of nomads and settled cultivators alike, who farmed and herded intensively just to sustain a basic living. More than just asserting their authority over established farms and farmers, Saudi rulers sought to tame the environment by actively facilitating agricultural expansion across the peninsula. This strategy thereby ensured that attempts to control the environment would play a key role in Saudi plans to deepen their power over the course of the twentieth century.[3]

The emergence of a modern state in the peninsula and the role of the environment in the process both served and reflected the power of the ruling elite. The story is not only a Saudi one, however. In addition to being politically vulnerable in the first half of the twentieth century, the Saudis also lacked the technical and material resources to master the environment, oil and water included. In the 1930s and 1940s they came to rely heavily on and collaborate with foreign experts, technical advisers, and an American oil conglomerate. The Saudis' goals were to simultaneously exploit their natural resources, engineer the environment, and strengthen centralized political authority. Their collaborators in these efforts helped to consolidate, institutionalize, and centralize Saudi political authority, as well as turn expertise and the environment itself into a source of royal power. In addition to helping establish centralized control over the environment—including natural resources such as oil and water, but also territory and people—these experts also helped build up the kingdom's administrative and governmental capacity, connecting bureaucratic power with environmental power. Their efforts brought millions of people into the state's emerging administrative order. They also assisted in securing the country's borders, created an entire new system of knowledge and information about the environment and society, directed and built the infrastructure that tied the far-flung provinces to authorities in Riyadh and Jidda, and spearheaded efforts to create a centrally controlled economy. In a place better known for the power of religion and religious scholars, it was the work of experts and their efforts to master the environment that sealed the political fortunes of the ruling elite.

While the kingdom relied heavily on foreign experts, the initiative to link the environment and the country's natural resources to power was driven by the Saudis. Saudi rulers increasingly sought and paid for information about territory, resources, and people from a variety of local and international sources. They well understood that their fortunes, like the fortunes of state builders and powerful elites everywhere, were connected to their ability to control and harness the power of nature. Environmental power was tantamount to power over people and their movements and also over commerce; ultimately environmental power was derived from and constituted the state's ability to control its own territory. And the need to control resources and space was dependent on the state's ability to catalogue and know about the environment over which it sought authority. This was especially important in the early stages of Saudi political development, when Saudi political power was tenuous and the state was only beginning to emerge. Information served the quest for control. Experts played an important role in building up knowledge about the environment and passing it along to central authorities, helping frame and shape the latter's decision making. But experts were not just compilers of data that central authorities used to know and oversee their dominion. They also influenced the very terms by which central authorities came to view the territory, resources, and people over which they sought command—shaping the terms of power and the nature of the relations through which power was enacted. Natural resources, territory, the environment more generally, and people emerged not just as things to control but also as obstacles to be overcome, projects to be developed, and subjects to be managed. Experts maintained that these objectives could only be achieved—and both the experts' and the government's interests served—through a strong central state. The result was that managerial ability and control over the environment were collapsed with political authority. Just as important, the political nature of the relationship was obscured, masked within the language of calculated, apolitical, detached science.

The Environment and the Saudi Imperial Will

The hard work of capturing, consolidating, and developing the Arabian Peninsula's environmental resources in the mid-to-late twentieth century—when the effort to do so was the most intense—was to some extent a continuation of Saudi efforts to link the environment with power in the first two decades of the century. Indeed, their experiences in central Arabia (Najd), the austere desert homeland of the Al Saud and their base of power from the late eighteenth through the twenty-first centuries, demonstrated to the Saudis that the expansion of agricultural production often proved a matter of life and death. The Najdi climate severely restricted the quantity and quality of agricultural production for those who lived there. Until the middle of the twentieth century, settled residents made do by tapping as much nutrition as they could from the region's limited environmental resources. Although the region was not particularly conducive to sustained agriculture, various forms of farming were nevertheless the central pillar of economic life. While Najdi farmers manipulated the limited resources available, they faced perennial hurdles in producing enough food to meet local needs. One historian has noted that "although every piece of cultivable land was used as intensively as traditional techniques allowed, Najdi towns were seldom self sufficient."[4] For the Saudis, who already constituted a ruling elite in the settled communities of Najd in the eighteenth century, environmental austerity and the limits it imposed on both economic and political power led them to look for ways to expand their reach into the more fertile and resource-rich regions of the Arabian Peninsula.

Indeed, well before they set out to use agriculture as a tool in consolidating their political power and defeating potential rivals in the twentieth century, the Al Saud sought to expand their sphere of influence in the eighteenth and nineteenth centuries by capturing rich agricultural resources beyond Najd. The eastern periphery of the peninsula was where they would find the most alluring prizes. The two oases of the Eastern Province, al-Hasa and Qatif, were particularly attractive.[5] In contrast to the deserts that dominated Najd, the oases of eastern Arabia were resource rich, awash in life-sustaining water and copiously stocked with lush palm groves and vegetable gardens. Hundreds of thousands of palm trees packed the eastern oases. Water, which flowed from artesian wells, streamed like veritable rivers through the region's gardens. Dates were the dominant crop. But Hasawi and Qatifi farmers also cultivated an array of fruits and vegetables, including pomegranates, apricots, peaches, figs, cucumbers, tomatoes, lemons, oranges, various melons, green beans, and even cotton.[6] Over the course of two centuries, Saudi leaders routinely strove to capture and control al-Hasa's and Qatif's abundant resources.[7] The Saudis coveted the produce, the wealth it generated—both al-Hasa and Qatif were connected to global trading networks that spread from the Persian Gulf to East Africa and South Asia—and the precious water and fertile soil that made it all possible. Their remote outpost in Najd was too isolated and possessed none of the potential for income so abundant in the east.

The Environmental Foundations of the Modern Saudi State

The same calculus was at work in the twentieth century, when in 1913 the Saudis would finally conquer the Eastern Province and set out to build a modern state. The fledgling Saudi state desperately needed considerable resources to survive. Through the first five decades of the century, it had access to limited revenue. In 1933, Riyadh entered an agreement with a consortium of American oil companies that would become the Arabian American Oil Company (Aramco) and began to rely heavily on them for loans drawn on future oil revenues. But oil production and the income that the sale of petroleum would deliver was limited until after World War II. Until the end of the war the main source of income for the central government was the tax revenues generated by the annual pilgrimage to Mecca (*Hajj*) by foreign visitors. Pilgrimage revenues were, however, unreliable in the long term. During times of global crisis, such as during the Great Depression or during World War II itself, when pilgrims faced financial or logistical obstacles to travel, the *Hajj* failed to bring in enough tax revenue to support the Saudi government or its ambitious schemes to establish hegemony across the very large Arabian Peninsula, which is roughly one-third the size of the United States. Control over the environment and agriculture and the search for new natural wealth was partly intended to help expand the Saudi purse, either through direct control or through the taxation of others. While tax revenue on farming would always remain limited, the Saudi state was hardly exceptional in its drive to drum up as much cash as possible. Eventually, oil would obviate the need for a systematic tax infrastructure. Beyond the issue of taxation, the Saudis also sought to integrate the peninsula's agricultural hinterlands into a centrally controlled economy. They did not fully succeed until midcentury, when efforts to co-opt merchants and farmers began to yield results, but the understanding that economic integration would strengthen Saudi power was accepted as necessary during the first few decades of the century.[8]

The most important aspect of the effort to control the environment and establish authority over the region's natural resources was political. Because most of the kingdom's subjects were engaged in some form of agriculture—thus engaged with or dependent on the environment—it made sense to target agriculture and the environment as objects of state power and control. Periodic surveys and studies revealed a consistent pattern over the course of the twentieth century. The United Nations Food and Agricultural Organization (FAO) documented in 1956 that 78 percent of the kingdom's citizens made their living from agriculture.[9] According to the FAO, only 22 percent of the national population was urban. Even as late as 1970, as much as half the population worked either as agricultural day laborers or on their own farms.[10] The Saudi Ministry of Agriculture reported in 1974 that out of a population of about seven million, around 45 percent of the entire labor force was in agriculture.[11] While much Saudi energy was spent on integrating farming communities into the kingdom's sphere of influence, most pastoralists were not permanently settled. No figures are available for the percentage of the population constituted by nomadic and semi-nomadic communities at the beginning of the century. But the FAO report claimed that at least 66 percent of the population continued to be nomadic as late as 1956. Although settled farming communities were not always easily pacified, the Al Saud and their supporters eventually quelled most into submission.

Rulers used water and agriculture as tools to subdue potential threats to their authority. Potential rivals included the settled farming and merchant communities that lived along the Arabian Peninsula's shores, including al-Hasa and the Hijaz in the west. These communities had much to lose financially and politically with the ascendance of the Saudis. And they would indeed eventually lose, although the Al Saud used a combination of incentives and penalties to compel the cooperation of the merchants. Most importantly, the Saudis and their backers used agriculture to rein in the tribal and Bedouin forces that threatened their newfound and still loose grip on power. The Saudi Arabian historian Abdulaziz al-Fahad has observed that "in writings about the country, the Saudi state is typically identified with the Bedouin, the tribe or nomads, and 'tribal values' are supposed to suffuse the state, at least at its inception. Such identification is difficult to sustain notwithstanding its prevalence, for this state had been (and continues to some extent to be) an exclusively *hadari* [settled] endeavor with profound anti-tribal and anti-Bedouin tendencies, and circumscribed roles for the Bedouins and their tribes."[12] Establishing the kingdom and successfully securing it depended on overcoming tribal tensions and defusing the threat posed by communities who had long enjoyed freedom of movement. Indeed, if the raiding (*ghazu*) that generated part of tribal income was allowed to continue, it would have represented a real threat to the integrity of the Saudi polity and the ability of the country's rulers to assert their power.

Almost immediately after grabbing control over Riyadh in 1902, the Saudis launched a two-pronged strategy to bring the Bedouin under control. The first tactic was to expose them to an intensive proselytizing and recruiting campaign, inculcating them with the Wahhabi spirit, thereby exploiting faith to build loyalty.[13] The second strategy was to settle them in permanent agricultural cooperative farms, a project that aimed to transform them into a *hadari* (settled) and, ultimately, warrior class answerable to centralized power in Najd.[14] The program proved very successful initially, and the newly settled Bedouin came to be known as the *ikhwan* (brothers). Their most important role was that they formed a mobile and rapid strike military force, one that proved instrumental in helping the Al Saud conquer the Arabian Peninsula.[15] Adept at war, the *ikhwan* proved less adept at farming, taking slowly if at all to agriculture.

Although the *ikhwan* did not take the agricultural imperative seriously, the strategic importance that the Saudis gave to the settlement project indicated a major turning point in the political history of the Arabian Peninsula. Promoting sedentarization and using settlements as instruments to overcome politically threatening raiding practices reflected a new strategic thinking on the part of Saudi leaders. With the aid of supportive religious scholars, who saw their own influence grow with the emerging power of the Al Saud, Saudi rulers partially justified their rule through the exploitation of the environment and through socio-environmental engineering.

The Al Saud learned to see water as strategically critical, especially when they sent their newly created warrior class the *ikhwan* on military missions. It made little sense to have settlements clustered tightly around the seat of power in Najd, although that would have made administrating and governing the communities considerably easier. Knowing about and establishing control over water wells spread across the peninsula made it possible to maintain military outposts at strategically vital locations, some closer to the Hijaz and others nearer al-Hasa.

Oil, Expertise, and Environmental Authority

With the discovery of oil in the late 1930s, a new era of environmental power emerged in Saudi Arabia. Before the discovery of oil and the "petrolization" of the Saudi state, however, the kingdom's leaders turned to the science of geology—and to American geologists in particular—in the hope of unearthing something of value from the arid landscape. In fact, the discovery of oil was the result of the government fully embracing environmental science and applying it in the consolidation of centralized authority. And the discovery of oil should furthermore be seen as a product of the continuation of the strategic thinking that first evolved while attempting to settle the Bedouin. Saudi leaders came to see that knowing their natural environment more systematically was a precursor to controlling the political one. The discovery of oil was the most important result of this determination, but it was not the only one. Oil would eventually strengthen Saudi power in ways previously unimaginable to the kingdom's rulers. So too would the work of American and other foreign experts who helped the kingdom locate and extract petroleum from the ground. But before the work of searching for and selling oil proceeded, the Saudis turned to Americans to help them find water.

The impact of the work carried out by American experts in support of the Al Saud was felt beyond Riyadh. Indeed, Saudi–American relations were first shaped in the 1930s by the work of American geologists, most of whom simultaneously served American and Saudi Arabian political and commercial interests. Since World War II, when the potential of the kingdom's bountiful oil reserves became well known, the U.S. government has prioritized the security and stability of the Saudi regime, no matter how dreadfully it has treated its own citizens. A stable tyrannical Saudi government beholden to American oil companies and to U.S. security assurances was far more preferable to a politically open state that would potentially prioritize its own citizens' needs and interests over that of American consumers or global energy markets. For American business—big oil as well as smaller independent consultants, scientists, and engineers who went to work for the Saudi government—helping safeguard the stability of the Saudi regime also meant ensuring access to some of the windfall generated by the sale of oil. The pursuit of profit and wealth went hand in hand with efforts to strengthen the capacity and reach of the central Saudi government.

Saudi leaders were attracted to the work of geologists and environmental experts in part because they held the scientific keys to expanding the state's knowledge of its natural resources. There were Saudis and other Arabs in the peninsula who would help in exploring and mapping the countryside, but the Saudis harbored concerns about relying on potential domestic rivals for collecting information. And while there were locals who would carry out the work on behalf of the authorities in Riyadh, their numbers were limited. In addition, the Americans offered something beyond political expediency and basic scientific ability. Equally important was the establishment of the connection between science, environmental expertise, and authority. Expertise itself became a measure of authority and this was something that state leaders would subsequently aspire to make a central part of their own ruling strategy.

The Saudi patriarch 'Abd al-'Aziz ibn Saud invited the first American geologist to survey his territory's natural resources several years before he became king. In 1930 he invited Charles R. Crane—a former U.S. representative to China, a philanthropist, and someone who became deeply involved in the political changes that swept the Middle East after World War I—to visit Arabia. The Saudi leader requested Crane's assistance in carrying out an inventory of the peninsula's water resources. Several years earlier Crane had undertaken a similar survey of Yemen. In Yemen, Crane turned to the American Karl Twitchell as his chief geologist and engineer, sending him on several tours to carry out the surveys as well as to complete other infrastructure projects. Twitchell looked into the possibility of building a Yemeni road network, establishing agricultural demonstration farms, installing water pumping windmills and irrigation networks, and erecting "the only steel truss highway bridge in Arabia."[16] According to the American geologist and engineer, who later recorded his experiences in a volume published in 1947, "reports of these unusual gifts reached Saudi Arabia," and "Mr. Crane accepted an invitation of King Ibn-Saud to visit him in Jidda" in order to explore potential ways the Americans might aid the Saudis.[17] "It soon appeared" to Twitchell that the Saudi king's "principal desire was to find ample water supplies, especially flowing artesian wells in the Hijaz and Najd."[18] Crane subsequently agreed to send Twitchell to Arabia to help the Saudis. Twitchell began a comprehensive survey, spanning 1,500 miles, of the Hijaz's water resources in April 1931. The survey returned disappointing results. Twitchell found "no geological evidence to justify the hope for flowing artesian wells."[19]

The Saudis remained undaunted. The absence of much-hoped-for natural riches in the Hijaz did not dampen their belief that the peninsula was home to mineral and natural wealth. Nor did the scarcity of water resources in the Hijaz undermine the Saudi belief that water was the key to political power elsewhere in the region. After his 1931 visit, Twitchell was subsequently contracted by 'Abd al-'Aziz to "advise him on the water resources and oil possibilities in his province of Hasa along the Persian Gulf. Although this would be a thousand mile trip over rough country, where no American had ever been, the invitation was readily accepted."[20] It was in al-Hasa that Twitchell and his fellow geologists would make their most significant discoveries. Twitchell spent several weeks studying al-Hasa's environment and geological features. He even sailed to Bahrain, a small island located a few miles off the eastern shore of the peninsula. Bahrain possessed similar characteristics to al-Hasa and exploratory drilling for oil was already under way on the small island. Twitchell believed that there were significant deposits of oil in Bahrain and on the peninsula. Even so, he advised caution. After returning to Jidda, Twitchell encouraged the Saudis to wait for the results of the Bahraini test drills before proceeding with any petroleum exploration plans in the east. The king heeded Twitchell's advice. Twitchell later recounted that 'Abd al-'Aziz made clear "that on account of the depression, with the lack of pilgrims and consequent fall in revenue, he could not afford to follow out the development previously planned and agreed upon. Furthermore he wished me to try to find capital to carry out the development previously discussed [more mining, surveys for water resources, and test drills for oil]."[21]

Twitchell returned to the United States in early 1932 to try to drum up support for oil exploratory work in Arabia. Motivated by the conviction that oil was there awaiting discovery, the geologist, once carrying out the work of a philanthropist, now saw an opportunity for personal gain. In July he began exploring for oil in the Arabian Peninsula. In spite of initial rejections by several mining and oil companies, Twitchell persisted and his efforts eventually led to the signing of Saudi Arabia's oil concession agreement with the Standard Oil Company of California (Socal) in May 1933, which formed an operating company that ultimately came to be known as the Arabian American Oil Company (or Aramco).[22] The Saudi Arabian government granted Socal the exclusive right to explore for and extract oil in al-Hasa (over an area of 318,000 square miles) in exchange for royalties if any was discovered in commercial quantities. Socal also secured a loan of £33,000 in gold sovereigns to be given to the kingdom in advance. Late in 1933, the first wave of oil company geologists landed and initiated their search for oil in al-Hasa. After five years of frustrating results, they struck commercially profitable oil at Jabal Dhahran in 1938. Within decades, Aramco would discover the largest oil field in the world at Ghawar, just west of the al-Hasa oasis, securing Saudi Arabia's place as the largest and most important oil producer on the planet.

The presence of oil in al-Hasa and its subsequent development proved to be the most important geological discovery in the kingdom's brief history. The wealth it eventually generated did more to shore up Saudi political authority than agriculture could have accomplished even in the best-case scenario. Saudi rulers appreciated this fact early on. Even so, it did not diminish their efforts to learn more about the still much-needed water and to pursue the intensification of agriculture. There was perhaps a simple reason for this. Oil generated income, but wealth alone was not sufficient to build power. It did not confer credibility and it did little to bring subjects directly into the orbit of the government. For these things to happen, oil wealth had to be spent. And it was through non-petroleum environmental projects that it would often be put to use. Most of the young kingdom's subjects continued to be engaged in agriculture and were hence dependent on water for their livelihoods. Even though the state did not look at its citizens as a source of revenue to be gained through taxes or other means, it continued to believe that the population needed to be productively engaged and that managing resources was a key to state oversight, administrative power, and security. Particular emphasis was placed on building an integrated administrative and economic network controlled or at least monitored by the central government. Moreover, aside from filling the Saudi purse with much-needed revenue, the discovery of oil also helped heighten both Saudi faith in science and their belief that it would be geologists who would help the kingdom locate and harness whatever resources remained undiscovered.

In 1940, King 'Abd al-'Aziz and Abdallah Sulaiman once again asked Karl Twitchell, who was visiting Riyadh, if he might be able to assist in locating natural resources in Najd and to introduce water drills, pumps, and other farming technology to the region. Twitchell was about to return to the United States to begin another search for partners who would be willing to sign on to carry out the work of surveying the Najdi environment. But before he left for the United States, the finance minister charged him with another task. Sulaiman asked Twitchell to travel south and undertake a study of the kingdom's southwestern province of Asir. Asir, home to rugged mountains, was difficult country to access. Because the Saudis hoped to ease their ability to access their southern reach, they sent Twitchell, along with Saudi mining engineer Ahmad Fakhry, to do the exhaustive work of surveying the landscape and making suggestions about how it might best be utilized. The two engineers, along with teams of local guides and assistants—whom Twitchell barely acknowledges in his account—were charged with several objectives. Sulaiman asked them to map the terrain, to measure and mark the steep mountain grades for a road network, to make initial determinations about the prospects for future mineral mining, and to catalogue the area's potential water resources and agricultural prospects. Twitchell, Fakhry, and their team spent several months making their way through Asir's passes. Twitchell corresponded regularly and directly with the king, providing detailed updates on his findings.

The team's work in Asir was particularly important to Saudi Arabia's rulers, both because of its potential environmental power and for strategic reasons. In fact, the two went hand in hand. Riyadh's grip on the region was still tenuous at best. The Saudis took control of the province in the mid-1920s. In 1934, they were forced to defend the possession from a rival claim leveled by forces in Yemen, on Asir's southern border. Because the Saudis valued Asir for its fertile soils and the potential it held for future mineral mining, they sent in Twitchell and Fakhry to carry out the preliminary work of fortifying the central government's presence. Not only did Twitchell and Fakhry investigate and inventory resources, but they also initiated the process of road building. The early roadwork carried out by the Twitchell-Fakhry team made passage as well as the extraction of resources and revenue much easier. More importantly, the new roads would also make it far easier for the central government to police and secure a vulnerable region.

Water and the Agricultural Imperative

Two years after completing his work in Asir, Twitchell returned as part of a group sponsored by the U.S. Department of Agriculture to carry out even more extensive surveys of the kingdom's natural resources. As was the case in Asir, the Americans were guided and influenced by local inhabitants who possessed their own knowledge and expertise of the region. The published report produced by the mission offered little insight into their role or their impact on the survey team's findings or experience. With the direct support of the U.S. government, the survey's mission was considerably expanded to cover much more of Saudi Arabia than just Najd. From May 15 to December 5, 1942, the team traveled eleven thousand miles by car, camel, and foot exploring and cataloguing the geological and agricultural possibilities in the kingdom.[23] The team's field research, while extensive, was still only a preliminary estimate of the water and agricultural potential of the kingdom. Other experts in subsequent years would carry out more exhaustive surveys that would yield greater insight and detail. Although it provided only a partial look, the survey itself, as well as its particular details and the suggestions it made, offered the first methodical portrayal of the region. It came to serve as a foundation for future efforts to engineer the environment and for more effective centralized oversight over the kingdom's vast territory and the people who lived there. Twitchell and his colleagues saw themselves as a force for progress, but their work also served specific political goals, especially the strengthening of the royal family and the polity.[24]

The U.S. agricultural mission's report was an inventory of water, land, and agricultural resources. In spite of the limitations imposed upon the survey team by the demands of time, poor transport, and the elements, the report is authoritative. The mission surveyed many of the communities and locations on the peninsula already actively engaged in settled farming, including al-Hasa, Najd, the Hijaz, and Asir in the southwestern corner of the kingdom. They inventoried a number of environmental metrics, including average annual rainfall, varieties of soil, quality of soil fertility, water resources, and kinds of crops being grown. They also offered some limited commentary on farming methods, especially at an experimental farm that had been set up in al-Kharj southeast of Riyadh.

Their primary objective was to outline where and how intensive efforts might expand the areas being cultivated. But the mission also helped establish the foundation for a new kind of political language and knowledge, one in which science overlapped with and reinforced geostrategic interests. In doing so, the U.S. agricultural mission contributed to the ongoing development of the Saudi strategic thinking that had emerged earlier in the century.

Twitchell argued that the kingdom's four regions—the territories conquered by the Saudis only a decade before that had historically maintained cultural, social, and political autonomy (al-Hasa, the Hijaz, Najd, Asir)—also served as a convenient classification system for describing the kingdom's environmental characteristics and water resources. Al-Hasa in the east was historically the most fertile area in the Arabian Peninsula and home to its richest underground water resources. Likewise, there are real topographical and geological features that distinguish the western provinces (the Hijaz and Asir) from the center (Najd) and the east, although they blend together at some point, making the actual distinctions between them somewhat arbitrary. But while the mission's classification system possessed geological and geographical plausibility, it also had political implications. The use of geology to justify the existing geopolitical classification reinforced to the Saudis that these areas were objects to be captured and exploited. This was further emphasized by the decision of the report's authors not to discuss people. Instead, they emphasized geology and water over actual human communities. The Americans created a powerful new knowledge system for the central government that prioritized descriptions and classifications of nature over descriptions of social and cultural life. This would continue to be a feature of state-sponsored agricultural work throughout the twentieth century.

Just as important as the knowledge system were the details it revealed about the location of the water resources being surveyed. Previously, various communities, including the Saudi family itself, had to compete for resources. The outcome of this competition was largely determined by knowledge of where things were and the ability to police them. The U.S. agricultural mission eliminated the need for Riyadh to bother with the process of "discovery."

The mission pinpointed with precision the location of some of the most important resources and provided valuable details about water depth, sustainability, and usability—all vital to a central government that considered control over such resources as important to its power. In addition, the report emphasized the very strategic logic that the Saudis had already begun developing themselves: because the center was natural-resource poor—the Najd was clearly the least fertile region and had the least water resources—capturing resources from the periphery was a key to power.

Thus, the U.S. agricultural mission sought to accomplish much more than simply help the central government capture resources from its provinces. In their final report the team provided a scientific framework, including a set of recommendations, that would help the central government more fully include far-flung areas within its sphere. It accomplished this by arguing that existing production levels of local agricultural areas were disappointing and by offering specific suggestions on how those levels could be expanded. The mission's aim was not to serve local cultivators, but rather the government that sought to increase revenue and establish its own presence. The Americans exhorted the Saudis to take on "reclamation" projects that would expand output throughout the kingdom. And their focus was on water—improving access to it, managing its use, and implementing extensive irrigation and pumping systems. The report pointed out that "it is [probable] that an inadequate water supply in all of its aspects, including quality, has caused the greatest damage" to agricultural production.[25] But even in spite of Arabia's scarce water resources, the report's authors were confident that better water management would yield much-improved returns. Improved irrigation would "conserve the precious gift of Allah—water—and will result in the springs and wells flowing for many years instead of flowing less and less each year until some day in the future they will stop flowing."[26]

Scientific management, the mission's report declared, offered the keys to accomplishing this. The report called for the introduction of drainage and irrigation networks all across the kingdom, in addition to ongoing efforts to find additional resources by continual test-drilling. The successful operation of an extensive national network of irrigation systems also required constant observation and data collection. Maintaining regular hydrological and climatic data served scientific ends, but it also offered an opportunity for central authorities to assert themselves and their interest over the environment more broadly. The report's authors even suggested combining data collection with the work of territorial occupation: "the work of collecting and recording these data could be assigned to the commandants of the various army posts throughout the country."[27] Of course, they envisioned these efforts to be in the service and under the control of the central government. The mission did not imagine that Riyadh could or would carry out all of the development work on its own. But the central authorities would have a guiding, managerial role to play, one in which they shaped practice, determined priorities, and pushed the populace toward outcomes useful to the state. Because the overwhelming majority of the country's subjects were engaged in some form of agriculture, the pursuit of expanded productivity would be best served by improving the farmers themselves, refining their methods of cultivation, and linking farmers in both material and less tangible ways to the state.

The Americans assumed that the kingdom's territory was unified, that it represented a single national space but simply lacked the infrastructure and systems to make it more efficient. The truth was that the kingdom was beset with internal rivalries and differences. In the 1940s, Saudi authority was feared by most of those who resided within the state's boundaries, but it was hardly accepted. It is unlikely that many cultivators from across the peninsula would have agreed with the U.S agricultural mission's objectives of expanding fertile areas in the interest of a national economy, although they may have appreciated the additional revenue generated by new fields. But this was not what the agricultural mission had in mind. Twitchell's team did not advocate greater privatization for local farmers nor the creation of "free markets." Instead, they called for more oversight and presence on the part of the central government, and they provided a detailed balance sheet about local resources to guide the government on where to concentrate its efforts. On the surface, the argument for greater centralized control over markets contradicted the capitalist free-trade model of development that was emerging in early twentieth-century American foreign policy and that would become a staple of the postwar period.[28] It is plausible that Twitchell and his colleagues did not consider the kingdom ready for the creation of open markets and that they believed the state was needed to lay the groundwork for a more mature national economy, though they did not remark openly on this.

In reality, the argument put forward by Twitchell served American interests—both government and business—particularly well, and the American preference for centralized Saudi control over its domestic market would indeed prove a cornerstone of U.S. foreign policy. The argument in favor of helping establish a strong government was tied directly to the American desire for political stability in Saudi Arabia. Even in the early 1940s, before the terms of American–Saudi relations had become totally clear, the United States understood that because of its rich oil deposits the kingdom would be a critical postwar partner and a key ally in the battle to control both the postwar global economy and the flow of energy resources.[29]

There was another powerful incentive for the Americans to help expand the kingdom's economy, encourage the development of its environment, and promote strong Saudi oversight. Often described as a philanthropist or as a scientist devoted to the principles of progress and to helping develop one of the United States' most important allies, Twitchell also sought to benefit materially from his ties to the kingdom and its growing oil revenues. When seen as someone on the payroll of the state, an entrepreneur in the pursuit of a small share of the oil bonanza, Twitchell's arguments for and influence on the creation of centralized Saudi power over a national economy undermine the claim that he was an objective expert. Instead, he was someone whose material interests almost certainly shaped the kind of expertise he provided. In the decade following the U.S. agricultural mission to the kingdom, Twitchell continued to serve the Saudi government as the vice president of the American Eastern Consortium (AEC), a business that provided expertise on mining and mineral extraction, and as a personal consultant to Abdullah Sulaiman. Even after completing the work of exploration, Twitchell turned to the work of extracting resources and taking on a more direct role in shaping the country's development planning capacity. He was committed to various mining activities in Najd and the Hijaz, and he devoted his personal energy to the mining of various minerals while working for the AEC. He also maintained an active presence in consulting for the government on agricultural, hydrological, and other geological matters.

In May 1949 at the request of Abdullah Sulaiman, Twitchell spent a great deal of time with the new minister of agriculture assessing the kingdom's agricultural strategy and offering his own insight on future efforts. He followed up the meeting with a personal letter to Sulaiman in which he outlined a long list of things the government should prioritize, from road building to dam building to the use of fertilizer and the creation of demonstration farms. A month later, again at the request of the finance minister, Twitchell even drew up a detailed three-year plan for development, a systematic blueprint advising the Saudis on how to best focus their energies on various matters, including agriculture, mining, transportation, education, communication, health care, and even Saudi Arabia's prisons. Twitchell wrote assertively on which regions needed most attention.[30]

Twitchell's expanded influence was based on his years of experience and his privileged access to Saudi power brokers. His access continued well after his extensive travels and surveys. Throughout the 1940s and well into the 1950s, Twitchell actively corresponded with Abdullah Sulaiman about mining, the environment, and agriculture, offering insight as to what he believed the country needed. He also used his access to cash in, serving as a purchasing agent, establishing relationships between the Saudi government and American suppliers, and ensuring that the Saudi and American markets were firmly interconnected. Mining work continued into the 1950s, although it is unclear how successful any of these efforts turned out to be. Beginning in the late 1940s Twitchell spent less time in the kingdom, offering advice mostly from afar. The Saudis never discovered large veins of gold nor did they turn up any mineral sources of revenue that rivaled oil. Failure of the mining operations to yield anything of value never diminished Twitchell's efforts to explore, and profit, further. The Saudis also remained committed to the search for new resources.

The Politics of Development and the Redistributive State

Karl Twitchell's failure to discover extensive new resource deposits did not alter the kingdom's environmental political imperative. Throughout the 1950s, 1960s, and 1970s, the state would continue to work to find new resources and to expand its control over the environment. And with rapidly expanding oil revenues from the 1950s on, the government became even more ambitious. In the middle decades of the century, the state began to spend billions of dollars on massive environmental development projects. These included dams, sprawling irrigation networks, agricultural research farms, and—perhaps most spectacular of all—dozens of expensive plants for the desalination of seawater. By the end of the 1970s, the Saudis were spending as much on desalination as they were on education; each plant cost hundreds of millions of dollars to build and operate.

By then the political logic that informed the kingdom's approach to the environment had undergone an important transformation. In the kingdom's early years, when the state was politically weak and vulnerable, establishing control over the environment was part of a program to build up the state's political and institutional capacity. By the 1950s and 1960s this had largely succeeded. The state remained susceptible to pressure, but it was in a considerably stronger position. Not all of the kingdom's subjects accepted the legitimacy of the ruling family, but the Al Saud were nevertheless firmly in control and in charge of an increasingly powerful centralized state. The Ministry of Agriculture and Water, an extensive agricultural loan program, and various development projects brought millions of Saudis directly into the material and administrative orbit of the state. By the late 1970s, the Saudis—now flush with billions of dollars in revenue as a result of the oil boom—were no longer pressed to build political capacity. Instead, the state was confronted with the challenge of redistributing some of its massive wealth. Subsidizing water, agriculture, and making the environment and the country's limited natural resources easier for citizens to access emerged as important parts of Saudi Arabia's post-boom redistributive political order, a system that used patronage more than coercion to ensure Saudi authority.

In the first half of the twentieth century, however, the Saudis could hardly have imagined the scope of the wealth they would eventually possess or the challenges of having too much money. Simply consolidating power was challenge enough. The Saudis though did have powerful patrons—including American oil interests, the American government, and American scientists and experts—who anticipated Arabia's environmental potential. Their efforts to explore for and extract water and oil helped the Saudis achieve their domestic political ambitions. Their efforts also helped secure the kingdom's place in the global economy and its centrality to the global energy regime dominated by oil.

Notes

1. This chapter is adapted from Toby Craig Jones, *Desert Kingdom: How Oil and Water Forged Modern Saudi Arabia*, (Cambridge, MA: Harvard University Press, 2010).

2. For more on the role of security in American foreign policy in the Persian Gulf, see F. Gregory Gause, III, *The International Relations of the Persian Gulf* (New York: Cambridge University Press, 2010).

3. In this way Saudi Arabia's modern political history was hardly exceptional. For more on the connections between state authority, space, technology, science, and the environment, see James C. Scott, *Seeing Like a State: How Certain Schemes to Improve the Human Condition Have Failed* (New Haven, CT: Yale University Press, 1998); Timothy Mitchell, *Rule of Experts: Egypt, Techno-Politics, Modernity* (Berkeley: University of California Press, 2002).

4. Guido Steinberg, "Ecology, Knowledge, and Trade in Central Arabia (*Najd*) during the Nineteenth and Early Twentieth Centuries," in *Counter-Narratives: History, Contemporary Society, and Politics in Saudi Arabia and Yemen*, ed. Madawi al-Rasheed and Robert Vitalis (New York: Palgrave Macmillan, 2004), 82.

5. Hamza al-Hassan, *al-Shi'a fi al-Mamlaka al-'Arabiyya al-Sa'udiyya*, vol. 2 (Mu'assasat al-Baqi li-Ihya' al-Turath, 1993), 293–294.

6. Muhammad Said al-Muslim, *Sahil al-Dhahab al-Aswad*, 2nd edn. (Beirut: Dar Maktabat al-Hayat, 1962), 205–206.

7. Alexei Vassiliev, *The History of Saudi Arabia* (London: Saqi Books, 1998), 225.

8. Kiren Aziz Chaudhry, *The Price of Wealth: Economies and Institutions in the Middle East* (Ithaca, NY: Cornell University Press, 1997); Daryl Champion, *The Paradoxical Kingdom: Saudi Arabia and the Momentum of Reform* (New York: Columbia University Press, 2003).

9. FAO, cited in Karl S. Twitchell, *Saudi Arabia: With an Account of the Development of its Natural Resources*, 3rd edn. (New York: Greenwood Press, 1969), 21.

10. Vassiliev, *History of Saudi Arabia*, 412–419; J. S. Birks and C. A. Sinclair, "The Domestic Political Economy of Development in Saudi Arabia," in *State, Society and Economy in Saudi Arabia*, ed. Timothy Niblock (New York: St. Martin's Press, 1982), 199–200. Vassiliev estimated the population in the 1960s to be between 3.5 and 4.5 million people. However, there was no census data and all figures must be viewed as loose estimates, as the wide range indicates.

11. Ministry of Agriculture and Water, *Seven Green Spikes* (Riyadh: Ministry of Agriculture and Water, 1974).

12. Abdulaziz H. al-Fahad, "The 'Imama vs. the 'Iqal: Hadari-Bedouin Conflict and the Formation of the Saudi State," in *Counter-Narratives: History, Contemporary Society, and Politics in Saudi Arabia and Yemen*, ed. Madawi al-Rasheed and Robert Vitalis (New York: Palgrave Macmillan, 2004), 35–36.

13. The Al Saud relied on the Wahhabi clergy, on whom it bestowed considerable power, to convince skeptics of the legitimacy of their message. To spread the faith, King 'Abd al-'Aziz dispatched religious missionaries known as *mutawwa'a*, religious figures who lacked the higher religious training of the more senior sheikhs but who made up for their lack of credentials with zealotry. In the twentieth century, the Al Saud helped connect the strict interpretative framework of Wahhabism—which viewed anyone who deviated from a set of narrow beliefs and rituals to be guilty of apostasy, and thereby worthy of either death or conquest—with their political aims of controlling the Arabian Peninsula.

14. The Al Saud relied on several incentives to induce permanent settlement. The *mutawwa'a* stressed that proper faith required sedentary living, a precept embodied in the name for the settlements, *hujjar* (sing. *hijra*). The choice of the word *hijra* for individual settlements was an intentional and symbolic choice, as it harkened back to the early Islamic era, specifically the Prophet Muhammad's migration from Mecca to Medina, a seminal event in the formation of Islamic belief and tradition. While Islam lent ideological substance to sedentarization, the location of *hujjar* near water resources and the potential for agriculture gave the Al Saud hope that the sedentarization program would be sustainable. In little over a decade the number of settlements jumped from one single settlement to over two hundred, with most located close to water resources spread across Najd. In addition to providing weapons for the *jihad* and building mosques and schools, the Al Saud gave the *ikhwan* money, seed, and the equipment necessary to engage in various kinds of settled agriculture.

15. John S. Habib, *Ibn Sa'ud's Warriors of Islam: The Ikhwan of Najd and their Role in the Creation of the Sa'udi Kingdom, 1910–1930* (Leiden: Brill, 1978), 16. Another significant objective was to harness but not eliminate the mobile-fighting power of the *ikhwan*. Habib notes that the military goal was to keep the *ikhwan* "mobile enough to cross the length and breadth of the peninsula, and sedentary enough to be in a specific locality when he [Ibn Sa'ud] needed them."

16. Twitchell, *Saudi Arabia*, 212.

17. Ibid.

18. Ibid.

19. Ibid., 213.

20. Ibid., 216.

21. Ibid., 219.

22. Shortly after obtaining the concession, a company for operating in Saudi Arabia was formed called the California Arabian Standard Oil Company. It would be renamed the Arabian American Oil Company in 1944.

23. *Report of the United States Agricultural Mission to Saudi Arabia* (Cairo: 1943), 1.

24. There is no indication in any of Twitchell's writings that he saw anything wrong with the expansion and consolidation of Saudi power. He certainly never wrote about local resistance to or animosity toward the Al Saud, which does not mean that he was unaware of such sentiment. His various accounts of his and the agricultural mission's work are apolitical in tone, but it is also clear from them that Twitchell understood whose political interests were being served.

25. *Report of the United States Agricultural Mission*, 46.

26. Ibid., 48.

27. Ibid., 58.

28. Michael Adas, *Dominance by Design: Technological Imperatives and America's Civilizing Mission* (Cambridge, MA: The Belknap Press of Harvard University Press, 2006).

29. Daniel Yergin, *The Prize: The Epic Quest for Oil, Money, and Power* (New York: Simon & Schuster, 1991); Nathan J. Citino, *From Arab Nationalism to OPEC: Eisenhower, King Sa'ūd, and the Making of U.S.-Saudi Relations* (Bloomington: Indiana University Press, 2002).

30. Letter to Sulayman al-Hamad, Assistant to the Minister of Finance, June 4, 1949. Karl S. Twitchell Papers. Public Policy Papers, Department of Rare Books and Special Collections, Princeton University Library.

II

Expanding the Nile's Watershed: The Science and Politics of Land Reclamation in Egypt

Jessica Barnes

A desert seems very different from a cultivated field. One is sandy, the other a swath of green. Yet the boundary between the desert and the sown shifts in time. Through a gradual process of watering, fertilizing, plowing, and planting, the desert can be brought into cultivation. This is the process of land reclamation. It is the conversion of a parched surface into a field of wheat, an orchard of oranges, a ground cover of clover. Egypt offers a valuable example of this landscape transformation. The stark contrast between the long-cultivated strip of the Nile Valley and Delta and the desert that abuts the fields on either side belies the transience of the border between the two. Over the last two centuries, land reclamation has reshaped the rural landscape of Egypt, increasing the territory of agricultural production.[1] Government agencies, farmers, international advisors, and foreign companies have worked to push out the boundaries of the land watered and drained by the world's longest river. They have come together in an ambitious endeavor to expand the Nile's watershed.

Histories of land reclamation in Egypt have charted the evolution of this program under different organizational regimes, highlighting the political motivations driving reclamation and the challenges reclamation projects have faced.[2] What has been largely missing from these histories, however, is the water that makes reclamation possible.[3] Water is not, of course, the only resource required to transform desert into field; other inputs are needed as well, like fertilizers and treatments to adjust soil texture. But whereas these inputs can be easily purchased, water cannot. The fortune of any attempt to reclaim new land lies, therefore, in the supply of water it receives from the Nile.[4] The channeling of water to the desert is what allows the conversion of sand to crop; the maintenance of that flow is what prevents the reversion of crop to sand.

This chapter places water at the forefront of the story of land reclamation in Egypt. Using the World Bank's New Land Development Project as a case study, it tracks how water, as both a physical resource and an object of scientific understanding, is central to reclamation efforts. This project, which started in 1980 and was completed in 1991, transformed twenty-four thousand feddans of desert west of the Nile delta into fields.[5] Yet a mere two years after it ended, many of the new fields lay fallow. Without the water necessary to sustain the land, production dropped significantly. The World Bank changed its assessment of the project from satisfactory to marginally satisfactory; it altered its evaluation of the project's sustainability to uncertain.[6]

The success of the project in creating productive lands in the desert was ultimately constrained by its failure to secure sufficient water resources to feed those lands. The roots of this failure lay in the assumptions that project managers made about where the water would come from and how it would reach the project area. The farmers who came to settle the newly reclaimed land no doubt had their own ideas about the project's progress and the water they received. Since these voices do not, however, feature in the archival records of the project, this chapter analyzes the project from the perspectives of the government officials, World Bank staff, and external consultants who managed the project and used their expertise to guide its activities. Correspondence, consultancy reports, and project documents from the World Bank archive offer insight into the types of water knowledge that formed the conceptual foundation of the project.[7] This project was not the biggest of the internationally funded reclamation programs implemented in Egypt in the late twentieth century. Its scale was small in comparison to the government's independent initiatives to transform the desert. But access to the project's internal records provides a unique opportunity to look at how staff from the World Bank and officials from the Ministries of Agriculture and Irrigation understood Egypt's water supply and the water requirements for reclamation.

There is a long history of water manipulation and agricultural development projects around the world that have failed to meet their ambitious goals. Visions of agricultural bounty and dreams of an endless water supply are frequently unfulfilled in reality.[8] The case of Egypt's New Land Development

Project demonstrates how a project's outcome is determined, in part, by the scientific understandings that underpin the project's activities. These understandings are the result of debates between expert groups, which hold different stakes in the project's outcomes. This is not just a matter of one group of experts being right and another group being wrong, but of each group framing the resource in question and its knowledge about that resource in a particular way. As the World Bank, other international donor agencies, and national governments continue to pursue water and agricultural development projects, this analysis shows how much can be learned about past and ongoing projects from their foundational assumptions.

In line with a body of scholarship within environmental history that has highlighted the role of nonhuman actors in influencing the course of historical events, this study also reveals how it was the flow of water—or rather the lack of flow—that ultimately determined the failure of this reclamation initiative.[9] The New Land Development Project dug canals, constructed pumping stations, and installed an irrigation network. But those canals ended up half empty, the stations operating at below capacity, the distribution ditches dry. Without water, the new infrastructure was futile. This provides a valuable lesson not only for rethinking the process of land reclamation, but for reconsidering the history of other farming communities in the region and beyond. Both water and the technologies for moving, blocking, storing, accessing, redirecting, and utilizing that water, from the scale of the river basin to the field, play a central role in the development of agrarian societies.

Land Reclamation in Egypt and the New Land Development Project

Land reclamation is not new in Egypt. The Ottoman state launched a number of projects to increase the amount of taxable agricultural land in Egypt, by expanding cultivation into barren and degraded land.[10] These efforts intensified over the course of the nineteenth century as the Ottoman governor general, Muhammad Ali, and his successors launched initiatives to drain waterlogged and salinized soils in the delta and offered grants of idle (ib'ādīya) land to prominent individuals, exempting them from taxes on the condition that they brought it into cultivation.[11] The building of the first Aswan Dam in 1902 made additional water available, opening up new possibilities for reclaiming desert land that had never previously been cultivated. It was in the wake of the 1952 revolution, however, that the government's

TABLE 11.1　Land Reclamation in Egypt since the 1950s

1950s	1953–67	*Egyptian-American Rural Improvement Services Project*	Reclaimed 37,100 feddans in Abis, Kom Oshim, and Quta.	• Reclaimed land distributed to landless farmers.
	1953–	*Tahrir Project*	Planned to reclaim 600,000 feddans west of the delta. By 1980 had reclaimed 122,000 feddans.	
	1956	Suez Crisis stalled progress.		
1960s	1960–65	*First Five-Year Plan*	Reclaimed 390,000 feddans.	• Private land reclamation companies nationalized.
	1965–70	*Second Five-Year Plan*	Reclaimed 300,000 feddans.	
	1967	Six Day War interrupted reclamation activities.		• Creation of state farms in reclamation areas.
1970s	1970–78	Low rates of growth	Less than 50,000 feddans reclaimed.	• Progress slow due to other budgetary priorities.
	1971	Completion of the Aswan High Dam	Increased the amount of water available for agricultural expansion.	
	1978	President Anwar al-Sadat launched a "Green Revolution" with the goal of developing 2.9 million feddans of agricultural land before the end of the century.		
1980s	1980–91	*New Land Development Project*	World Bank-funded project reclaimed 24,000 feddans west of the Delta.	• Sale of state land to private investors for reclamation.
	1981	Public Law No. 143	Removed public sector's legal monopoly on reclamation, opening up reclamation to the private sector.	• Privatization of state farms. • Increasing skepticism among international donors about the economic viability of reclamation.
	1983–88	*Third Five-Year Plan*	Reclaimed 189,000 feddans.	
	1987–	*Mubarak Project for Developing and Serving the Land Allocated to Youth Graduates*	Started to distribute land in parcels of five feddans to high school or college graduates and beneficiaries (veterans and those who lost land in the 1992 tenure reform).	
	1988–93	*Fourth Five-Year Plan*	Reclaimed 656,000 feddans.	
1990s	1993–97	*Fifth Five-Year Plan*	Reclaimed 469,000 feddans.	• Increasing private sector involvement.
	1997–17	*Thirty-Year Strategy*	3.4 million feddans to be reclaimed by 2017.	

TABLE 11.1 *(Continued)*

2000s	2009–	*Agricultural Strategy Towards 2030*	1.25 million feddans to be reclaimed by 2017 and 3.1 million feddans by 2030.	• Private-public partnerships in the new lands.

Data Sources: Jon Alterman, *Egypt and American Foreign Assistance 1952–1956: Hopes Dashed* (New York: Palgrave Macmillan, 2002); Sayed Hussein et al., *Study of New Land Allocation Policy in Egypt*, Report No. 65, Agricultural Policy Reform Program (Cairo: Ministry of Agriculture and Land Reclamation, 1999); Pamela Johnson et al., *Egypt: The Egyptian American Rural Improvement Service, a Point Four Project, 1952–63*, AID Project Impact Evaluation No. 43 (Washington, DC: USAID, 1983); Günter Meyer, "Economic Changes in the Newly Reclaimed Lands: From State Farms to Small Holdings and Private Agricultural Enterprises," in *Directions of Change in Rural Egypt*, ed. Nicholas Hopkins and Kirsten Westergaard (Cairo: American University in Cairo Press, 1998); A. Nyberg, S. Barghouti, and S. Rehman, *Arab Republic of Egypt Land Reclamation Subsector Review*, Report No. 8047 (Washington, DC: World Bank, 1990); Robert Springborg, "Patrimonialism and Policy Making in Egypt: Nasser and Sadat and the Tenure Policy for Reclaimed Lands," *Middle Eastern Studies* 15 (1979): 49–69; Sarah Voll "Egyptian Land Reclamation Since the Revolution," *Middle East Journal* 34 (1980): 127–148 ; T. Zalla et al., *Availability and Quality of Agricultural Data for the New Lands in Egypt*, Impact Assessment Report No. 12, Agricultural Policy Reform Project (Cairo: Ministry of Agriculture and Land Reclamation, 2000).

land reclamation program took off. Driven by an aspiration to expand the cultivated area, increase agricultural production, and create new jobs for farmers and laborers, the government launched a set of ambitious five-year plans. Construction of the Aswan High Dam in the 1960s further increased the amount of water available for irrigation, generating the potential for even more expansion.

While the governments of Anwar al-Sadat (1970–1981) and Hosni Mubarak (1981–2011) continued to promote land reclamation, the 1970s and early 1980s were a time of increasing doubt within the international donor community, which provided an important source of funding to Egypt's agricultural sector, about the viability of transforming desert into fields. These concerns were bolstered by a study commissioned in 1980 by the U.S. Agency for International Development (USAID), which concluded that "large-scale reclamation of new desert lands in Egypt can take place only at considerable cost to the economy."[12] The authors found that the reclamation of land where farmers must pump the water up more than twenty meters had a negative rate of return, due to the high energy costs of lifting the water. In fact, their models only produced positive economic results "under the most heroic assumptions regarding yields."[13] The report was highly controversial; according to one source, it "soured relations with the MOLR [Ministry of Land Reclamation]," which was "extremely displeased."[14]

In October 1977, the World Bank sent a reconnaissance mission to Egypt to look for new lending opportunities in the agricultural sector. The bank was keen to increase its lending to Egypt as a mechanism to gain leverage in Egypt's political economy and push for liberal reforms.[15] The mission found that reclamation was one of the government's top agricultural priorities. Indeed President Sadat was on the eve of launching his bold "Green Revolution," designed to reclaim 2.9 million feddans of desert land by the end of the century.[16] Up until this point, the bank's position had been that the Egyptian government should focus on improving the situation in existing lands rather than expanding into new ones. The mission recognized, however, that reclamation was an area where the government would welcome bank funding. After much debate, senior bank officials came to a decision that they should reverse their "hands off stance" on new land projects.[17] A number of staff remained skeptical, but they realized that whatever their position, the Egyptian government would press ahead with this program. As one bank official commented, "New land development is going to take place in Egypt, whether we like it or not. And sooner or later the bank is bound to be involved in such projects. If we participate early we may help the Egyptians to avoid costly mistakes."[18] This change of position was supported by the bank's assessment that reclamation could be economically viable. Having conducted their own studies, bank staff came to the conclusion that the USAID study was "excessively pessimistic" and that their proposed reclamation project would have a positive rate of economic return.[19]

The New Land Development Project marked the bank's first foray into irrigation and land reclamation work in Egypt. The project's primary goal was to develop twenty-four thousand feddans of land west of the Nile through a reclamation and smallholder resettlement project.[20] Project implementation met with problems from the outset; what was meant to be a six-year project ended up taking eleven. The project's goal to develop a training farm failed, in part because the land that the government allocated to the farm did not have a source of irrigation water.[21] The agricultural extension program, which was to provide settlers with information on the most suitable crop rotations, never materialized.[22] Despite these problems, by 1991 the project had succeeded in its goal of reclaiming around twenty-four thousand feddans. "The area has changed from a barren desert into an oasis," the project completion report concluded.[23] The government's evaluation was even more effusive. "The project is a success in all aspects," the borrower's response section of the completion report stated. "The project is considered as one of the most successful projects in Egypt and ever in the whole region. It is one of the ideal projects in the field of land reclamation."[24]

However, a critical weakness at the heart of the project was soon revealed. There was not enough water to irrigate the new land. When bank staff carried out a performance audit mission in 1993 they found extensive areas of project land left fallow. The cropping intensity in the summer was low. Production

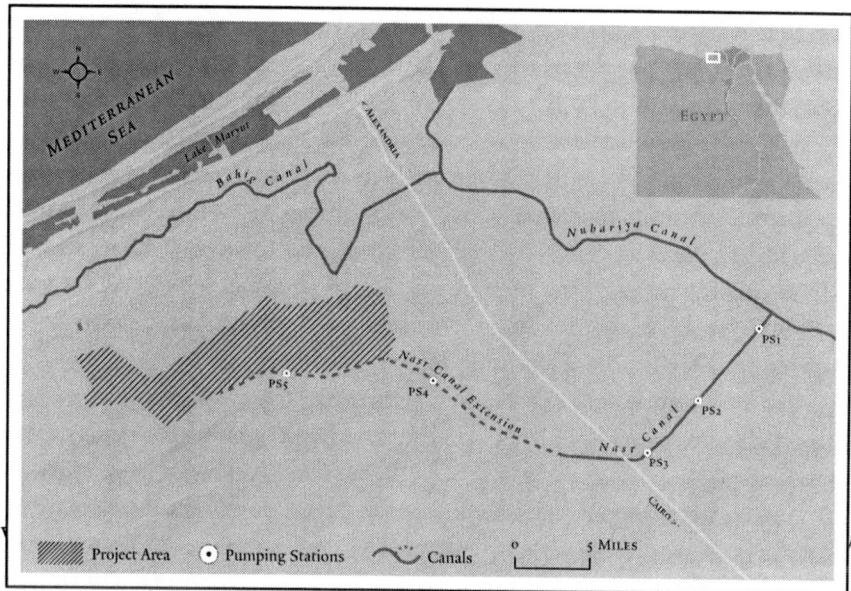

MAP 11.1. New Land Development Project

water supplies are reducing benefits," the evaluation team stated unequivo-cally.[25] The origins of this problem, which were manifest a mere two years after the project came to a close, can be found in three of the project's key assumptions. The first was an assumption about how much water the new lands would require; the second was an assumption about where that water would come from; and the third was an assumption about how much water was available. In forging these assumptions, the Egyptian and expatriate experts produced and contested a set of scientific understandings at the level of the feddan, canal, and basin about how water would be used for desert transfor-mation. These understandings were not just about contrasting scientific judg-ments but reflected the broader political and economic stakes that the different experts held in the project's success. The project's failure to attain sustainabil-ity ultimately lay in the limitations of these negotiated understandings.

Calculating Demand: How Much Water Does a Desert Feddan Need?

The plan to reclaim twenty-four thousand feddans was underpinned by a calcula-tion of how much water each new feddan would require. This was partly a ques-tion of how much water would be needed for the reclamation process. To turn desert land into cultivable soil, water must be washed through the soil to leach salts out of the top layers and reduce them to tolerable salinity levels. Preliminary field trials, conducted by a British consultancy firm in 1979 and 1980, found that the soils in the area were easily leachable and that high yields could be obtained from the first year of cropping. Based on these results, the project staff assumed a primary leaching of one meter, which meant that each new field would be inundated up to a meter in depth as part of the reclamation process.[26] They predicted that in subsequent years the deep percolation of about 20 percent of the water applied to the soil would be sufficient to remove accumulated salts, so no further applications specifically for leaching would be required.[27]

The next step was to assess how much water the crops would need once those soils were planted. This was done by ascertaining how much water a "reference crop" would evapotranspire (conduct to the atmosphere by evaporation from plant surfaces and transpiration through the leaves) under the climatic conditions in that location, and then multiplying that figure by a crop factor to represent the crops that would actually be planted. The British consultants preparing the feasibility study used a theoretical method known as the Blaney Criddle, as modified by the Food and Agriculture Organization (FAO) of the United Nations, to calculate reference crop evapotranspiration with data from the South Tahrir climate station.[28] They then adjusted this value based on a coefficient to reflect crop characteristics. Since the farmers within the project were going to be allowed to choose what to cultivate—which meant that cropping patterns were unknown—the consultants applied a mean crop factor of 1.05 based on data from the FAO. This figure was an average coefficient that reflected the most commonly grown crops' peak water requirements in July.[29]

Finally, the project planners had to account for the fact that some water would be lost as it passed through both the distribution network and the fields. The consultants estimated that the efficiency of the conveyance system would be quite high at 0.93, meaning that only 7 percent of the water that entered the canals would be lost to evaporation or leakage before reaching the fields. In terms of the field-level irrigation efficiency, this depended on the method of irrigation farmers would use. The Ministry of Irrigation had a policy that farmers use sprinkler or drip irrigation on all new lands, but initial field trials had indicated that the more commonly used technique of flood irrigation would actually be more appropriate in the project area. Not only are sprinklers expensive to install, operate, and maintain, but few of the smallholder settlers would be familiar with their use. In addition, the trials showed that contrary to expectations, the efficiency of water use was lower with sprinkler irrigation. On fields where they used sprinklers, consultants found that a soil crust developed which caused up to 40 percent of the water

applied to runoff over the surface.[30] Combined with high rates of evapotranspiration losses, the trials concluded that sprinkler irrigation's efficiency was
a mere 0.5 in comparison to flood irrigation's 0.65.[31] Government officials
remained skeptical that the method which they understood to be more efficient could, in this case, be less so. But the bank staff were adamant that
flood irrigation be used and adopted the figure of 0.65 as their estimate of
field-level efficiency.

Combining these three factors, the feasibility study therefore determined
that each feddan would need a peak of fifty-four cubic meters of water a day
in July.[32] This was substantially more than Egyptian officials thought a feddan
needed. Indeed government engineers had designed the Nasr Canal, from
which the project was to draw its water, to irrigate 300,000 feddans based
on an assumption that the peak water requirement would be only thirty-three
cubic meters of water per feddan each day.[33] It was not surprising, therefore,
that the calculation of water duty should be hotly contested.

A few months after the consultancy firm submitted its feasibility study, the
irrigation subcommittee for the project, comprising two senior officials from
the Ministry of Land Reclamation, an irrigation engineer from the World Bank,
and an external consultant, met to revise the water duty to be "less conserva-
tive."[34] First, they adopted the original rather than the modified Blaney Criddle
method to calculate reference evapotranspiration. The Egyptian members
of the committee asserted that based on experiments carried out at a nearby
research station, this method would give more realistic (and lower) results.
Second, they argued that the climate data from South Tahrir was too extreme.
Located fifty kilometers southwest of the project, the station is in a hotter environment where crop water requirements, as a result, are higher. The committee recommended using figures from North Tahrir station, situated close to
the project area, averaged with data from the Borg al-'Arab station, located in
a slightly milder environment on the Mediterranean coast. Third, the committee increased the conveyance efficiency to 0.94, on the basis that not only the
main canals but also the tertiary canals would be lined. By making these adjustments, the committee brought the figure of water duty down to forty-eight
cubic meters per feddan each day.[35]

Furthermore, committee members identified two characteristics of the
project area that they argued would reduce the water duty. The first was the
fact that part of the project lands would be irrigated with sprinklers, which—
despite the field trial results—committee members assumed to have a lower
water requirement (of 36.4 cubic meters per feddan each day). The second
was the fact that some of the project land would be planted with trees, which
require less water than field crops (typically only thirty-one cubic meters per

feddan each day). By taking these factors into consideration, they calculated an average water duty of just forty-six cubic meters per feddan each day, in contrast to the initial fifty-four cubic meters. In addition, they pointed out that about a quarter of the land would be under villages, roads, canals, and drains, and thus not require any water at all; thus, the water duty for the project as a whole would be only 34.4 cubic meters per feddan each day. By changing the method of calculation and the assumptions on which that calculation was based, this group of Egyptian and foreign engineers therefore concluded that each feddan would need only 64 percent as much water as the British consultants originally anticipated.[36]

The bank was only willing, however, to accept a reduction in the water duty up to forty-eight cubic meters per feddan each day. Bank staff rejected further revisions since they said that no provisions had been made in the project for tree crop cultivation or sprinkler irrigation and that non-agricultural areas would be excluded from the project area.[37] A figure of forty-eight cubic meters per feddan each day was therefore adopted as the project design guideline. Although the Ministry of Irrigation gave an assurance at the time of project negotiations that it would supply this water, many government officials remained skeptical that this volume was really necessary. One bank consultant noted that "Mr Makhlouf [Undersecretary] of the Ministry of Irrigation has still some reservations about this demand."[38] Other questioning voices came from the Ministry of Land Reclamation, where engineers challenged the use of FAO data on crop water requirements. Arguing that crop coefficients should be based on local experience instead, they recalculated peak water requirements as being only 43.4 cubic meters per feddan each day, based on the inclusion of the local 130-day maize variety, rather than the 100-day variety included in the initial estimates.[39]

This difference in opinion between the government officials and international consultants can be explained by the two groups' contrasting stakes in the project. To the government officials, the decision about how much water was needed to reclaim the project lands carried with it a much broader significance. It helped to define the feasibility of the government's plans to expand the nation's cultivated area by hundreds of thousands of feddans. If they agreed that more water was needed to reclaim each feddan, it would mean that the government's ambitious reclamation plans would require a far greater volume of water than what was available, thereby undermining their vision of national progress through agricultural expansion. These issues may not have been openly discussed in meetings between government officials and project staff, but they were acknowledged in internal project documents. In a "back-to-office report," for example, the bank official cited above noted that the undersecretary's reservations about raising the water requirements for reclamation were "due obviously to political reasons." He added, by means of explanation, that "the government several times has announced an ambitious plan for expansion of irrigation, based on a much lower water demand per feddan."[40] For the international consultants, on the other hand, their key concern was the productivity of the twenty-four thousand feddans within the project area. If those feddans flourished, it would reflect well on them and on the bank's development program. From their perspective, therefore, it was better to plan to accommodate the maximum volume of water that the land would need during the time of peak water requirements.

Government officials were hence caught between meeting the demands of the project and fulfilling the vision of the politicians. In light of this fact, it is not altogether surprising that they ultimately failed to fulfill their commitment to supply the project lands with the volume of water set by the bank staff at the outset.

Tracing Flows: Where Would the Water Come From?

The water that the government assured the project consultants it would supply each year was to come from a twenty-eight-kilometer extension to the existing Nasr Canal. For the water to reach the project site, two new pumping stations would be needed to lift water from the canal up forty-three to fifty-five meters to the higher desert land. Even before the project started, the government, keen to develop new lands in this area, had begun to extend the canal and install the pumping stations. Although there were a number of delays in the construction work, the extension was complete by late 1983 and the final pumping station started operating in 1986.[41]

The project consultants were concerned, however, about the poor state of repair of the first twenty-two-kilometer section of the Nasr Canal. Built in 1971, the canal's walls were deteriorating and its concrete lining was broken. The Ministry of Irrigation promised that the rehabilitation would be completed by the end of 1983, but as it had not begun this work when the project negotiations were underway, some within the bank were dubious. If the canal collapsed, no water would be able to flow to the project site and reclamation would be impossible. Bank staff called for the signing of the project agreement to be postponed until the ministry had found contractors to repair the canal.[42] The ministry put out a tender for repair work and received two bids, but it rejected both on account of their "unreasonably high cost."[43] In the end, the bank capitulated, agreeing in October 1981 to move forward with the project based only on government assurance that the repair work would be carried out.[44]

At the same time, bank staff working on the project called for intervention at the highest levels to pressure the government into proceeding with the work. In November 1981, the director of the bank's department for the Middle East region sent a letter to the minister of irrigation about the progressive deterioration of the canal. "If serious damage and deterioration of the canal have occurred with the present regime," he wrote, "one can easily imagine what disastrous effect the higher flow and velocity [required as new lands are developed along the canal] will have on the banks of the canal already damaged by the forces of erosion. Our considered judgment on the matter is that a collapse of the canal cannot be ruled out. This would have serious financial and social consequences which, I am sure, you are keen to avert."[45]

Despite the bank's efforts to galvanize swift action on this matter, it was not until the middle of 1983 that the government finally authorized two contracts for the repairs.[46] Work started in early 1985, but the contractors met a number of technical difficulties that delayed their progress. By April 1987, one contract was 40 percent complete, the second only 24 percent.[47] Even when the project came to a close, sections of the canal were still in a poor state of repair.[48] The project completion report highlighted the canal's damaged condition as the largest risk to the project's sustainability. "Part of the Nasr Canal, the lifeline of the project, remained in need of urgent repair," the report stated. It continued, "Although some rehabilitation work was started, much more work remains to be done to ensure adequate irrigation water delivery to the project area in the future."[49]

The failure to mend the canal lay partly in the challenge of carrying out complex repair work while not interrupting the flow of water through the canal. But it also lay in the fact that the Egyptian engineers differed from the expatriate engineers in their evaluation of the urgency of this maintenance work. When the minister of irrigation visited Washington during the early stages of the project, he maintained that "even without repairs, the canal could carry a sufficient amount of water to service the New Lands Project." He argued that the first priority was to complete the extension of the canal.[50] The bank official overseeing the project had a different view. "To my mind," he wrote, "there is no need to complete the end of the canal when there is a risk that nothing will pass through the first part."[51] As the project progressed, government officials continued to express ambivalence about the necessity of this repair work. While this position may in part have been due to a judgment that the canal's status posed no threat, it could also have reflected the budgetary constraints that the ministry faced during this period as the economic situation worsened in Egypt.[52] In a time of limited funds (but still ambitious goals), it is not surprising that the ministry was unwilling to spend its money on repair work that it did not deem to be absolutely essential. The government's response to the project completion report rejected the bank's concerns, asserting that "efforts have been directed to ensure the repair of any damage to the lining of this canal and to assure its full maintenance...minimizing the risk of water supply failure from this canal and hence the sustainability of the project is guaranteed."[53]

Although the disagreement over the significance of the repair works remained unresolved right up to the end of the project, by the time of the follow-up mission in 1993, the level of water in the canal was so low that its poor state of repair posed little problem.

Spreading Water Thinly: How Much Water Was Available?

According to its design specifications, the Nasr Canal could supply 118 cubic meters of water a second. This volume would be sufficient to irrigate from 270,000 feddans to over 370,000 feddans, depending on how much water each feddan required.[54] Since at the time of project development the government had only committed to irrigating 218,000 feddans along the canal, including the new project, neither bank staff nor government officials anticipated water shortage problems.[55] The project's feasibility study stated unequivocally that "the availability of water is not a constraint to the development of irrigated agriculture in the project area."[56]

But water availability is not merely a function of the size of the canal that supplies the water. It is also a question of how much water government officials choose to direct into that part of the water distribution system. For 118 cubic meters of water to flow into the Nasr Canal each second, government engineers had to make particular decisions about water distribution at a number of points farther up the irrigation network. The Nasr Canal draws from the Nubariya Canal, which in turn draws from the Rayah al-Behera and Rayah al-Nasseri, which lead off from the Nile just upstream of the Delta Barrages (see figure 11.1). At each of these junctions in the irrigation network, officials from the Ministry of Irrigation determine how much water should go into each branch of the system, adjusting the gates of regulators or weirs accordingly. Tracing the flow back farther, the amount of water available in the Nile at the Delta Barrages is linked to how much water farmers draw off upstream, the government's releases from the Aswan High Dam, and the amount of rainfall in the East African source regions of the Nile. So the presumption that the Nasr Canal

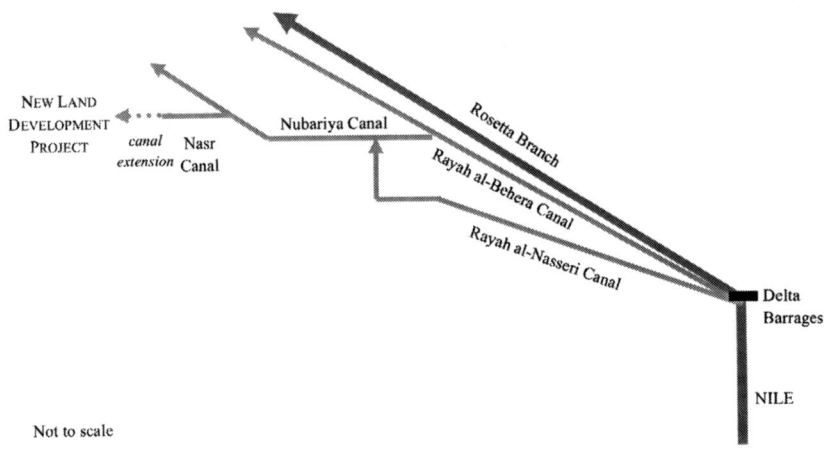

FIGURE II.I. New Land Development Project Schematic Diagram

would always be able to operate at its maximum capacity, constantly delivering 118 cubic meters of water per second, was quite optimistic.

The bank staff recognized that part of the project's success in securing an adequate irrigation water supply hinged on increasing the flow through the Nubariya Canal.[57] This was something that the Ministry of Irrigation was already in the process of doing, as it knew that there would be an increasing demand for water in the area served by the canal. In 1980, the flow through the canal was 12.5 million cubic meters per day, just over half its design discharge of 22.5 million cubic meters per day. To increase the canal's inflow to fifteen million cubic meters per day, the ministry was working to enlarge the Rayah al-Nasseri.[58] By the middle of 1982, twelve dredgers were at work and had expanded the canal along 45 percent of its length; the remainder was expected to be complete by June 1983.[59] There is no reference in the archival record to suggest that this work was delayed. Thus it was most likely finished by 1983, before water started flowing to project lands.

However, acknowledgment of the need to increase the flow through the Nubariya Canal was the only indication in the project preparation documents that something would need to change within the broader system of water distribution to secure water for new land development. Indeed, the project was founded on the idea that there was in fact *too much* water. As one bank irrigation specialist wrote, "There are millions of feddans of undeveloped land in Egypt which should (and can) be developed to utilize surplus waters now available in the Nile River to provide a livelihood for a growing population."[60] Thus, far from anticipating water shortages, the project forecast that the supply would increase in the future, meaning that the government could pursue additional reclamation initiatives along the Nasr Canal in later years. Based on this prediction, project consultants recommended that the extension to the Nasr Canal be constructed at a capacity greater than necessary for the current project, to allow for subsequent expansion.[61]

Over the course of the 1980s, though, there were growing concerns among bank staff that there might not be enough water for further reclamation. These concerns were exacerbated by a string of low rainfall years from the late 1970s through the mid-1980s in the source regions of the Nile, which led to reduced Nile flows and mounting pressure on the national water supply.[62] In 1983, a bank report came to the conclusion that the availability of water would soon become a constraint in the new lands. The report judged that the government's plans to develop new lands were "overambitious, not to say unrealistic."[63] An internal bank memorandum from 1984 noted that "although it is not possible to define precisely when water will become a limiting factor in absolute terms [for new land development], two phases can be anticipated: a shortage of water during the peak demand period in the summer caused by conveyance constraints in the Nile and canal distribution system, to be followed later by an overall shortage of water."[64] Bank officials tried to raise these concerns with Egyptian officials but found them generally unwilling to heed their words of warning. Government officials were, of course, acutely aware of the fact that Nile flows had been low for a number of years and that storage supplies in their main reservoir of Lake Nasser were nearing a critical point. However, publicly acknowledging this serious point of vulnerability was politically unacceptable.

In the later years of the project, some water shortages became apparent in the newly developed lands. In 1986 and 1987, the level of water in the Nasr Canal fell, leaving parts of the canal system nearly dry. A report from July 1987 stated that "serious crop losses have resulted."[65] During this period, a number of farmers who had settled in the project area traveled to the nearby city to complain to the resident engineer about the inadequate water supply.[66] But bank staff judged these water shortages to be due more to problems of water distribution than fundamental resource insufficiency. They recommended modifications to the weirs and offtakes to ensure that each part of the distribution system would receive water even when the supply was low. These adjustments seem to have been successful since the final years of the project saw no further complaints of shortages.[67]

Just two years after the project came to an end in 1991, though, the availability of water was in doubt. When bank staff visited the region to conduct a performance audit, they found that the flow of water through the Nubariya Canal was insufficient during the summer to allow the Nasr Canal to operate at its design capacity. Summer supplies were so low that irrigation managers were only able to provide water to the canals in the project area once every two weeks rather than once a week. The water level in the canals was up to ninety centimeters below the design level, meaning that engineers had to reconstruct the offtakes so that water would still be able to flow out of the canals to the fields. This shortage cannot be explained by the status of the broader Nile system; by the early 1990s rainfall had increased in the river's source regions and flows had rebounded. Rather, what had evidently happened was that government officials had stopped channeling the same volume of water into these branches of the irrigation network as they had previously. The bank report noted that "the New Lands project has been undermined by the government's policy of spreading water thinly (for social and political reasons)." It continued with its prognosis: "The long range water duty figure used by the government is well below the figure used to design the project, and therefore the current realistic prospect is that the scheme will have to adjust for the foreseeable future to water supplies much below the level expected."[68]

Watering the Desert

Thus the confluence of three assumptions—about how much water was needed, where the water would come from, and how much water was available—led to a situation, shortly after the project came to an end, in which much of the newly developed land lay fallow. Without water, the new land could not flourish.

These three assumptions were the outcome of heated debates between Egyptian and expatriate experts, waged in project memoranda, reports, and letters. The debates revolved around the question of accessing water for new land development. They focused on the science of how much water was required to reclaim the desert. Yet they were about more than just water. They were about the relations between an international funding institution that wanted to push a particular reform agenda and a national government that needed funds to finance its activities but had its own set of interests, between politicians who saw a vision for a future greener Egypt and government officials who were tasked with implementing those visions, and between international consultants whose immediate concern was to ensure their project's success and ministry staff who had to meet their government's political and budgetary priorities.

The case of the New Land Development Project, therefore, offers useful insights into ongoing efforts to expand the Nile's watershed. Although the international donors have largely withdrawn from active engagement in new land development, the Egyptian government continues its reclamation program. In the south of the country, the Toshka Scheme channels water fifty kilometers from Lake Nasser into the Western Desert to cultivate what the government hopes will eventually be over a half million feddans. In the north, the El-Salam Canal channels water into the Sinai Peninsula with the goal of reclaiming 620,000 feddans of desert. All along the boundaries of the agricultural zone, through both government projects and independent initiatives, farmers are extending their lands out into the desert. But where will the water come from to irrigate all this new land? How much land can the Nile irrigate? Who gets to make this decision? How do recent theories about climate change and its impact on East African rainfall patterns alter the debates over water availability and the potential for desert development? What will happen in the future if other Nile Basin countries build large dams and modify the volume of water Egypt receives from the river?

As the Egyptian government looks to expand the Nile's watershed even farther—by 3.1 million feddans by the year 2030—questions about where the water will come from to feed that expansion are paramount.[69] Understanding how water flows through canals and soil will be critical if that ambitious goal is to become a reality, and most significantly, if those fields are to be sustained into the future.

Acknowledgment

I thank the staff of the World Bank archives for helping me access records from the New Land Development Project. I also thank David Kneas, Alan Mikhail, and two anonymous reviewers for their insightful comments on drafts of this chapter.

Notes

1. While the term *land reclamation* (*istiṣlāḥ*) can refer to a range of activities, including draining marshlands or transforming saline soils, land reclamation in Egypt since the middle of the twentieth century has been primarily about cultivating the desert. In this sense, the term is something of a misnomer, since it is not a process of re-claiming something that has been lost but of creating something new.

2. J. Tony Allan, "Some Phases in Extending the Cultivated Area in the Nineteenth and Twentieth Centuries in Egypt," *Middle Eastern Studies* 19 (1983): 470–481; Jon Alterman, *Egypt and American Foreign Assistance 1952–1956: Hopes Dashed* (New York: Palgrave Macmillan, 2002), 63–95; Günter Meyer, "Economic Changes

in the Newly Reclaimed Lands: From State Farms to Small Holdings and Private Agricultural Enterprises," in *Directions of Change in Rural Egypt*, ed. Nicholas Hopkins and Kirsten Westergaard (Cairo: American University in Cairo Press, 1998), 334–357; Robert Springborg, "Patrimonialism and Policy Making in Egypt: Nasser and Sadat and the Tenure Policy for Reclaimed Lands," *Middle Eastern Studies* 15 (1979): 49–69; Sarah Voll, "Egyptian Land Reclamation Since the Revolution," *Middle East Journal* 34 (1980): 127–148; William Willcocks, *Egyptian Irrigation*, 2nd edn. (London: E. & F. N. Spon, 1899), 238–254.

3. One exception is Jeannie Sowers's analysis, which discusses the "ecological critiques" of land reclamation regarding whether or not there is actually a sufficient quantity and quality of water available from the Nile for agricultural expansion. Jeannie Sowers, "Remapping the Nation, Critiquing the State: Environmental Narratives and Desert Land Reclamation in Egypt," in *Environmental Imaginaries of the Middle East and North Africa*, ed. Diana K. Davis and Edmund Burke III (Athens: Ohio University Press, 2011), 158–191.

4. Irrigation is the only source of water for agriculture, since rainfall is so low throughout Egypt (with the exception of a small area along the Mediterranean Coast where rain-fed agriculture is possible). Reclamation also takes place in the Western Desert and Sinai Peninsula, drawing on deep groundwater, but this chapter focuses on reclamation based on Nile water.

5. The feddan is the unit of area measurement in Egypt. One feddan is equal to 0.42 hectare or 1.04 acres.

6. Robert Picciotto and H. Eberhard Köpp, *New Land Development Project, Egypt: Performance Audit Report*, Report No. 13275 (Washington, DC: World Bank, 1994), xvii–xviii.

7. All archival references are from the World Bank Group Archives (WBGA) in Washington, DC. The analysis draws from four boxes within the WB IBRD/IDA 01 Country Operation Files 1946–1998. For brevity, the fond is omitted and reference is made only to the box number (which is followed by the letter "B"). Three types of records are used: correspondence records (from files titled "Credit 1083, New Land Development Project"), monthly reports (from files titled "New Land Development Project-MR"), and quarterly reports (from files titled "New Land Development Project-QR"). These sources are referred to, respectively, as CR, MR, and QR with the relevant file number.

8. See, for instance, James C. Scott's analysis of the limited success of efforts to introduce a model of modern, scientific agriculture in the developing world. James C. Scott, *Seeing Like a State: How Certain Schemes to Improve the Human Condition Have Failed* (New Haven, CT: Yale University Press, 1998), 262–306. For an example of the ecological and social costs of attempts to manipulate rivers for irrigation and urban use, see Donald Worster, *Rivers of Empire: Water, Aridity, and the Growth of the American West* (New York: Pantheon Books, 1985).

9. On the significance of nonhuman actors in historical narratives, see, for example, Timothy Mitchell's analysis of the role of mosquitoes, fertilizers, and dams in mid-twentieth-century Egyptian development. Timothy Mitchell, *Rule of Experts: Egypt, Techno-Politics, Modernity* (Berkeley: University of California Press, 2002),

19–53. See also Nancy Reynolds's account in this volume of how the physical nature of stone influenced the construction of the Aswan High Dam.

10. On late seventeenth- and eighteenth-century land reclamation in Egypt, see Alan Mikhail, *Nature and Empire in Ottoman Egypt: An Environmental History* (Cambridge: Cambridge University Press, 2011), 66–71.

11. On nineteenth-century land reclamation in Egypt, see Allan, "Some Phases in Extending the Cultivated Area"; Gabriel Baer, *A History of Landownership in Modern Egypt, 1800–1950* (London: Oxford University Press, 1962); Kenneth M. Cuno, *The Pasha's Peasants: Land, Society, and Economy in Lower Egypt, 1740–1858* (Cambridge: Cambridge University Press, 1992); Willcocks, *Egyptian Irrigation,* 238–254.

12. Leon Hesser et al., *New Lands Productivity in Egypt: Technical and Economic Feasibility Study* (Washington, DC: Pacific Consultants, 1980), i.

13. Ibid., 23.

14. Jennifer Bremer, *New Lands Concept Paper II: Rethinking an AID Assistance Strategy for the New Lands of Egypt* (Cairo: Development Alternatives, Inc., 1981), 8.

15. Robert Springborg, *Mubarak's Egypt: Fragmentation of the Political Order* (Boulder, CO: Westview Press), 256.

16. Meyer, "Economic Changes in the Newly Reclaimed Lands," 334.

17. WBGA, 125456B, CR 901228, "Investment Possibilities in the 'New Lands,'" Memo Donovan to Files, December 22, 1977.

18. WBGA, 125456B, CR 901230, "New Land Development," Memo Maiss to Köpp, April 1, 1980.

19. WBGA, 125456B, CR 901231, "New Lands Project," Memo Naylor to Haynes, June 17, 1980.

20. The project also included a bilharzia control program, which this chapter does not discuss.

21. World Bank, *New Land Development Project: Project Completion Report*, Report No. 10631 (Washington, DC: World Bank, 1992), 4.

22. Ibid., 5.

23. Ibid., 6.

24. Ibid., 24.

25. Picciotto and Köpp, *Performance Audit Report*, cover memorandum.

26. WBGA, 72424B, 84541, "Feasibility Study of West Nubariya Extension Reclamation and Settlement Project—Final Report," Volume 3, Annexes 3 and 4, ULG Consultants Limited, August 1979, 19.

27. Ibid., 22.

28. Ibid., 12.

29. WBGA, 66892B, MR 1227169, "Progress Report for the Period 26 April–26 May 1980," Halcrow-ULG Ltd.

30. WBGA, 72424B, 84541, "Feasibility Study," Volume 3, 29.

31. WBGA, 125456B, CR 901232, Back-to-Office Report, Economides, September 16, 1980, 6.

32. WBGA, 72424B, 84541, "Feasibility Study," Volume 3, 5.

33. Ibid., 3.

34. WBGA, 125456B, CR 901230, "West Nubariya Extension: Reclamation and Settlement Project Feasibility Study," Memo ULG Consultants Ltd, October 1, 1979, 4.

35. Ibid.

36. Ibid.

37. WBGA, 125456B, CR 901231, Letter Naylor to Lloyd, June 12, 1980.

38. WBGA, 125456B, CR 901232, Back-to-Office Report, Economides, September 16, 1980, 1.

39. WBGA, 66892B, MR 1227169, "Progress Report for the Period January 1982," Halcrow-ULG Ltd, B2.

40. WBGA, 125456B, CR 901232, Back-to-Office Report, Economides, September 16, 1980, 1.

41. WBGA, 125456B, CR 901241, Letter Lloyd to van Tuijl, September 5, 1986.

42. WBGA, 125456B, CR 901231, Telex Naylor to Makhlouf, August 4, 1980.

43. WBGA, 125456B, CR 901232, Back-to-Office Report, Economides, January 12, 1981.

44. WBGA, 125456B, CR 901233, "NLDP: Declaration of Effectiveness," Memo Karaosmanoglu to Köpp, October 1, 1981.

45. WBGA, 125456B, CR 901233, Letter Picciotto to Samaha, December 22, 1981.

46. WBGA, 125456B, CR 901237, Supervision Report, Economides, January 10, 1984.

47. WBGA, 125456B, CR 901242, Project Execution Report, van Tuijl and Fuad, April 1987.

48. A number of recent works of environmental history have highlighted the often overlooked role of the laborers who build and maintain irrigation canals. Alan Mikhail, for example, provides a detailed analysis of how the organization of repair works on irrigation canals in Egypt changed between the late seventeenth and early nineteenth centuries, linking this to the emergence of new conceptions of population and society. Mikhail, *Nature and Empire in Ottoman Egypt*, 170–200. See also Julie Greene, *The Canal Builders: Making America's Empire at the Panama Canal* (New York: Penguin Press, 2009) for an interesting account of the multinational workforce that built the Panama Canal. In the case of the canal extension and repair works carried out for the New Land Development Project, however, labor does not seem to have been a key issue. Indeed there is no mention within the archival record of who actually implemented the work on the canals. Instead, it is clear from the records that the disagreement between bank and government officials hinged on the selection of the contractors, whom they ultimately held responsible for the quality of the work, rather than the laborers per se.

49. World Bank, *Project Completion Report*, 9.

50. WBGA, 125456B, CR 901233, "Meetings with Eng. Samaha, Minister of Irrigation and Sudan Affairs, November 16 and 17, 1981," Memo Swayze to Files, November 20, 1981, 2.

51. WBGA, 125456B, CR 901233, "Briefing Note," Memo Naylor to Picciotto, November 17, 1981, 2.

52. Nadia Farah, *Egypt's Political Economy: Power Relations in Development* (Cairo: American University in Cairo Press, 2009), 80.

53. World Bank, *Project Completion Report*, 20.

54. WBGA, 66892B, MR 1227169, "Progress Report for the Period November 1982," Halcrow-ULG Ltd, B19.

55. P. Economides et al., *New Land Development Project: Staff Appraisal Report*, Report No. 3010 (Washington, DC: World Bank, 1980), 7.

56. WBGA, 72424B, 84521, "Feasibility Study," Volume 1, Main Report and Summary, 34.

57. WBGA, 125456B, CR 901230, "New Lands Development Project—Issues Paper," Memo Economides et al. to Naylor, February 22, 1980.

58. P. Economides et al., *Staff Appraisal Report*, 7.

59. WBGA, 125456B, CR 901233, Supervision Report, Economides, May 28, 1982.

60. WBGA, 125456B, CR 901231, "NLDP Yellow Cover Review Comments," Memo Hotes to Baum, June 19, 1980.

61. Economides et al., *Staff Appraisal Report*, 17.

62. Declan Conway and Mike Hulme, "Recent Fluctuations in Precipitation and Runoff Over the Nile Sub-Basins and their Impact on Main Nile Discharge," *Climatic Change* 25 (1993): 127–151.

63. World Bank, *Selected Issues in Agriculture, Irrigation, and Land Reclamation*, Report No. 4013 (Washington, DC: World Bank, 1983), 19.

64. WBGA, 125456B, CR 901237, "5 Year Plan Investment Review in Land Reclamation," Memo Quicke to Ramasubbo, February 10, 1984.

65. WBGA, 66891B, QR 901225, "Quarterly Report for the Three Months Ending 31 July, 1987," Halcrow-ULG Ltd.

66. WBGA, 66891B, QR 901225, "Quarterly Report for the Period Ending 30 April, 1986," Halcrow-ULG Ltd.

67. WBGA, 66891B, QR 901225, "Progress Report 10," Halcrow-ULG Ltd, March 31, 1988, 9.

68. Picciotto and Köpp, *Performance Audit Report*, xvii.

69. Ministry of Agriculture and Land Reclamation, *Sustainable Agricultural Development Strategy Towards 2030* (Cairo: MALR Committee on Agricultural Research and Development, 2009).

Bibliography

Archival sources, interviews, websites, manuscripts, governmental reports, and other unpublished materials are not included in this bibliography.

'Abd al-Ghanī, Aḥmad Shalabī Ibn. *Awḍaḥ al-Ishārāt fī man Tawallā Miṣr al-Qāhira min al-Wuzarā' wa al-Bāshāt*, edited by 'Abd al-Raḥīm 'Abd al-Raḥman 'Abd al-Raḥīm. Cairo: Tawzī' Maktabat al-Khānjī, 1978.

'Abdülkâdir Efendi. *Topçular Kâtibi 'Abdülkâdir Efendi Tarihi*, edited by Ziya Yılmazer. Ankara: Türk Tarih Kurumu Yayınları, 2003.

Aberth, John. *The Black Death: The Great Mortality of 1348–1350, A Brief History with Documents*. New York: Palgrave Macmillan, 2005.

Abi-Mershed, Osama. *Apostles of Modernity: Saint-Simonians and the Civilizing Mission in Algeria*. Stanford, CA: Stanford University Press, 2010.

Abu al-Izz, M. S. *Landforms of Egypt*, translated by Y. A. Fayid. Cairo: American University in Cairo Press, 1971.

Abu-Lughod, Janet L. *Before European Hegemony: The World System A.D. 1250–1350*. New York: Oxford University Press, 1989.

Adams, Robert McC. *Land Behind Baghdad: A History of Settlement on the Diyala Plains*. Chicago: University of Chicago Press, 1965.

Adams, William M. *Green Development: Environment and Sustainability in a Developing World*. 3rd edn. London: Routledge, 2009.

Adas, Michael. *Dominance by Design: Technological Imperatives and America's Civilizing Mission*. Cambridge, MA: The Belknap Press of Harvard University Press, 2006.

Afyoncu, Erhan. "Türkiye'de Tahrir Defterlerine Dayalı Olarak Hazırlanmış Çalışmalar Hakkında Bazı Görüşler." *Türkiye Araştırmaları Literatür Dergisi* 1 (2003): 267–286.

Ágoston, Gábor. *Guns for the Sultan: Technology, Industry, and Military Power in the Ottoman Empire*. New York: Cambridge University Press, 2004.

Agrawal, Arun, and K. Sivaramakrishnan, eds. *Agrarian Environments: Resources, Representations, and Rule in India*. Durham, NC: Duke University Press, 2000.

Aḥmad, Laylā ʿAbd al-Laṭīf. *Al-Idāra fī Miṣr fī al-ʿAṣr al-ʿUthmānī*. Cairo: Maṭbaʿat Jāmiʿat ʿAyn Shams, 1978.

Ahmed, Hasan Bey-Zâde. *Hasan Bey-Zâde Târîhi*, edited by Şevki Nezihi. Ankara: Türk Tarihi Kurumu Yayınevi, 2004.

Akdağ, Mustafa. *Celâlî İsyanları (1550–1603)*. Ankara: Ankara Üniversitesi Basımevi, 1963.

Akdağ, Mustafa. "Celâli İsyanlarından Büyük Kaçgunluk." *Tarih Araştırmaları Dergisi* 2 (1964): 1–49.

Akdağ, Mustafa. *Türk Halkının Dirlik ve Düzenlik Kavgası*. Ankara: Bilgi Yayınevi, 1975.

Akdağ, Mustafa. *Türkiye'nin İktisadi ve İçtimai Tarihi*. Ankara: Türk Tarih Kurumu Basımevi, 1971.

Akkemik, Ünal, et al. "Tree-Ring Reconstructions of Precipitation and Streamflow for Northwestern Turkey." *International Journal of Climatology* 28 (2008): 173–183.

Alam, Muzaffar, and Sanjay Subrahmanyam. *Indo-Persian Travels in the Age of Discoveries, 1400–1800*. Cambridge: Cambridge University Press, 2007.

Alatout, Samer. "Narratives of Power: Territory, Population, and Environmental Narratives in Palestine and Israel." In *Palestinian and Israeli Environmental Narratives*, edited by Stuart Schoenfeld, 299–315. Toronto: The York Centre for International and Security Studies, 2005.

Alatout, Samer. "Towards a Bio-Territorial Conception of Power: Territory, Population, and Environmental Narratives in Palestine and Israel." *Political Geography* 25 (2006): 601–621.

Albert, Jeff, Magnus Bernhardsson, and Roger Kenna, eds. *Transformations of Middle Eastern Natural Environments: Legacies and Lessons*. Bulletin Series, no. 103. New Haven, CT: Yale School of Forestry and Environmental Sciences, 1998.

Alder, Garry. "The Origins of the 'Pusa experiment': The East India Company and Horse Breeding in Bengal, 1793–1808." *Bengal Past and Present* 98 (1979): 10–12.

Âlî, Gelibolu Mustafa. *Künhü'l-Ahbâr*, edited by Faris Çerçi. Kayseri: Erciyes Üniversitesi Yayınları, 2000.

ʿAli, Idris. *Dongola: A Novel of Nubia*, translated by Peter C. Theroux. Fayetteville: University of Arkansas Press, 1998.

ʿAli, Idris. *Poor*, translated by Elliott Colla. Cairo: American University in Cairo Press, 2007.

Ali, Siham. "La Revalorisation du SMIG au Maroc ne satisfait pas les salaries." *Maghrebia*. July 8, 2009.

Allan, J. A. *The Middle East Water Question: Hydropolitics and the Global Economy*. London: I. B. Tauris, 2002.

Allan, J. Tony. "Some Phases in Extending the Cultivated Area in the Nineteenth and Twentieth Centuries in Egypt." *Middle Eastern Studies* 19 (1983): 470–481.

Allsen, Thomas T. *The Royal Hunt in Eurasian History*. Philadelphia: University of Pennsylvania Press, 2006.

Alterman, Jon. *Egypt and American Foreign Assistance 1952–1956: Hopes Dashed.* New York: Palgrave Macmillan, 2002.

Ambraseys, N. N., and C. F. Finkel. *The Seismicity of Turkey and Adjacent Areas: A Historical Review, 1500–1800.* Istanbul: Eren, 1995.

Amery, Hussein A., and Aaron T. Wolf, eds. *Water in the Middle East: A Geography of Peace.* Austin: University of Texas Press, 2000.

Amery, H. A., and A. A. Kubursi. "The Litani River Basin: The Politics and Economics of Water." *The Beirut Review: A Journal on Lebanon and the Middle East* 1 (1992): 95–107.

Antes, John. *Observations on the Manners and Customs of the Egyptians, the Overflowing of the Nile and Its Effects; with Remarks on the Plague and Other Subjects. Written During a Residence of Twelve Years in Cairo and its Vicinity.* London: Printed for J. Stockdale, 1800.

Anthony, David. *The Horse, the Wheel and Language: How Bronze-Age Riders from the Eurasian Steppes Shaped the Modern World.* Princeton, NJ: Princeton University Press, 2007.

Anthony, Robert. *Like Froth Floating on the Sea: The World of Pirates and Seafarers in Late Imperial South China.* Berkeley, CA: Institute of East Asian Studies, 2003.

Arslan, Hüseyin. *16. yy. Osmanlı Toplumunda Yönetim, Nüfus, İskân, Göç ve Sürgün.* Istanbul: Kaknüs Yayınları, 2001.

Arzoumanian, Ara Alain. "Activist: Top Leaders Don't Care about Environment." *Daily Star.* June 21, 2003.

Ashtiyani, Mirza Mahmud Taqi. *'Ibratnama. Khatirati az Dawran-i Pas az Jangha-yi Herat va Marv [c. 1278–1288/1860–1870],* edited by Husayn 'Imadi Ashtiyani. Tehran: Nashr-i Markaz, 2003.

Ashtor, Eliyahu. "The Economic Decline of the Middle East During the Later Middle Ages: An Outline." *Asian and African Studies* 15 (1981): 253–286.

Asmar, Basel N. "The Science and Politics of the Dead Sea: Red Sea Canal or Pipeline." *Journal of Environment and Development* 12 (2003): 325–339.

Attallah, Pierre. *Al-Nifayat al-Sama fi Bilad al-Arz: al-Qissa al-Kamila.* Beirut: Dar al-Nahar, 1998.

Atwell, William S. "Volcanism and Short-Term Climatic Change in East Asian and World History c.1200–1699." *Journal of World History* 12 (2001): 29–98.

Baer, Gabriel. *A History of Landownership in Modern Egypt, 1800–1950.* London: Oxford University Press, 1962.

Baer, Jean. "Aperçu Historique de la Protection de la Nature." *Biological Conservation* 1 (1968): 7–12.

Bagis, Ali Ihsan. "Turkey's Hydropolitics of the Euphrates-Tigris Basin." *International Journal of Water Resources Development* 13 (1997): 567–582.

Bahadir Han, Ebülgâzî. *Histoire des Mogols et des Tatares,* translated by Le Baron Desmaisons. 2 vols. St. Pétersbourg: Imprimerie de l'Académie Impériale des sciences, 1871–1874.

Bahjat, Ahmad. "Wajhun fi'l-Ziham." *al-Ahram.* January 8, 1960.

al-Bakrī, Muḥammad Ibn Abī al-Surūr. *Al-Nuzha al-Zahiyya fi Dhikr Wulāt Miṣr wa al-Qāhira al-Mu'izziyya,* edited by 'Abd al-Rāziq 'Abd al-Rāziq 'Īsā. Cairo: al-'Arabī lil-Nashr wa al-Tawzī', 1998.

Ball, John. *A Description of the First or Aswan Cataract of the Nile*. Cairo: National Printing Department for the Ministry of Finance Survey Department, 1907.

Barfield, Thomas J. *The Perilous Frontier: Nomadic Empires and China*. Cambridge: Basil Blackwell, 1989.

Barkan, Ömer Lütfi. "Edirne Askeri Kassam'ına ait Tereke Defterleri (1545–1659)." *Belgeler* III/5–6 (1966): 1–479.

Barkan, Ömer Lütfi. "Essai sur les données statistiques des registres de recensement dans l'Empire ottoman aux XVᵉ et XVIᵉ siècles." *Journal of the Economic and Social History of the Orient* 1 (1958): 9–36.

Barkan, Ömer Lütfi. "İstanbul Saraylarına ait Muhasebe Defterleri." *Belgeler* IX/13 (1979): 1–380.

Barkan, Ömer Lütfi. "The Price Revolution of the Sixteenth Century: A Turning Point in the Economic History of the Near East." *International Journal of Middle East Studies* 6 (1975): 3–28.

Barkan, Ömer Lütfi. *Türkiye'de Toprak Meselesi*. Istanbul: Gözlem Yayınları, 1980.

Barkey, Karen. *Bandits and Bureaucrats: The Ottoman Route to State Centralization*. Ithaca, NY: Cornell University Press, 1994.

Bartol'd, V. V. *Mīr 'alī-Shīr: A History of the Turkman People*. Vol. 3 of *Four Studies on the History of Central Asia*, translated by V. and T. Minorsky. Leiden: Brill, 1962.

Başaran, Betül. "The 1829 Census and Istanbul's Population during the Late 18th and Early 19th Centuries." In *Studies on İstanbul and Beyond, The Freely Papers*, edited by Robert G. Ousterhout, 53–71. Vol. 1. Philadelphia: University of Pennsylvania Press, 2007.

Bayly, C. A. *Empire and Information: Intelligence Gathering and Social Communication in India, 1780–1870*. Cambridge: Cambridge University Press, 1996.

Beaumont, Peter. "Water Resource Development in Iran." *The Geographical Journal* 140 (1974): 418–431.

Beckwith, Christopher I. *Empires of the Silk Road: A History of Central Eurasia from the Bronze Age to the Present*. Princeton, NJ: Princeton University Press, 2009.

Beinart, William, and Lotte Hughes. *Environment and Empire*. In *The Oxford History of the British Empire Companion Series*, edited by Wm. Roger Louis. Oxford: Oxford University Press, 2007.

Beinin, Joel. *Workers and Peasants in the Modern Middle East*. Cambridge: Cambridge University Press, 2001.

Belon du Mans, Pierre. *Voyage au Levant (1553): Les Observations de Pierre Belon du Mans*, edited and introduced by Alexandra Merle. Paris: Chandeigne, 2001.

Benkheira, Mohamed Hocine, Catherine Mayeur-Jaouen, and Jacqueline Sublet. *L'animal en islam*. Paris: Indes savantes, 2005.

Beydoun, Ahmad. "Lebanon's Sects and the Difficult Road to a Unifying Identity." *The Beirut Review: A Journal on Lebanon and the Middle East* 6 (1993): 16.

Bilgin, Arif. *Osmanlı Saray Mutfağı*. Istanbul: Kitabevi, 2004.

Binark, Ismet, et al., eds. *Bolu, Kastamonu, Kengırı ve Koca-ili Livâları: Dizin ve Tıpkıbasım*. Vol. 2 of *438 Numaralı Muhâsebe-i Vilâyet-i Anadolu Defteri (937/1530)*. Ankara: Başbakanlık Devlet Arşivleri Genel Müdürlüğü, 1994.

Birks, J. S., and C. A. Sinclair. "The Domestic Political Economy of Development in Saudi Arabia." In *State, Society and Economy in Saudi Arabia*, edited by Timothy Niblock, 198–213. New York: St. Martin's Press, 1982.

Bishop, Elizabeth. "Talking Shop: Egyptian Engineers and Soviet Specialists at the Aswan High Dam." Ph.D. diss., University of Chicago, 1997.

Bistami, Muhammad Tahir. *Futuhat-i Fariduniyah: Sharh-i Jang'ha-yi Faridun Khan Charkas Amir al-Umara-yi Shah 'Abbas-i Avval*, edited by Mir Muhammad Sadiq and Muhammad Nadir Nasiri Muqaddam. Tehran: Nuqtah, 2001.

Blackbourn, David. *The Conquest of Nature: Water, Landscape, and the Making of Modern Germany*. New York: W. W. Norton, 2006.

Blackmar, Elizabeth. "Contemplating the Force of Nature." *Radical Historians Newsletter* 70 (1994): 4.

Blench, Roger. *"You Can't Go Home Again." Extensive Pastoral Livestock Systems: Issues and Options for the Future*. London: Overseas Development Institute, 2000.

Bloom, Jonathan. *Paper Before Print: The History and Impact of Paper in the Islamic World*. New Haven, CT: Yale University Press, 2001.

Bloxam, Elizabeth. "QuarryScapes Report: The Assessment of Significance of Ancient Quarry Landscapes—Problems and Possible Solutions. The Case of the Aswan West Bank." No. 5. Trondheim: QuarryScapes Project, 2007.

Bonneval, M. le Général de. *Cahiers du centenaire de l'Algérie*. Vol. 7 of *L'Algérie touristique*, edited by Comité National Métropolitain du Centenaire de l'Algérie. 12 vols. Orléans: Imprimerie A. Pigelet et Cie., 1930.

Borsch, Stuart J. *The Black Death in Egypt and England: A Comparative Study*. Austin: University of Texas Press, 2005.

Bostan, İdris. *Osmanlı Bahriye Teşkilâtı: XVII. Yüzyılda Tersâne-i Âmire*. Ankara: Türk Tarih Kurumu Basımevi, 1992.

Boudy, Paul. *Économie forestière nord-africaine: Description forestière de l'Algérie et de la Tunisie*. Vol. 4. Paris: Editions Larose, 1955.

Boudy, Paul. *Économie forestière nord-africaine: Milieu physique et milieu humain*. Vol. 1. Paris: Éditions Larose, 1948.

Boujrouf, Said, Mireille Bruston, Philippe Duhamel, et al. "Les conditions de la mise en tourisme de la haute montagne et ses effets sur le territoire." *Revue de Géographie Alpine* 86 (1998): 67–82.

Bournoutian, George A., trans. *The History of Vardapet Aṛak'el of Tabriz*. Costa Mesa, CA: Mazda Publishers, 2005.

Bousquet, Bernard. *Guide Des Parcs Nationaux d'Afrique: Afrique du Nord, Afrique de l'Ouest*. Paris: Delachaux et Niestlé, 1992.

Boutros-Ghali, Boutros. *Egypt's Road to Jerusalem: A Diplomat's Story of the Struggle for Peace in the Middle East*. New York: Random House, 1997.

Braudel, Fernand. *The Mediterranean and the Mediterranean World in the Age of Philip II*, translated by Siân Reynolds. 2 vols. Berkeley: University of California Press, 1995.

Brázdil, Rudolf, et al. "Historical Climatology in Europe—the State of the Art." *Climatic Change* 70 (2005): 363–430.

Bregel, Yuri. "Uzbeks, Qazaqs, and Turkmen." In *The Cambridge History of Inner Asia: The Chinggisid Age*, edited by Nicola Di Cosmo, Allen J. Frank, and Peter B. Golden, 221–236. Cambridge: Cambridge University Press, 2009.

Bremer, Jennifer. *New Lands Concept Paper II: Rethinking an AID Assistance Strategy for the New Lands of Egypt*. Cairo: Development Alternatives, Inc., 1981.

Brice, William C., ed. *The Environmental History of the Near and Middle East Since the Last Ice Age*. London: Academic Press, 1978.

Brookfield, Michael. "The Desertification of the Egyptian Sahara during the Holocene (the Last 10,000 Years) and Its Influence on the Rise of Egyptian Civilization." In *Landscapes and Societies: Selected Cases*, edited by I. Peter Martini and Ward Chesworth, 91–108. Dordrecht: Springer, 2010.

Brower, Benjamin Claude. *A Desert Named Peace: The Violence of France's Empire in the Algerian Sahara, 1844–1902*. New York: Columbia University Press, 2009.

Bukhari, Mir 'Abd al-Karim. *Histoire de l'Asie Centrale (Afghanistan, Boukhara, Khiva, Khoqand) depuis les dernières années de règne de Nadir Chah (1153)*, edited by Charles Schefer. Paris: E. Leroux, 1876.

Bulliet, Richard W. *The Camel and the Wheel*. Cambridge, MA: Harvard University Press, 1975.

Bulliet, Richard W. *The Case for Islamo-Christian Civilization*. New York: Columbia University Press, 2004.

Bulliet, Richard W. *Conversion to Islam in the Medieval Period: An Essay in Quantitative History*. Cambridge, MA: Harvard University Press, 1979.

Bulliet, Richard W. *Cotton, Climate, and Camels in Early Islamic Iran: A Moment in World History*. New York: Columbia University Press, 2009.

Bulliet, Richard W. *Hunters, Herders, and Hamburgers: The Past and Future of Human-Animal Relationships*. New York: Columbia University Press, 2005.

Bulliet, Richard W. *Islam: The View from the Edge*. New York: Columbia University Press, 1994.

Bulliet, Richard W. *Kicked to Death by a Camel*. New York: Harper and Row, 1973.

Bulliet, Richard W. *The One-Donkey Solution: A Satire*. Bloomington: iUniverse, 2011.

Bulliet, Richard W. *The Patricians of Nishapur: A Study in Medieval Islamic Social History*. Cambridge, MA: Harvard University Press, 1972.

Bulliet, Richard W., Pamela Kyle Crossley, Daniel R. Headrick, Steven W. Hirsch, Lyman L. Johnson, and David Northrup. *The Earth and Its Peoples: A Global History*. 5th edn. Boston: Wadsworth, 2010.

Burke, Edmund, III. "The Big Story: Human History, Energy Regimes, and the Environment." In *The Environment and World History*, edited by Edmund Burke III and Kenneth Pomeranz, 33–53. Berkeley: University of California Press, 2009.

Burke, Edmund, III. "The Transformation of the Middle Eastern Environment, 1500 B.C.E.–2000 C.E." In *The Environment and World History*, edited by Edmund Burke III and Kenneth Pomeranz, 81–117. Berkeley: University of California Press, 2009.

Burnes, Alexander. *Travels into Bokhara; Being the Account of a Journey from India to Cabool, Tartary, and Persia*. 3 vols. London: J. Murray, 1834.

Burns, G., T. C. Billard, and K. M. Matsui. "Salinity Threat to Upper Egypt." *Nature* 344 (March 1, 1990): 25.

Butzer, Karl W. *Early Hydraulic Civilization in Egypt: A Study in Cultural Ecology.* Chicago: University of Chicago Press, 1976.

Carle, Georges, and Jean Gattefossé. "Réserves naturelles et parc chérifiens." *Revue Scientifique Illustrée* 17 (1933): 622–628.

Carpentier, Elisabeth. "Autour de la Peste Noire: famines et épidémies dans l'histoire du XIVe siècle." *Annales* 17 (1962): 1062–1092.

Carruthers, Jane. *The Kruger National Park: A Social and Political History.* Pietermaritzburg: University of Natal Press, 1995.

Cartier, E. "A National Park in the Belgian Congo." *Science* 16 (1925): 623–624.

Chamieh, Naji, and Jihad Issa. "Policy Paper on the Environment in Lebanon." Beirut: Lebanese Center for Policy Studies, 1996.

Champion, Daryl. *The Paradoxical Kingdom: Saudi Arabia and the Momentum of Reform.* New York: Columbia University Press, 2003.

Chase, Kenneth. *Firearms: A Global History to 1700.* Cambridge: Cambridge University Press, 2003.

Chatty, Dawn. "Enclosures and Exclusions: Conserving Wildlife in Pastoral Areas of the Middle East." *Anthropology Today* 14 (1998): 2–7.

Chaudhry, Kiren Aziz. *The Price of Wealth: Economies and Institutions in the Middle East.* Ithaca, NY: Cornell University Press, 1997.

Christensen, Peter. *The Decline of Iranshahr: Irrigation and Environments in the History of the Middle East, 500 B.C. to A.D. 1500.* Copenhagen: Museum Tusculanum Press, 1993.

Christian, David. "Silk Roads or Steppe Roads? The Silk Roads in World History." *Journal of World History* 11 (2000): 1–26.

Cioc, Mark. *The Game of Conservation: International Treaties to Protect the World's Migratory Animals.* Athens: Ohio University Press, 2009.

Citino, Nathan J. *From Arab Nationalism to OPEC: Eisenhower, King Sa'ūd, and the Making of U.S.–Saudi Relations.* Bloomington: Indiana University Press, 2002.

Clancy-Smith, Julia A. *Mediterraneans: North Africa and Europe in an Age of Migration, c. 1800–1900.* Berkeley: University of California Press, 2011.

Clapp, Gordon R. "Iran: A TVA for the Khuzestan Region." *Middle East Journal* 11 (1957): 1–11.

Clarence-Smith, William Gervase, and Steven Topik. *The Global Coffee Economy in Africa, Asia, and Latin America, 1500–1989.* Cambridge: Cambridge University Press, 2003.

Cohen, Shaul Ephraim. *The Politics of Planting: Israeli-Palestinian Competition for Control of Land in the Jerusalem Periphery.* Chicago: University of Chicago Press, 1993.

Collins, Robert O. *The Nile.* New Haven, CT: Yale University Press, 2002.

A Common Trust: The Preservation of the Ancient Monuments of Nubia. Paris: UNESCO, 1960.

Conant, Louis C. "Review: Geology of North Africa." *Science.* New Series, 140 (April 5, 1963): 41.

Conrad, Lawrence I. "Arabic Plague Chronologies and Treatises: Social and Historical Factors in the Formation of a Literary Genre." *Studia Islamica* 54 (1981): 51–93.

Conrad, Lawrence I. "The Biblical Tradition for the Plague of the Philistines." *Journal of the American Oriental Society* 104 (1984): 281–287.

Conrad, Lawrence I. "The Plague in the Early Medieval Near East." Ph.D. diss.,
    Princeton University, 1981.

Conway, Declan, and Mike Hulme. "Recent Fluctuations in Precipitation and Runoff
    Over the Nile Sub-Basins and Their Impact on Main Nile Discharge." *Climatic
    Change* 25 (1993): 127–151.

Cook, Michael. *Population Pressure in Rural Anatolia, 1450–1600.* New York: Oxford
    University Press, 1972.

Cooper, David E., and Simon P. James. *Buddhism, Virtue and Environment.* Aldershot,
    UK: Ashgate, 2005.

Copponi, Niccolò. *Victory of the West: The Great Christian-Muslim Clash at the Battle of
    Lepanto.* Cambridge: Da Capo Press, 2007.

Cordova, Carlos E. *Millennial Landscape Change in Jordan: Geoarchaeology and Cultural
    Ecology.* Tucson: University of Arizona Press, 2007.

Di Cosmo, Nicola. *Ancient China and Its Enemies: The Rise of Nomadic Power in East
    Asian History.* Cambridge: Cambridge University Press, 2002.

Di Cosmo, Nicola, Allen J. Frank, and Peter B. Golden. "Introduction." In *The
    Cambridge History of Inner Asia: The Chinggisid Age,* edited by Nicola Di Cosmo,
    Allen J. Frank, and Peter B. Golden, 1–6. Cambridge: Cambridge University
    Press, 2009.

Couliboeuf de Blocqueville, Henri de. "Quatorze mois de captivite, chez les Turcomans
    aux frontieres du Turkestan et de la Perse, 1860–1861 (Frontières du Turkestan et
    de la Perse)." In *Le Tour du Monde,* 225–272. Paris: Librairie de L. Hachette, 1866.

Cronon, William. *Changes in the Land: Indians, Colonists, and the Ecology of New
    England.* 1st rev. edn. New York: Hill and Wang, 2003.

Cronon, William. "A Place for Stories: Nature, History, and Narrative." *Journal of
    American History* 78 (1992): 1347–1376.

Crosby, Alfred W. *Ecological Imperialism: The Biological Expansion of Europe, 900–1900.*
    Cambridge: Cambridge University Press, 2004.

Crossley, Pamela Kyle. *Orphan Warriors: Three Manchu Generations and the End of the
    Qing World.* Princeton, NJ: Princeton University Press, 1990.

Cullen, Karen J. *Famine in Scotland: The 'Ill Years' of the 1690s.* Edinburgh: Edinburgh
    University Press, 2010.

Cuno, Kenneth M. "Commercial Relations between Town and Village in Eighteenth-
    and Early Nineteenth-Century Egypt." *Annales Islamologiques* 24 (1988): 111–135.

Cuno, Kenneth M. *The Pasha's Peasants: Land, Society, and Economy in Lower Egypt,
    1740–1858.* Cambridge: Cambridge University Press, 1992.

Çarkoğlu, Ali, and Mine Eder. "Development *alla Turca*: The Southeastern Anatolia
    Development Project (GAP)." In *Environmentalism in Turkey: Between Democracy
    and Development?* edited by Fikret Adaman and Murat Arsel, 167–184. Aldershot,
    UK: Ashgate, 2005.

Çarkoğlu, Ali, and Mine Eder. "Domestic Concerns and the Water Conflict over the
    Euphrates-Tigris River Basin." *Middle Eastern Studies* 37 (2001): 41–71.

Dale, Stephen Frederic. *Indian Merchants and Eurasian Trade, 1600–1750.* Cambridge:
    Cambridge University Press, 1994.

al-Damurdāshī Katkhudā 'Azabān, Aḥmad. *Kitāb al-Durra al-Muṣāna fī Akhbār al-Kināna*, edited by 'Abd al-Raḥīm 'Abd al-Raḥman 'Abd al-Raḥīm. Cairo: Institut français d'archéologie orientale, 1989.

Darling, Linda T. *Revenue-Raising and Legitimacy: Tax Collection and Finance Administration in the Ottoman Empire, 1560–1660*. Leiden: Brill, 1996.

Davis, Diana K. "Environmentalism as Social Control? An Exploration of the Transformation of Pastoral Nomadic Societies in French Colonial North Africa." *The Arab World Geographer* 3 (2000): 182–198.

Davis, Diana K. "Potential Forests: Degradation Narratives, Science, and Environmental Policy in Protectorate Morocco, 1912–1956." *Environmental History* 10 (2005): 211–238.

Davis, Diana K. "Power, Knowledge, and Environmental History in the Middle East and North Africa." *International Journal of Middle East Studies* 42 (2010): 657–659.

Davis, Diana K. *Resurrecting the Granary of Rome: Environmental History and French Colonial Expansion in North Africa*. Athens: Ohio University Press, 2007.

Davis, Diana K., and Edmund Burke III, eds. *Environmental Imaginaries of the Middle East and North Africa*. Athens: Ohio University Press, 2011.

Davis, Diana K., and Denys Frappier. "The Social Context of Working Equines in the Urban Middle East: The Example of Fez Medina." In *The Walled Arab City in Literature, Architecture and History: The Living Medina in the Maghrib*, edited by Susan Slyomovics, 51–68. London: Frank Cass Publishers, 2001.

Davis, Mike. *Late Victorian Holocausts: El Niño Famines and the Making of the Third World*. London: Verso, 2001.

Decker, Michael. "Plants and Progress: Rethinking the Islamic Agricultural Revolution." *Journal of World History* 20 (2009): 187–206.

Deeb, Lara. *An Enchanted Modern: Gender and Public Piety in Shi'i Lebanon*. Princeton, NJ: Princeton University Press, 2006.

Delaye, Théophile. "Le Parc National du Toubkal." *Revue de Géographie Marocaine* 28 (1944): 3–14.

Desroches-Noblecourt, Christiane, and Georg Gerster. *The World Saves Abu Simbel*. Vienna: Verlag, 1968.

Deveciyan, Karekin. *Türkiye'de Balık ve Balıkçılık*, translated by Erol Üyepazarcı. Istanbul: Aras Yayınları, 2011.

Dewald, Jonathan, Geoffrey Parker, Michael Marmé, and J. B. Shank. "AHR Forum: The General Crisis of the Seventeenth Century Revisited." *American Historical Review* 113 (2008): 1029–1099.

Dictionary of American Family Names, s.v. "Tahan."

Digby, Simon. *War-Horse and Elephant in the Delhi Sultanate: A Study of Military Supplies*. Oxford: Orient Monographs, 1971.

Dimishquieh, Nadim. *Mahatat fi Hayati al-Diblomassiyah: Dhikriatun fi al-Siyassa wa al-'Alaqat al Dawaliyya*. Beirut: Dar al-Nahar, 1995.

Djoundourian, Salpie. "Environmental Movement in Lebanon." *Environment, Development and Sustainability* 11 (2009): 427–438.

Dols, Michael W. *The Black Death in the Middle East*. Princeton, NJ: Princeton University Press, 1977.

Dols, Michael W. "The General Mortality of the Black Death in the Mamluk Empire." In *The Islamic Middle East, 700–1900: Studies in Social and Economic History*, edited by Abraham L. Udovitch, 397–428. Princeton, NJ: Darwin Press, 1981.

Dols, Michael W. "Ibn al-Wardī's *Risālah al-Naba' 'an al-Waba'*, A Translation of a Major Source for the History of the Black Death in the Middle East." In *Near Eastern Numismatics, Iconography, Epigraphy and History: Studies in Honor of George C. Miles*, edited by Dickran K. Kouymjian, 443–455. Beirut: American University of Beirut, 1974.

Dols, Michael W. "al-Manbijī's 'Report of the Plague:' A Treatise on the Plague of 764–765/1362–1364 in the Middle East." In *The Black Death: The Impact of the Fourteenth-Century Plague*, edited by Daniel Williman, 65–75. Papers of the Eleventh Annual Conference of the Center for Medieval and Early Renaissance Studies. Binghamton, NY: Center for Medieval and Early Renaissance Studies, 1982.

Dols, Michael W. "Plague in Early Islamic History." *Journal of the American Oriental Society* 94 (1974): 371–383.

Dols, Michael W. "The Second Plague Pandemic and Its Recurrences in the Middle East: 1347–1894." *Journal of the Economic and Social History of the Orient* 22 (1979): 162–189.

Doumani, Beshara. *Rediscovering Palestine: Merchants and Peasants in Jabal Nablus, 1700–1900*. Berkeley: University of California Press, 1995.

Dukes, J. S. "Burning Buried Sunshine: Human Consumption of Ancient Solar Energy." *Climatic Change* 61 (2003): 31–44.

Dursun, Selçuk. "Forest and the State: History of Forestry and Forest in the Ottoman Empire." Ph.D. diss., Sabancı University, 2007.

Eaton, Richard M. "Islamic History as Global History." In *Islamic and European Expansion: The Forging of a Global Order*, edited by Michael Adas, 1–36. Philadelphia: Temple University Press, 1993.

Eckersley, Robyn. *Environmentalism and Political Theory: Toward an Ecocentric Approach*. Albany: State University of New York Press, 1992.

Economides, P., et al. *New Land Development Project: Staff Appraisal Report*. Report No. 3010. Washington, DC: World Bank, 1980.

Elliott, Mark C. *The Manchu Way: The Eight Banners and Ethnic Identity in Late Imperial China*. Stanford, CA: Stanford University Press, 2001.

Elmusa, Sharif S., ed. *Culture and the Natural Environment: Ancient and Modern Middle Eastern Texts*. Vol. 26, no. 1 of Cairo Papers in Social Science. Cairo: American University in Cairo Press, 2003.

Elver, Hilal. *Peaceful Uses of International Rivers: The Case of the Euphrates and Tigris Rivers Dispute*. Ardsley, NY: Transnational, 2002.

Elver, Hilal. "Turkey's Rivers of Dispute." *Middle East Report* 254 (2010): 14–18.

Elvin, Mark. *The Pattern of the Chinese Past*. Stanford, CA: Stanford University Press, 1973.

Elvin, Mark. *The Retreat of the Elephants: An Environmental History of China*. New Haven, CT: Yale University Press, 2004.

Elvin, Mark, and Liu Ts'ui-jung, eds. *Sediments of Time: Environment and Society in Chinese History*. Cambridge: Cambridge University Press, 1998.

The Encyclopedia of Religion and Nature. London: Thoemmes Continuum, 2005.

Erder, Leila, and Suraiya Faroqhi. "Population Rise and Fall in Anatolia 1550–1620." *Middle East Studies* 15 (1979): 322–345.

Erler, Mehmet Yavuz. *Osmanlı Devleti'nde Kuraklık ve Kıtlık Olayları, 1800–1880*. Istanbul: Libra Kitap, 2010.

Estes, J. Worth, and LaVerne Kuhnke. "French Observations of Disease and Drug Use in Late Eighteenth-Century Cairo." *Journal of the History of Medicine and Allied Sciences* 39 (1984): 121–152.

Evenden, Matthew D. *Fish versus Power: An Environmental History of the Fraser River*. Cambridge: Cambridge University Press, 2004.

Evliya Çelebi bin Derviş Mehemmed Zılli. *Evliya Çelebi Seyahatnâmesi, Topkapı Sarayı Bağdat 304 Yazmasının Transkripsyonu—Dizini*, edited by Robert Dankoff, Seyit Ali Kahraman, and Yücel Dağlı. Istanbul: Yapı Kredi Yayınları, 2006.

Fagan, Brian. *The Little Ice Age*. New York: Basic Books, 2000.

al-Fahad, Abdulaziz H. "The 'Imama vs. the 'Iqal: Hadari-Bedouin Conflict and the Formation of the Saudi State." In *Counter-Narratives: History, Contemporary Society, and Politics in Saudi Arabia and Yemen*, edited by Madawi al-Rasheed and Robert Vitalis, 35–75. New York: Palgrave Macmillan, 2004.

Fahim, Hussein M. *Dams, People and Development: The Aswan High Dam Case*. New York: Pergamon Press, 1981.

Fahim, Hussein M. *Egyptian Nubians: Resettlement and Years of Coping*. Salt Lake City: University of Utah Press, 1983.

Farah, Nadia. *Egypt's Political Economy: Power Relations in Development*. Cairo: American University in Cairo Press, 2009.

Faramarzi, 'Ali Sultani Gird, ed. *Du Farasnama-yi Manthur va Manzum*. Tehran: University of Tehran, 1987.

Farnie, D. A. *East and West of Suez: The Suez Canal in History, 1854–1956*. Oxford: Clarendon Press, 1969.

Faroqhi, Suraiya, ed. *Animals and People in the Ottoman Empire*. Istanbul: Eren, 2010.

Faroqhi, Suraiya. *Approaching Ottoman History: An Introduction to the Sources*. Cambridge: Cambridge University Press, 1999.

Faroqhi, Suraiya. *Artisans of Empire: Crafts and Craftspeople under the Ottomans*. London: I. B. Tauris, 2009.

Faroqhi, Suraiya. "Camels, Wagons, and the Ottoman State in the Sixteenth and Seventeenth Centuries." *International Journal of Middle East Studies* 14 (1982): 523–539.

Faroqhi, Suraiya. "Introduction." In *Animals and People in the Ottoman Empire*, edited by Suraiya Faroqhi, 11–54. Istanbul: Eren, 2010.

Faroqhi, Suraiya. *Kultur und Alltag im Osmanischen Reich: Vom Mittelalter bis zum Anfang des 20. Jahrhunderts*. Munich: Verlag C. H. Beck, 1995.

Faroqhi, Suraiya, ed. *The Later Ottoman Empire, 1603–1839*. Vol. 3 of *The Cambridge History of Turkey*. Cambridge: Cambridge University Press, 2006.

Faroqhi, Suraiya. *Men of Modest Substance: House Owners and House Property in Seventeenth-Century Ankara and Kayseri*. New York: Cambridge University Press, 1987.

Faroqhi, Suraiya. *The Ottoman Empire and the World Around It*. London: I. B. Tauris, 2004.

Faroqhi, Suraiya. "Ottoman Peasants and Rural Life: The Historiography of the Twentieth Century." *Archivum Ottomanicum* 18 (2000): 153–182.

Faroqhi, Suraiya. "The Peasants of Saideli in the Late Sixteenth Century." *Archivum Ottomanicum* 8 (1983): 215–250.

Faroqhi, Suraiya. *Towns and Townsmen of Ottoman Anatolia: Trade, Crafts, and Food Production in an Urban Setting, 1520–1650*. New York: Cambridge University Press, 1984.

Faroqhi, Suraiya, and Huri İslamoğlu. "Crop Patterns and Agricultural Production Trends in Sixteenth-Century Anatolia." *Review* 2 (1979): 401–436.

Fawal, Ibrahim. *Youssef Chahine*. London: British Film Institute, 2001.

Fawaz, Leila Tarazi. *An Occasion for War: Civil Conflict in Lebanon and Damascus in 1860*. Berkeley: University of California Press, 1994.

Fawaz, Mona. "Neoliberal Urbanity and the Right to the City: A View from Beirut's Periphery." *Development and Change* 40 (2009): 827–852.

Fernea, Elizabeth Warnock, and Robert A. Fernea, with Aleya Rouchdy. *Nubian Ethnographies*. Prospect Heights, IL: Waveland Press, 1991.

Fernea, Robert A. *Nubians in Egypt: Peaceful People*. Austin: University of Texas Press, 1973.

Fernea, Robert A., and Aleya Rouchdy. "Contemporary Egyptian Nubians." In *Nubian Ethnographies*, edited by Elizabeth Warnock Fernea and Robert A. Fernea, with Aleya Rouchdy, 183–202. Prospect Heights, IL: Waveland Press, 1991.

Ferrier, J. P. *Caravan Journeys and Wanderings in Persia, Afghanistan, Turkistan, and Beloochistan*. London: J. Murray, 1856.

Finkel, Caroline. *The Administration of Warfare: The Ottoman Military Campaigns in Hungary 1593–1606*. Vienna: VWGÖ, 1988.

Flinn, Michael. *Scottish Population History from the Seventeenth Century to the 1930s*. Cambridge: Cambridge University Press, 1977.

Foltz, Richard. *Animals in Islamic Tradition and Muslim Cultures*. Oxford: Oneworld, 2006.

Foltz, Richard., ed. *Environmentalism in the Muslim World*. New York: Nova Science Publishers, 2005.

Foltz, Richard. "Is There an Islamic Environmentalism?" *Environmental Ethics* 22 (2000): 63–72.

Foltz, Richard C., Frederick M. Denny, and Azizan Baharuddin, eds. *Islam and Ecology: A Bestowed Trust*. Cambridge, MA: Harvard University Press, 2003.

Ford, Caroline. "Reforestation, Landscape Conservation, and the Anxieties of Empire in French Colonial Algeria." *American Historical Review* 113 (2008): 341–362.

Franklin, James L. *Pompeii: The "Casa del Marinaio" and Its History.* Rome: "L'Erma" di Bretschneider, 1990.

Fraser, James Baillie. *Narrative of a Journey into Khorasān, in the Years 1821 and 1822.* London: Longman, Hurst, Rees, Orme, Brown, and Green, 1825.

Friedman, Thomas L. *The World Is Flat 3.0: A Brief History of the Twenty-First Century.* New York: Picador, 2007.

Gadgil, Madhav, and Ramachandra Guha. *This Fissured Land: An Ecological History of India.* Berkeley: University of California Press, 1993.

Garnier, Emmanuel. *Les dérangements du temps: 500 ans de chaud et de froid en Europe.* Paris: Plon 2010.

Gause, F. Gregory, III. *The International Relations of the Persian Gulf.* New York: Cambridge University Press, 2010.

Gerbault, Pascale, et al. "Evolution of Lactase Persistence: An Example of Human Niche Construction." *Philosophical Transactions of the Royal Society B* 366 (2011): 863–877.

Gerber, Haim. *Economy and Society in an Ottoman City: Bursa 1600–1700.* Jerusalem: Hebrew University of Jerusalem, 1988.

Gies, Frances and Joseph. *Cathedral, Forge, and Waterwheel: Technology and Invention in the Middle Ages.* New York: HarperCollins, 1994.

Glacken, Clarence J. *Traces on the Rhodian Shore: Nature and Culture in Western Thought from Ancient Times to the End of the Eighteenth Century.* Berkeley: University of California Press, 1967.

Glaser, Rüdiger. *Klimageschichte Mitteleuropas: 1200 Jahre Wetter, Klima, Katastrophen.* Darmstadt: Primus, 2008.

Glick, Thomas F. *Irrigation and Hydraulic Technology: Medieval Spain and Its Legacy.* Brookfield, VT: Variorum, 1996.

Glick, Thomas F. *Irrigation and Society in Medieval Valencia.* Cambridge, MA: Harvard University Press, 1970.

Goldsmith, Edward, and Nicholas Hildyard. *The Social and Environmental Effects of Large Dams.* San Francisco: Sierra Club Books, 1984.

Goldstone, Jack A. *Revolution and Rebellion in the Early Modern World.* Berkeley: University of California Press, 1991.

Gommans, Jos. "The Horse Trade in Eighteenth-Century South Asia." *Journal of the Economic and Social History of the Orient* 37 (1994): 228–250.

Gommans, Jos. *Mughal Warfare: Indian Frontiers and Highroads to Empire.* New York: Routledge, 2002.

Gommans, Jos. *The Rise of the Indo-Afghan Empire, c. 1710–1780.* Leiden: Brill, 1995.

Gommans, Jos. "Warhorse and Post-Nomadic Empire in Asia, c. 1000–1800." *Journal of Global History* 2 (2007): 1–21.

Gordon, Joel. *Revolutionary Melodrama: Popular Film and Civic Identity in Nasser's Egypt.* Chicago: Middle East Documentation Center, 2002.

Gould, Andrew Gordon. "Pashas and Brigands: Ottoman Provincial Reform and Its Impact on the Nomadic Tribes of Southern Anatolia, 1840–1885." Ph.D. diss., University of California, Los Angeles, 1973.

Gradeva, Rossitsa. "Ottoman Policy towards Christian Church Buildings." In *Rumeli under the Ottomans, 15th–18th Centuries: Institutions and Communities*, edited by Rossitsa Gradeva, 339–368. Istanbul: Isis, 2004.

Green, Nile. *Bombay Islam: The Religious Economy of the West Indian Ocean, 1840–1915*. Cambridge: Cambridge University Press, 2011.

Greene, Julie. *The Canal Builders: Making America's Empire at the Panama Canal*. New York: Penguin Press, 2009.

Greenwood, Antony. "Istanbul's Meat Provisioning: A Study of the *Celepkeşan* System." Ph.D. diss., University of Chicago, 1988.

Griswold, William. "Climatic Change: A Possible Factor in the Social Unrest of Seventeenth Century Anatolia." In *Humanist and Scholar: Essays in Honor of Andreas Tietze*, edited by Heath W. Lowry and Donald Quataert, 37–57. Istanbul: Isis Press, 1993.

Griswold, William. *The Great Anatolian Rebellion, 1000–1020/1591–1611*. Berlin: Klaus Schwarz Verlag, 1983.

Grousset, René. *The Empire of the Steppes: A History of Central Asia*, translated by Naomi Walford. New Brunswick, NJ: Rutgers University Press, 1970.

Grove, A. T., and Oliver Rackham. *The Nature of Mediterranean Europe: An Ecological History*. New Haven, CT: Yale University Press, 2001.

Grove, Jean M., and Annalisa Conterio. "Climate in the Eastern and Central Mediterranean, 1675 to 1715." In *Climatic Trends and Anomalies in Europe 1675–1715: High Resolution Spatio-Temporal Reconstructions from Direct Meteorological Observations and Proxy Data, Methods and Results*, edited by Burkhard Frenzel, Christian Pfister, and Birgit Gläser, 275–286. New York: G. Fischer, 1994.

Grove, Richard H., Vinita Damodaran, and Satpal Sangwan, eds. *Nature and the Orient: The Environmental History of South and Southeast Asia*. Delhi: Oxford University Press, 1998.

Guha, Ramachandra. *Environmentalism: A Global History*. New York: Longman, 2000.

Guha, Ramachandra. *The Unquiet Woods: Ecological Change and Peasant Resistance in the Himalaya*. Berkeley: University of California Press, 1990.

Guha, Ramachandra, and J. Martinez-Alier. *Varieties of Environmentalism: Essays North and South*. London: Earthscan Publications, 1997.

Guli, Amin. *Tarikh-i Siyasi va Ijtima'i-i Turkman'ha*. Tehran: Nashr-i 'Ilm, 1987.

Gunn, Joel D., ed. *The Years without Summer: Tracing AD 536 and Its Aftermath*. Oxford: Archaeopress, 2000.

Güçer, Lütfi. *Osmanlı İmparatorluğunda Hububat Meselesi ve Hububattan Alınan Vergiler*. Istanbul: İstanbul Üniversitesi İktisat Fakültesi, 1964.

Gümüşçü, Osman. *Tarihî Coğrafya Açısından Bir Araştırma: XVI. Yüzyıl Larende (Karaman) Kazasında Yerleşme ve Nüfus*. Ankara: Türk Tarihi Kurumu Basımevi, 2001.

Habib, John S. *Ibn Sa'ud's Warriors of Islam: The Ikhwan of Najd and Their Role in the Creation of the Sa'udi Kingdom, 1910–1930*. Leiden: Brill, 1978.

Halawi, Majed. *A Lebanon Defied: Musa al-Sadr and the Shi'a Community*. Boulder, CO: Westview Press, 1992.

Haleem, Harifyah Abdel, ed. *Islam and the Environment*. London: Ta-Ha Publishers, 1998.

Hall, Charles, et al. "Hydrocarbons and the Evolution of Human Culture." *Nature* 426 (2003): 318–322.

Hämäläinen, Pekka. *The Comanche Empire.* New Haven, CT: Yale University Press, 2008.

Hambly, Gavin R. G. "Iran during the Reigns of Fath 'Ali Shah and Muhammad Shah." In *From Nadir Shah to the Islamic Republic.* Vol. 7 of *The Cambridge History of Iran,* edited by Peter Avery, Gavin Hambly, and Charles Melville, 144–173. Cambridge: Cambridge University Press, 1991.

Hamdan, Fouad. "Toxic Attack Against Lebanon, Case One: Toxic Waste from Italy, A Chronology." Beirut: Greenpeace, 1997.

Hamzeh, A. Nizar. "Clientalism, Lebanon: Roots and Trends." *Middle Eastern Studies* 37 (2001): 167–178.

Hanioğlu, M. Şükrü. *A Brief History of the Late Ottoman Empire.* Princeton, NJ: Princeton University Press, 2008.

Hanjra, Sadaquat H., Bakht B. Khan, A. R. Barque, M. Tufail, and M. Aftab Khan. "Comparative Efficiency of Draught Animals." *Journal of Animal Sciences* 11 (1980): 79–84.

Harris, Leila. "Postcolonialism, Postdevelopment, and Ambivalent Spaces of Difference in Southeastern Turkey." *Geoforum* 39 (2008): 1698–1708.

Harris, Leila. "Water and Conflict Geographies of the Southeastern Anatolia Project." *Society and Natural Resources* 15 (2002): 743–759.

Harroy, Jean-Paul. "L'Union Internationale pour la Conservation de la Nature et de ses Ressources: Origine et Constitution." *Biological Conservation* 1 (1969): 106–110.

Hartline, Beverly Karplus. "Irrigation Threatens Egyptian Temples." *Science* 209 (August 15, 1980): 796.

Hasbini, Tala. "Environmental NGOs as Mediators in the Policy-Making Process: Exploring the Establishment of Nature Reserves in Lebanon." M.A. thesis, American University of Beirut, 2009.

"Hashimi," Ibn Sayyid Abu'l Husain. *The Faras-Nama of Hashimi,* edited by D. C. Phillott. Calcutta: Asiatic Society, 1910.

Hassan, Fekri. "Historical Nile Floods and Their Implications for Climatic Change." *Science* 212 (1981): 1142–1145.

al-Hassan, Hamza. *Al-Shi'a fi al-Mamlaka al-'Arabiyya al-Sa'udiyya.* Vol. 2. Mu'assasat al-Baqi li-Ihya' al-Turath, 1993.

Hattox, Ralph. *Coffee and Coffeehouses: The Origins of a Social Beverage in the Medieval Near East.* Seattle: University of Washington Press, 1985.

Hayward, G. S. W. "Route from Jellalabad to Yarkand Through Chitral, Badakhshan, and Pamir Steppe, Given by Mahomed Amin of Yarkand." *Proceedings of the Royal Geographical Society of London* 13 (1868–1869): 122–130.

Heldal, Tom, and Per Storemyr. "The Quarries at the Aswan West Bank." In *Quarryscapes Report: Characterisation of Complex Quarry Landscapes, an Example from the West Bank Quarries, Aswan,* edited by Elizabeth Bloxam, Tom Heldal, and Per Storemyr, 69–140. No. 4. Trondheim: QuarryScapes Project, 2007.

Herodotus. *The Persian War.* New York: Modern Library, 1942.

Hesser, Leon, et al. *New Lands Productivity in Egypt: Technical and Economic Feasibility Study.* Washington, DC: Pacific Consultants, 1980.

Heston, Alan, H. Hasnain, S. Z. Hussain, and R. N. Khan. "The Economics of Camel Transportation in Pakistan." *Economic Development and Cultural Change* 34 (1985): 121–141.

Hidayat, Riza Quli Khan. *Rawzat al-Safa-yi Nasiri*. Vol. 9. Edited by Jamshid Kiyanfar. Tehran: Asatir, 2001.

Hidayat, Riza Quli Khan. *Sifaratnama-yi Khvarazm*, edited by Charles Schefer. Paris: Ernest Leroux, 1876.

Hijazi, S. S., A. Abulaban, Z. Ammarin, and G. Flatz. "Distribution of Adult Lactase Phenotypes in Bedouins and in Urban and Agricultural Populations of Jordan." *Tropical & Geographical Medicine* 35 (1983): 157–161.

Hirst, L. Fabian. *The Conquest of Plague: A Study of the Evolution of Epidemiology*. Oxford: Clarendon Press, 1953.

Hobsbawm, E. J. "The Crisis of the 17th Century—II." *Past and Present* 6 (1954): 44–64.

Hobsbawm, E. J. "The General Crisis of the European Economy in the 17th Century." *Past and Present* 5 (1954): 33–53.

Hodge, Adam R. "Pestilence and Power: The Smallpox Epidemic of 1780–1782 and Intertribal Relations on the Northern Great Plains." *The Historian* 72 (2010): 543–567.

Hoffman, Richard C., Nancy Langston, James C. McCann, Peter C. Perdue, and Lise Sedrez. "AHR Conversation: Environmental Historians and Environmental Crisis." *American Historical Review* 113 (2008): 1431–1465.

Holden, Clare, and Ruth Mace. "Phylogenetic Analysis of the Evolution of Lactose Digestion in Adults." *Human Biology* 81 (2009): 597–617.

Holden, Stacy E. "Famine's Fortune: The Pre-Colonial Mechanization of Moroccan Flour Production." *Journal of North African Studies* 15 (2010): 71–84.

Holt, Richard. *The Mills of Medieval England*. Oxford: Blackwell, 1988.

Homewood, Katherine M. *Livestock Economy and Ecology in El Kala, Algeria: Evaluating Ecological and Economic Costs and Benefits in Pastoralist Systems*. Pastoral Development Network Paper, no. 35. London: Overseas Development Institute, 1993.

Hopwood, Derek. *Egypt, Politics and Society, 1945–1981*. London: Allen and Unwin, 1982.

Hourani, Albert. "Lebanon from Feudalism to Modern State." *Middle Eastern Studies* 2 (1966): 256–263.

Hourani, Albert. "Ottoman Reform and the Politics of the Notables." In *The Modern Middle East: A Reader*, edited by Albert Hourani, Philip S. Khoury, and Mary C. Wilson, 83–109. Berkeley: University of California Press, 1994.

Howard, David. "Ottoman Historiography and the Literature of 'Decline' in the Sixteenth and Seventeenth Centuries." *Journal of Asian History* 22 (1988): 52–76.

Hughes, J. Donald. *Ecology in Ancient Civilizations*. Albuquerque: University of New Mexico Press, 1975.

Hussein, Sayed, et al. *Study of New Land Allocation Policy in Egypt*. Report No. 65 of the Agricultural Policy Reform Program. Cairo: Ministry of Agriculture and Land Reclamation, 1999.

Hütteroth, Wolf-Dieter. "Ecology of the Ottoman Lands." In *The Cambridge History of Turkey, Volume 3: The Later Ottoman Empire, 1603–1839*, edited by Suraiya N. Faroqhi, 18–43. Cambridge: Cambridge University Press, 2006.

Hütteroth, Wolf-Dieter. *Laendliche Siedlungen im südlichen Inneranatolien in den Letzen vierhundert Jahren*. Göttingen: Göttingen Universität Geographischen Institut, 1968.

Hütteroth, Wolf-Dieter, and Kamal Abdulfattah. *Historical Geography of Palestine, Transjordan and Southern Syria in the Late Sixteenth Century*. Erlangen: Fränkische Geographische Ges., 1977.

Ibn al-Ḥājj Ibrāhīm Ṭābiʻ al-Marḥūm Ḥasan Aghā ʻAzabān al-Damurdāshī, Muṣṭafā. *Tārīkh Waqāʼiʻ Miṣr al-Qāhira al-Maḥrūsa Kinānat Allah fī Arḍihi*, edited by Ṣalāḥ Aḥmad Harīdī ʻAlī. 2nd edn. Cairo: Dār al-Kutub wa al-Wathāʼiq al-Qawmiyya, 2002.

Ibn Khaldun. *The Muqaddimah: An Introduction to History*, translated by Franz Rosenthal. Princeton, NJ: Princeton University Press, 1967.

Ibrāhīm, Nāṣir Aḥmad. *Al-Azamāt al-Ijtimāʻiyya fī Miṣr fī al-Qarn al-Sābiʻ ʻAshar*. Cairo: Dār al-Āfāq al-ʻArabiyya, 1998.

Imber, Colin. *The Ottoman Empire, 1350–1650: The Structure of Power*. New York: Palgrave Macmillan, 2002.

Imber, Colin. "The Reconstruction of the Ottoman Fleet After the Battle of Lepanto, 1571–1572." In *Studies in Ottoman History and Law*, edited by Colin Imber, 85–101. Istanbul: Isis, 1996.

Issar, Arie S. *Water Shall Flow from the Rock: Hydrogeology and Climate in the Lands of the Bible*. Berlin: Springer-Verlag, 1990.

Issar, Arie S., and Mattanyah Zohar. *Climate Change—Environment and Civilization in the Middle East*. Berlin: Springer, 2004.

Issawi, Bahay. "Nubia Sandstone: A Discussion." *American Association of Petroleum Geologists Bulletin* 55 (June 1971): 885–887.

Iʻtimad al-Saltana, Muhammad Hasan Khan. *Mirʼat al-Buldan*, edited by ʻAbdul Husayn Navaʼi. Tehran: University of Tehran, 1988.

Izzat-Allah, Sayyid. *Travels in Central Asia by Meer Izzut-Oollah in the Years 1812–1813*, translated by Captain Henderson. Calcutta: Foreign Department Press, 1872.

Izzi Dien, Mawil. *The Environmental Dimensions of Islam*. Cambridge: Lutterworth Press, 2000.

Izzi Dien, Mawil. "Islam and the Environment: Theory and Practice." *Journal of Beliefs and Values* 18 (1997): 47–57.

İnalcık, Halil. "Istanbul." *Encyclopaedia of Islam*. 2nd edn. Leiden: Brill, 2006.

İnalcık, Halil. "The Origins of the Ottoman-Russian Rivalry and the Don-Volga Canal 1569." *Les annales de l'Université d'Ankara* 1 (1946–1947): 47–106.

İnalcık, Halil. *The Ottoman Empire: The Classical Age*, translated by Norman Itzkowitz and Colin Imber. New York: Praeger Publishers, 1973.

İnalcık, Halil. "The Socio-Political Effects of the Diffusion of Firearms in the Middle East." In *War, Technology and Society in the Middle East*, edited by V. J. Parry and M. E. Yapp, 195–217. New York: Oxford University Press, 1975.

İnalcık, Halil, with Donald Quataert. *An Economic and Social History of the Ottoman Empire.* Vol. 1. New York: Cambridge University Press, 1994.

İslamoğlu-İnan, Huri. *State and Peasant in the Ottoman Empire: Agrarian Power Relations and Regional Economic Development in Ottoman Anatolia during the Sixteenth Century.* Leiden: Brill, 1994.

İz, Fahir. "XVII. Yüzyılda Halk Dili ile Yazılmış bir Tarih Kitabı: Hüseyin Tuği 'Vak'a-i Sultan Osman Han.'" *Türk Dili Araştırmaları Yıllığı* (1967): 119–155.

al-Jabartī, 'Abd al-Raḥman. *'Ajā'ib al-Āthār fī al-Tarājim wa al-Akhbār,* edited by Ḥasan Muḥammad Jawhar, 'Abd al-Fattāḥ al-Saranjāwī, 'Umar al-Dasūqī, and al-Sayyid Ibrāhīm Sālim. 7 vols. Cairo: Lajnat al-Bayān al-'Arabī, 1958–1967.

Jacobsen, Thorkild. *Salinity and Irrigation Agriculture in Antiquity: Diyala Basin Archaeological Report on Essential Results, 1957–58.* Bibliotheca Mesopotamica. Vol. 14. Malibu: Undena Publications, 1982.

Jacobsen, Thorkild, and Robert M. Adams. "Salt and Silt in Ancient Mesopotamian Agriculture." *Science* 128 (1958): 1251–1258.

Jacoby, Karl. *Crimes against Nature: Squatters, Poachers, Thieves, and the Hidden History of American Conservation.* Berkeley: University of California Press, 2001.

Johnson, Douglas H. "Tribe or Nationality? The Sudanese Diaspora and the Kenyan Nubis." *Journal of Eastern African Studies* 3 (2009): 112–131.

Johnson, Pamela, et al. *Egypt: The Egyptian American Rural Improvement Service, a Point Four Project, 1952–63.* AID Project Impact Evaluation No. 43. Washington, DC: USAID, 1983.

Jones, Toby Craig. *Desert Kingdom: How Oil and Water Forged Modern Saudi Arabia.* Cambridge, MA: Harvard University Press, 2010.

Julien, Charles-André. *Histoire de l'Algérie contemporaine: La conquete et les débuts de la colonisation (1827–1871).* Paris: Presses Universitaires de France, 1964.

Jung, Abdallah Khan Firoze. *A Treatise on Horses, Entitled Saloter, or, A Complete System of Indian Farriery,* translated by Joseph Earles. Calcutta: George Gordon, 1788.

Jutikkala, Eino. "The Great Finnish Famine in 1696–1697." *Scandinavian Economic History Review* 3 (1955): 48–63.

Kafadar, Cemal. "The Question of Ottoman Decline." *Harvard Middle Eastern and Islamic Review* 4 (1997–98): 30–75.

Kaplan, Steven Laurence. *Provisioning Paris: Merchants and Millers in the Grain and Flour Trade During the Eighteenth Century.* Ithaca, NY: Cornell University Press, 1984.

Karam, Karam. "Revendiquer, Mobiliser, Participer: Les Associations Civiles dans le Liban de l'Apres-Guerre." Ph.D. thesis, Aix Marseille III, 2004.

Karnouk, Liliane. *Contemporary Egyptian Art.* Cairo: American University in Cairo Press, 1995.

Kasaba, Reşat. *A Moveable Empire: Ottoman Nomads, Migrants, and Refugees.* Seattle: University of Washington Press, 2009.

Kassler, P. "The Structural and Geomorphic Evolution of the Persian Gulf." In *The Persian Gulf: Holocene Carbonate Sedimentation and Diagenesis in a Shallow Epicontinental Sea,* edited by B. H. Purser, 11–32. Berlin: Springer-Verlag, 1973.

Kazemi, Farhad, and John Waterbury, eds. *Peasants and Politics in the Modern Middle East*. Miami: Florida International University Press, 1991.

Kelekna, Pita. *The Horse in Human History*. Cambridge: Cambridge University Press, 2009.

Khalaf, Samir. "Changing Forms of Political Patronage in Lebanon." In *Patrons and Clients in Mediterranean Societies*, edited by Ernest Gellner and John Waterbury, 185–205. London: Duckworth, 1977.

Khalaf, Samir. *Persistence and Change in 19th Century Lebanon: A Sociological Essay*. Beirut: American University of Beirut, 1979.

Khalid, Fazlun M., with Joanne O'Brien, eds. *Islam and Ecology*. New York: Cassell, 1992.

Khan, Ibrahim. "Route of Ibrahim Khan from Kashmir through Yassin to Yarkand in 1870." *Proceedings of the Royal Geographical Society of London* 15 (1870–71): 387–392.

Khan, Muhammad Fazil. *Tarikh-i Manazili Bukhara*, translated by Iqtidar Husain Siddiqui. Srinagar: Centre of Central Asian Studies for the University of Kashmir, 1981.

al-Khashshāb, Ismāʿīl Ibn Saʿd. *Akhbār Ahl al-Qarn al-Thānī ʿAshar: Tārīkh al-Mamālīk fī al-Qāhira*, edited by ʿAbd al-ʿAzīz Jamāl al-Dīn and ʿImād Abū Ghāzī. Cairo: al-ʿArabī lil-Nashr wa al-Tawzīʿ, 1990.

al-Khashshāb, Ismāʿīl Ibn Saʿd. *Khulāṣat mā Yurād min Akhbār al-Amīr Murād*, edited and translated by Hamza ʿAbd al-ʿAzīz Badr and Daniel Crecelius. Cairo: al-ʿArabī lil-Nashr wa al-Tawzīʿ, 1992.

el-Khazen, Farid. *The Breakdown of the State in Lebanon: 1967–1976*. Cambridge, MA: Harvard University Press, 2000.

el-Khazen, Farid. "Lebanon's Communal Elite-Mass Politics: The Institutionalization of Disintegration." *The Beirut Review: A Journal on Lebanon and the Middle East* 3 (1992): 61.

Khazeni, Arash. *Tribes and Empire on the Margins of Nineteenth-Century Iran*. Seattle: University of Washington Press, 2010.

Kılıç, Orhan. "Osmanlı Devleti'nde Meydana Gelen Kıtlıklar." *Türkler* 10 (2002): 718–730.

King, Brian. "Conservation Geographies in Sub-Saharan Africa: The Politics of National Parks, Community Conservation and Peace Parks." *Geography Compass* 4 (2009): 14–27.

Kingston, Paul. "Patron, Clients, and Civil Society in Lebanon: A Case Study of Environmental Politics in Post-War Lebanon." *Arab Studies Quarterly* 23 (2001): 55–72.

Kiwan, Fadia. "The Formation of Lebanese Civil Society." *The Beirut Review: A Journal on Lebanon and the Middle East* 6 (1993): 72–73.

Klemm, Rosemarie, and Dietrich D. Klemm. *Stones and Quarries in Ancient Egypt*. London: British Museum Press, 2008.

Klitzsch, E., and Heinz Schandelmeier. "South Western Desert." In *The Geology of Egypt*, edited by Rushdi Said, 249–258. Rotterdam: A. A. Balkema, 1990.

Kolars, John F., and William A. Mitchell. *The Euphrates River and the Southeast Anatolia Development Project*. Carbondale: Southern Illinois University Press, 1991.

Konchine, M. "La Question de l'Oxus." *Annales de Géographie* 5 (1895–96): 496–504.

Kömürcüyan, Eremya Çelebi. *İstanbul Tarihi: XVII. Asırda İstanbul*, edited by Hrand D. Andreasyan, with additions by Kevork Pamukciyan. 2nd edn. Istanbul: Eren, 1988.

Krane, Jim. *City of Gold: Dubai and the Dream of Capitalism*. New York: St. Martin's Press, 2009.

Kuhnke, LaVerne. *Lives at Risk: Public Health in Nineteenth-Century Egypt*. Berkeley: University of California Press, 1990.

Kumar, Deepak, Vinita Damodaran, and Rohan D'Souza, eds. *The British Empire and the Natural World: Environmental Encounters in South Asia*. Oxford: Oxford University Press, 2011.

Kunniholm, Peter. "Archaeological Evidence and Non-Evidence for Climate Change." *Philosophical Transactions of the Royal Society of London* 330 (1990): 645–655.

Kunt, Metin. "State and Sultan up to the Age of Süleyman: Frontier Principality to World Empire." In *Süleyman the Magnificent and His Age: The Ottoman Empire in the Early Modern World*, edited by Metin Kunt and Christine Woodhead, 3–19. London: Longman, 1995.

Kupferschmidt, Hugo. *Die Epidemiologie der Pest: Der Konzeptwandel in der Erforschung der Infektionsketten seit der Entdeckung des Pesterregers im Jahre 1894*. Aarau: Sauerländer, 1993.

Kurat, A. N. "The Turkish Expedition to Astrakhan and the Problem of the Don-Volga Canal." *Slavonic and East European Review* 40 (1961): 7–23.

Kuru, Mehmet. "The Relations between Ottoman Corsairs and the Imperial Navy in the 16th Century." Ph.D. diss., Sabancı University, 2009.

Kütükoğlu, Mübahat. *Osmanlılarda Narh Müessesesi ve 1640 Tarihli Narh Defteri*. Istanbul: Enderun Kitabevi, 1983.

Lal, Mohan. *Travels in the Panjab, Afghanistan, Turkistan, to Balk, Bokhara, and Herat*. London: W. H. Allen, 1846.

Lamb, H. F., U. Eichner, and V. R. Switsur. "An 18,000-Year Record of Vegetation, Lake-level and Climate Change from Tigalmamine, Middle Atlas, Morocco." *Journal of Biogeography* 1 (1989): 65–74.

Lamb, H. H. "Volcanic Dust in the Atmosphere; with a Chronology and Assessment of Its Meteorological Significance." *Philosophical Transactions of the Royal Society of London (Series A., Mathematical and Physical Sciences)* 266 (1970): 425–533.

Lane, Edward William. *Description of Egypt: Notes and Views in Egypt and Nubia, Made during the Years 1825, 26, 27, and 28*, edited and introduced by Jason Thompson. Cairo: American University in Cairo Press, 2000.

Lane, Kris. *Pillaging the Empire: Piracy in the Americas, 1500–1750*. Armonk, NY: M. E. Sharpe, 1998.

Langdon, John. "The Economics of Horses and Oxen in Medieval England." *Agricultural History Review* 30 (1982): 31–40.

Lapidus, Ira M. "The Grain Economy of Mamluk Egypt." *Journal of the Economic and Social History of the Orient* 12 (1969): 1–15.

Larèrre, Raphael, Bernadette Lizet, and Martine Berlan-Dargué. *Histoire des parcs nationaux: Comment prendre soin de la nature?* Paris: Éditions Quai, 2009.

Lavauden, Louis. "La Tunisie et les Reserves Naturelles." In *Contribution à l'Étude des Réserves Naturelles et des Parcs Nationaux*, edited by Société de Biogéographie, 139–150. Paris: Paul Lechevalier, 1937.

Le Roy Ladurie, Emmanuel. *Histoire humaine et comparée du climat.* 3 vols. Paris: Fayard, 2004–2009.

Le Strange, G. *The Lands of the Eastern Caliphate: Mesopotamia, Persia, and Central Asia from the Moslem Conquest to the Time of Timur.* Cambridge: University Press, 1905.

Létolle, René. "Histoire de l'Ouzboi, cours fossil de l'Amou Darya: synthese et elements nouveaux." *Studia Iranica* 29 (2002): 195–240.

Létolle, René, Philip Micklin, Nikolay Aladin, and Igor Plotnikov. "Uzboy and the Aral Regressions: A Hydrological Approach." *Quaternary International* 173–174 (2007): 125–136.

Levi, Scott C. *The Indian Diaspora in Central Asia and Its Trade, 1550–1900.* Leiden: Brill, 2002.

Lindner, Rudi. "Nomadism, Horses, and Huns." *Past and Present* 92 (1981): 3–20.

Lindner, Rudi. *Nomads and Ottomans in Medieval Anatolia.* Bloomington: Indiana University Press, 1983.

Little, Tom. *High Dam at Aswan: The Subjugation of the Nile.* New York: John Day, 1965.

Louis, Wm. Roger, and Roger Owen, eds. *Suez 1956: The Crisis and Its Consequences.* Oxford: Clarendon Press, 1989.

Lowi, Miriam R. *Water and Power: The Politics of a Scarce Resource in the Jordan River Basin.* Cambridge: Cambridge University Press, 1993.

Lucas, Adam Robert. "Industrial Milling in the Ancient and Medieval Worlds: A Survey of the Evidence for an Industrial Revolution in Medieval Europe." *Technology and Culture* 46 (2005): 1–30.

Ludwig, Emil. *The Nile: The Life-Story of a River*, translated by Mary H. Lindsay. New York: Viking Press, 1937.

Luterbacher, J., et al. "The Late Maunder Minimum—A Key Period for Studying Decadal Climate Change in Europe." *Climatic Change* 49 (2001): 441–462.

Luterbacher, J., et al. "Mediterranean Climate Variability over the Last Centuries: A Review." In *The Mediterranean Climate: An Overview of the Main Characteristics and Issues*, edited by P. Lionello, P. Malanotte-Rizzoli, and R. Boscolo, 27–148. Amsterdam: Elsevier, 2006.

Lydon, Ghislaine. *On Trans-Saharan Trails: Islamic Law, Trade Networks, and Cross-Cultural Exchange in Nineteenth-Century Western Africa.* Cambridge: Cambridge University Press, 2009.

Lydon, Ghislaine. "Writing Trans-Saharan History: Methods, Sources and Interpretations Across the African Divide." *Journal of North African Studies* 10 (2005): 293–324.

Ma'alouf, Habib. *'Al al-Haffa: Madkhal ila al-Falsafa al-Bi'aya.* Beirut: Markaz al-Thaqafi al-'Arabi, 2002.

MacGregor, Charles Metcalfe. *Narrative of a Journey Through the Province of Khorassan and on the N.W. Frontier of Afghanistan in 1875.* 2 vols. London: W. H. Allen, 1879.

Maciejewski, E., et al. "Morocco: Selected Issues." Washington, DC: International Monetary fund, 1997.

Magny, M., B. Vanniere, G. Zanchetta, and E. Fouache. "Possible Complexity of the Climatic Event around 4200–4000 cal. BP in the Central and Western Mediterranean." *The Holocene* 19 (2009): 823–833.

Mahdavi, 'Abd al-Husayn, ed. *Ganjina-yi Baharistan: Majmu'a-yi 6 Farasnama*. Tehran: Majlis-i Shuray-i Islami, 2008.

Makdisi, Karim. "Towards a Human Rights Approach to Water in Lebanon: Implementation beyond 'Reform.'" *International Journal of Water Resources Development* 23 (2007): 369–390..

Makdisi, Karim. "Trapped Between Sovereignty and Globalization: Implementing International Environmental and Natural Resource Treaties in Developing Countries, The Case of Lebanon." Ph.D. diss., Tufts University, 2001.

Makdisi, Ussama. *The Culture of Sectarianism: Community, History, and Violence in Nineteenth-Century Ottoman Lebanon*. Berkeley: University of California Press, 2000.

Makdisi, Ussama. "Reconstructing the Nation-State: The Modernity of Secularism in Lebanon." *Middle East Report* 200 (1996): 23–26 and 30.

Malcolm, John. *The History of Persia, from the Most Early Period to the Present Time*. 2 vols. London: J. Murray, 1815.

Mallat, Hyam. *Le droit de l'urbanisme, de la construction, de l'environnement et de l'eau au Liban*. Paris: Delta, 1997.

Mallat, Hyam. "Water Laws in Lebanon." In *Water in the Middle East: Legal, Political, and Commercial Implications*, edited by J. A. Allan and Chibli Mallat, with Shai Wade and Jonathan Wild, 151–176. London: I. B. Tauris, 1995.

Mantran, Robert. *Istanbul dans la seconde moitié du xvi'e siècle*. Paris: Maisonneuve, 1962.

Marc, Henri. *Notes sur les forêts de l'Algérie, Collection du centenaire de l'Algérie, 1830–1930*. Paris: Librairie Larose, 1930.

Marcus, Abraham. *The Middle East on the Eve of Modernity: Aleppo in the Eighteenth Century*. New York: Columbia University Press, 1989.

Markovits, Claude, Jacques Pouchepadass, and Sanjay Subrahmanyam, eds. *Society and Circulation: Mobile People and Itinerant Cultures in South Asia, 1750–1950*. Delhi: Permanent Black, 2003.

Marlowe, John. *World Ditch: The Making of the Suez Canal*. New York: Macmillan, 1964.

Marvin, Charles. *Merv, the Queen of the World; and the Scourge of the Man-Stealing Turcomans*. London: W. H. Allen, 1881.

Masri, Basheer Ahmad. *Animal Welfare in Islam*. Markfield: The Islamic Foundation, 2007.

Masri, Rania. "Environmental Challenges in Lebanon." *Journal of Developing Societies* 13 (1997): 73–115.

Mauelshagen, Franz. *Klimageschichte der Neuzeit, 1500–1900*. Darmstadt: Wissenschaftliche Buchgesellschaft, 2010.

McCann, James C. *Maize and Grace: Africa's Encounter with a New World Crop, 1500–2000*. Cambridge, MA: Harvard University Press, 2007.

McNeill, J. R. *Atlantic Empires of France and Spain: Louisbourg and Havana, 1700–1763*. Chapel Hill: University of North Carolina Press, 1985.

McNeill, J. R. "Chinese Environmental History in World Perspective." In *Sediments of Time: Environment and Society in Chinese History*, edited by Mark Elvin and Liu Ts'ui-jung, 31–52. Cambridge: Cambridge University Press, 1998.

McNeill, J. R. "Ecology and Strategy in the Mediterranean." In *Naval Policy and Strategy in the Mediterranean: Past, Present, and Future*, edited by John Hattendorf, 374–392. London: Frank Cass, 2000.

McNeill, J. R. "The First Hundred Thousand Years." In *The Turning Points of Environmental History*, edited by Frank Uekoetter, 13–28. Pittsburgh: University of Pittsburgh Press, 2010.

McNeill, J. R. *Mosquito Empires: Ecology and War in the Greater Caribbean, 1620–1914*. New York: Cambridge University Press, 2010.

McNeill, J. R. *The Mountains of the Mediterranean World: An Environmental History*. Cambridge: Cambridge University Press, 1992.

McNeill, J. R. "Observations on the Nature and Culture of Environmental History." *History and Theory* 42 (2003): 5–43.

McNeill, J. R. *Something New under the Sun: An Environmental History of the Twentieth-Century World*. New York: W. W. Norton, 2000.

McNeill, J. R., and William H. McNeill. *The Human Web: A Bird's-Eye View of World History*. New York: W. W. Norton, 2003.

McNeill, William H. *Plagues and Peoples*. Garden City, NY: Anchor Press/Doubleday, 1976.

Mehmed Ağa, Silahdar Fındıklı. *Silahdar Tarihi*. Vol. 2. Istanbul: Devlet Matbaası, 1928.

Meiggs, Russell. *Trees and Timber in the Ancient Mediterranean World*. Oxford: Clarendon Press, 1982.

Melhaoui, Mohammed. *Peste, contagion et martyre: Histoire du fléau en Occident musulman médiéval*. Paris: Publisud, 2005.

Merchant, Carolyn. "Shades of Darkness: Race and Environmental History." *Environmental History* 8 (2003): 380–394.

Meyer, Günter. "Economic Changes in the Newly Reclaimed Lands: From State Farms to Small Holdings and Private Agricultural Enterprises." In *Directions of Change in Rural Egypt*, edited by Nicholas Hopkins and Kirsten Westergaard, 334–357. Cairo: American University in Cairo Press, 1998.

Meyerhof, Max. "La Peste en Égypte à la fin du XVIII siècle et le Mèdecin Enrico di Wolmar." *La Revue Médicale d'Égypte* 1 (1913): 1–13.

Mikhail, Alan. "Animals as Property in Early Modern Ottoman Egypt." *Journal of the Economic and Social History of the Orient* 53 (2010): 621–652.

Mikhail, Alan. "Global Implications of the Middle Eastern Environment." *History Compass* 9 (2011): 952–970.

Mikhail, Alan. "An Irrigated Empire: The View from Ottoman Fayyum." *International Journal of Middle East Studies* 42 (2010): 569–590.

Mikhail, Alan. *Nature and Empire in Ottoman Egypt: An Environmental History*. New York: Cambridge University Press, 2011.

Mikhail, Alan. "The Nature of Plague in Late Eighteenth-Century Egypt." *Bulletin of the History of Medicine* 82 (2008): 249–275.

Ministry of Agriculture and Land Reclamation. *Sustainable Agricultural Development Strategy Towards 2030*. Cairo: MALR Committee on Agricultural Research and Development, 2009.

Ministry of Agriculture and Water. *Seven Green Spikes*. Riyadh: Ministry of Agriculture and Water, 1974.

Mirpanja, Sarhang Isma'il. *Khatirat-i Asarat: Ruznama-yi Safar-i Khvarazm va Khiva [1280/1862]*, edited by Safa al-Din Tabarrayan. Tehran: Mu'assasa-yi Pajhuhish va Mutala'at-i Farhangi, 1991.

Mitchell, Timothy. "Are Environmental Imaginaries Culturally Constructed?" In *Environmental Imaginaries of the Middle East and North Africa*, edited by Diana K. Davis and Edmund Burke III, 265–273. Athens: Ohio University Press, 2011.

Mitchell, Timothy. "Can the Mosquito Speak?" In *Rule of Experts: Egypt, Techo-Politics, Modernity*, 19–53. Berkeley: University of California Press, 2002.

Mitchell, Timothy. *Carbon Democracy: Political Power in the Age of Oil*. London: Verso, 2011.

Mitchell, Timothy. *Colonising Egypt*. Berkeley: University of California Press, 1991.

Mitchell, Timothy. *Rule of Experts: Egypt, Techno-Politics, Modernity*. Berkeley: University of California Press, 2002.

Montgomerie, T. G. "A Havildar's Journey Through Chitral to Faizabad in 1870." *Journal of the Royal Geographical Society of London* 42 (1872): 180–201.

Montgomerie, T. G. "A Havildar's Journey Through Chitral to Faizabad in 1870." *Proceedings of the Royal Geographical Society of London* 16 (1871–72): 253–261.

Montgomerie, T. G. "Journey to Shigatz, in Tibet, and Return by Dingri-Maidan into Nepaul, in 1871, by the Native Explorer No. 9." *Journal of the Royal Geographical Society of London* 45 (1875): 330–349.

Montgomerie, T. G. "Report of the Mirza's Exploration of the Route from Caubul to Kashgar." *Proceedings of the Royal Geographical Society of London* 15 (1870–71): 181–204.

Montgomerie, T. G. "Report of 'The Mirza's' Exploration from Caubul to Kashgar." *Journal of the Royal Geographical Society of London* 41 (1871): 132–193.

Montgomerie, T. G. "Report on the Trans-Himalayan Explorations, in Connexion with the Great Trigonometrical Survey of India, during 1856–7: Route-Survey made by Pundit -----." *Proceedings of the Royal Geographical Society of London* 12 (1867–68): 146–175.

Moorcroft, William. *Travels in the Himalayan Provinces of Hindustan and the Panjab*. 2 vols. London: J. Murray, 1841.

Moore, Clement Henry. *Images of Development: Egyptian Engineers in Search of Industry*. 2nd edn. Cairo: American University in Cairo Press, 1994.

Moser, Henri. *À travers l'Asie Centrale: la Steppe Kirghize, le Turkestan russe, Boukhara, Khiva, le pays des Turcomans et la Perse, impressions de voyage*. Paris: Librairie Plon, 1885.

Mudawar, Kamal. *Al-Intihar aw Hadm al-Bi'ah*. Beirut: Dar al-Abjadiyya, 1983.

Munro, John. "Industrial Energy from Water-Mills in the European Economy, Fifth to Eighteenth Centuries: The Limitations of Power." Working Paper No. 16 (April 4, 2002). Department of Economics and Institute for Policy Analysis, University of Toronto.

Murphey, Rhoads. "The Decline of North Africa Since the Roman Occupation: Climatic or Human?" *Annals of the Association of American Geographers* 41 (1951): 116–132.

Murphey, Rhoads. *Ottoman Warfare, 1500–1700*. New Brunswick, NJ: Rutgers University Press, 1999.

Murphey, Rhoads. "Provisioning Istanbul: The State and Subsistence in the Early Modern Middle East." *Food and Foodways* 2 (1988): 217–263.

al-Muslim, Muhammad Said. *Sahil al-Dhahab al-Aswad*. 2nd edn. Beirut: Dar Maktabat al-Hayat, 1962.

Naff, Thomas, and Ruth C. Matson, eds. *Water in the Middle East: Conflict or Cooperation?* Boulder, CO: Westview Press, 1984.

Najmi, Nasir. *Iran dar Miyan-i Tufan ya Zindigani-yi 'Abbas Mirza*. Tehran: Kanun-i Ma'rifat, 1957.

Nategh, Homa. "'Abbas Mirza va Turkamanan-i Khurasan." *Nigin* 10, no. 112 (September 22, 1974): 13–17.

Needham, Joseph. *Mechanical Engineering*. Part 2 of *Physics and Physical Technology*. Vol. 4 of *Science and Civilization in China*. Cambridge: Cambridge University Press, 1965.

Neumann, Roderick P. "Dukes, Earls, and Ersatz Edens: Aristocratic Nature Preservationists in Colonial Africa." *Environment and Planning D: Society and Space* 14 (1996): 79–98.

Neumann, Roderick P. *Imposing Wilderness: Struggles over Livelihood and Nature Preservation in Africa*. Berkeley: University of California Press, 1998.

Neustadt (Ayalon), David. "The Plague and Its Effects upon the Mamlûk Army." *Journal of the Royal Asiatic Society of Great Britain and Ireland* (1946): 67–73.

Norton, Richard Augustus. *Hezbollah: A Short History*. Princeton, NJ: Princeton University Press, 2007.

Nouschi, André. *Enquête sur le niveau de vie des populations rurales constantinois de la conquête jusqu'en 1919*. Paris: Presses Universitaires de France, 1961.

Nyberg, A., S. Barghouti, and S. Rehman. *Arab Republic of Egypt Land Reclamation Subsector Review*. Report No. 8047. Washington, DC: World Bank, 1990.

O'Donovan, Edmund. *The Merv Oasis: Travels and Adventures East of the Caspian during the Years 1879–80–81, Including Five Months' Residence among the Tekkés of Merv*. 2 vols. London: Smith Elder, 1882.

Ofeish, Sami. "Lebanon's Second Republic: Secular Talk, Sectarian Application." *Arab Studies Quarterly* 21 (1999): 97–116.

Olivier, Antonie. *18. Yüzyılda Türkiye ve İstanbul*, translated by Aloda Kaplan. Istanbul: Kesit, 2007.

Oman, Luke, Alan Robock, Georgiy L. Stenchikov, and Thorvaldur Thordarson. "High-Latitude Eruptions Cast Shadow over the African Monsoon and the Flow of the Nile." *Geophysical Research Letters* 33 (2006): L18711.

Orhonlu, Cengiz. *Osmanlı İmparatorluğunda Aşiretleri İskân Teşebbüsü, 1691–96*. Istanbul: Edebiyat Fakültesi Basımevi, 1963.

Ouelmouhoub, Samir. "Gestion multi-usage et conservation du patrimoine forestier: Cas des subéraies du Parc National d'El Kala (Algérie)." Master of Science, CIHEAM-IAMM, 2005.

Owen, E. R. J. *Cotton and the Egyptian Economy, 1820–1914.* Oxford: Clarendon Press, 1969.

Özcan, Abdülkadir, ed. *Zübde-i Vekayiât.* Ankara: Türk Tarih Kurumu Basımevi, 1995.

Özel, Oktay. "Population Changes in Ottoman Anatolia during the 16th and 17th Centuries: The 'Demographic Crisis' Reconsidered." *International Journal of Middle East Studies* 36 (2004): 183–205.

Pakravan, Emineh. *Abbas Mirza.* Paris: Buchet/Chastel, 1973.

Pamuk, Şevket. "The Price Revolution in the Ottoman Empire Reconsidered." *International Journal of Middle East Studies* 33 (2001): 69–89.

Panzac, Daniel. *La peste dans l'Empire Ottoman, 1700–1850.* Leuven: Editions Peeters, 1985.

Panzac, Daniel. *Quarantaines et lazarets: l'Europe et la peste d'Orient (XVIIe-XXe siècles).* Aix-en-Provence: Édisud, 1986.

Parc national de Théniet El Had and Direction Générale des Forêts. *Atlas des parcs nationaux algériens.* Cité administrative de Théniet El Had, Algérie: Parc national de Théniet El Had, 2006.

Parker, Geoffrey. "Crisis and Catastrophe: The Global Crisis of the Seventeenth Century Reconsidered." *American Historical Review* 113 (2008): 1053–1079.

Parker, Geoffrey. *The Military Revolution: Military Innovation and the Rise of the West, 1500–1800.* New York: Cambridge University Press, 1996.

Parker, Geoffrey, and Lesley M. Smith, eds. *The General Crisis of the Seventeenth Century.* 2nd edn. London: Routledge, 1997.

Parks, Peggy. *The Aswan High Dam.* New York: Blackbirch Press, 2004.

Peçevi, İbrahim. *Peçevî Tarihi,* edited by Murat Uraz. Istanbul: Neşriyat, 1968.

Peirce, Leslie. *Morality Tales: Law and Gender in the Ottoman Court of Aintab.* Berkeley: University of California Press, 2003.

Peluso, Nancy. "Coercing Conservation?: The Politics of State Resource Control." *Global Environmental Change* 3 (1993): 199–217.

Perdue, Peter C. *China Marches West: The Qing Conquest of Central Asia.* Cambridge, MA: Harvard University Press, 2005.

Perdue, Peter C. "Fate and Fortune in Central Eurasian Warfare: Three Qing Emperors and Their Mongol Rivals." In *Warfare in Inner Asian History,* edited by Nicola Di Cosmo, 369–404. Leiden: Brill, 2002.

Perevolotsky, Avi, and No'am Seligman. "Role of Grazing in Mediterranean Rangeland Ecosystems." *BioScience* 48 (1998): 1007–1017.

Périer, J.-A.-N. *Exploration scientifique de l'Algérie: Sciences médicales: de l'hygiène en Algérie.* 2 vols. Paris: Imprimerie Royale, 1847.

Perry, John. "Forced Migration in Iran During the Seventeenth and Eighteenth Centuries." *Iranian Studies* 8 (1975): 199–215.

de Peyerimhoff, P. "Les 'Parcs Nationaux' d'Algérie." In *Contribution à l'étude des reserves naturelles et des parcs nationaux,* edited by Société de Biogéographie, 127–138. Paris: Paul Lechevalier, 1937.

Picard, Elizabeth. *Lebanon, A Shattered Country: Myths and Realities of the Wars in Lebanon.* New York: Holmes & Meier, 1996.

Picciotto, Robert, and H. Eberhard Köpp. *New Land Development Project, Egypt: Performance Audit Report.* Report No. 13275. Washington, DC: World Bank, 1994.

Planhol, Xavier. "Les nomades, la steppe, et la foret en Anatolie." *Geographische Zeitschrift* 52 (1965): 101–116.

Pohl, Hans, Henry Wasserman, and Zeev Barkai. "Sugar Industry and Trade." *Encyclopaedia Judaica*. 2nd edn. Detroit: Macmillan Reference, 2007.

Pollitzer, Robert. *Plague*. Geneva: World Health Organization, 1954.

Popper, William. *The Cairo Nilometer: Studies in Ibn Taghrî Birdî's Chronicles of Egypt, I*. Berkeley: University of California Press, 1951.

Postel, Sandra. *Pillar of Sand: Can the Irrigation Miracle Last?* New York: W. W. Norton, 1999.

Prochaska, David. "Fire on the Mountain: Resisting Colonialism in Algeria." In *Banditry, Rebellion and Social Protest in Africa*, edited by Donald Crummey, 229–252. London: James Currey, 1986.

Pumpelly, Raphael, ed. *Explorations in Turkestan, Expedition of 1904: Prehistoric Civilizations of Anau: Origins, Growth, and Influence of Environment*. 2 vols. Washington, DC: Carnegie Institution of Washington, 1908.

Pyne, Stephen J. *Vestal Fire: An Environmental History, Told through Fire, of Europe and Europe's Encounter with the World*. Seattle: University of Washington Press, 1997.

Pyne, Stephen J. *World Fire: The Culture of Fire on Earth*. Seattle: University of Washington Press, 1997.

al-Qaradawi, Yusuf. *Ri'ayat al-Bi'ah fi Shari'at al-Islam*. Cairo: Dar al-Shuruq, 2001.

Qaragazlu, Hajji 'Abdullah Khan. *Majmu'a-yi Athar*, edited by 'Inayatullah Majidi. Tehran: Miras Makub, 2003.

Quataert, Donald. *The Ottoman Empire, 1700–1922*. 2nd edn. Cambridge: Cambridge University Press, 2005.

Racz, Lajos. "Variations of Climate in Hungary, 1540–1779." In *European Climate Reconstructed from Documentary Data: Methods and Results*, edited by Burkhard Frenzel, Christian Pfister, and Birgit Gläser, 125–136. New York: G. Fisher, 1992.

Ramzī, Muḥammad. *Al-Qāmūs al-Jughrāfī lil-Bilād al-Miṣriyya min 'Ahd Qudamā' al-Miṣriyyīn ilā Sanat 1945*. 6 vols in 2 parts. Cairo: al-Hay'a al-Miṣriyya al-'Āmma lil-Kitāb, 1994.

Rangarajan, Mahesh, ed. *Environmental Issues in India: A Reader*. New Delhi: Pearson Education, 2009.

Rangarajan, Mahesh, and K. Sivaramakrishnan, eds. *India's Environmental History*. 2 vols. Ranikhet: Permanent Black, 2012.

Raymond, André. *Artisans et commerçants au Caire au XVIIIe siècle*. 2 vols. Damascus: Institut Français de Damas, 1973–1974.

Raymond, André. "Les Grandes Épidémies de peste au Caire aux XVIIe et XVIIIe siècles." *Bulletin d'Études Orientales* 25 (1973): 203–210.

Raymond, André. *Grandes villes arabes à l'époque ottomane*. Paris: Sindbad, 1985.

Raymond, André. "La population du Caire et de l'Égypte à l'époque ottomane et sous Muḥammad 'Alî." In *Mémorial Ömer Lûtfi Barkan*, 169–178. Paris: Librairie d'Amérique et d'Orient Adrien Maisonneuve, 1980.

Redhouse, James W. *A Turkish and English Lexicon*. Istanbul: H. Matteosian, 1921.

Refik, Ahmet. *Anadolu'da Türk Aşiretleri (966–1200)*. 2nd edn. Istanbul: Enderun Kitabevi, 1989.

Reindl-Kiel, Hedda. "The Chickens of Paradise: Official Meals in the Seventeenth-Century Ottoman Palace." In *The Illuminated Table, the Prosperous House: Food and Shelter in Ottoman Material Culture*, edited by Suraiya Faroqhi and Christoph Neumann, 59–88. Istanbul: Orient-Institut, 2003.

Report of the United States Agricultural Mission to Saudi Arabia. Cairo: 1943.

Republic of Lebanon Ministry of Environment, United Nations Development Programme, and ECODIT. *State and Trends of the Lebanese Environment 2010*. Beirut: Ministry of Environment, 2011.

Reynolds, Terry S. *Stronger than a Hundred Men: A History of the Vertical Water Wheel*. Baltimore: Johns Hopkins University Press, 1983.

Richards, John F. *The Unending Frontier: An Environmental History of the Early Modern World*. Berkeley: University of California Press, 2003.

Riehl, Simone. "Climate and Agriculture in the Ancient Near East: A Synthesis of the Archaeobotanical and Stable Carbon Isotope Evidence." *Vegetation History and Archaeobotany* 17 (2008): 43–51.

Rivlin, Helen Anne B. *The Agricultural Policy of Muḥammad 'Alī in Egypt*. Cambridge, MA: Harvard University Press, 1961.

Robbins, Paul. *Political Ecology*. Oxford: Blackwell Publishing, 2004.

Roberts, J. M. *A Short History of the World*. New York: Oxford University Press, 1993.

Rosen, Arlene M. *Civilizing Climate: Social Responses to Climate Change in the Ancient Near East*. Lanham, MD: Altamira Press, 2007.

Ruet, Commandant. *La Transhumance dans le moyen Atlas et la haute Moulouya*. CHEAM Unpublished Report, 1952.

Russell, Josiah C. "Population in Europe." In *The Middle Ages*. Vol. 1 of *The Fontana Economic History of Europe*, edited by Carlo M. Cipolla, 25–71. London: Collins/ Fontana, 1972.

Russell, Josiah C. "That Earlier Plague." *Demography* 5 (1968): 174–184.

Sa'adat Yar Khan. *The Faras-Nama-e Rangin or The Book of the Horse*, translated by D. C. Phillott. London: Bernard Quaritch, 1911.

Saad-Ghorayeb, Amal. *Hizbu'llah: Politics and Religion*. London: Pluto Press, 2002.

Sadr, Karim. *The Development of Nomadism in Ancient Northeast Africa*. Philadelphia: University of Pennsylvania Press, 1991.

Sahli, Mohammed. "Protection de la nature et développement: Cas du Parc national du Belezma." *New Medit. A Mediterranean Journal of Economics, Agriculture and Environment* 3 (2004): 38–43.

Said, Edward W. *Orientalism*. New York: Pantheon Books, 1978.

Said, Rushdi. *The Geology of Egypt*. New York: Elsevier, 1962.

Said, Rushdi, ed. *The Geology of Egypt*. Rotterdam: A. A. Balkema, 1990.

Said, Rushdi. *The River Nile: Geology, Hydrology, and Utilization*. New York: Pergamon Press, 1993.

Sa'id, Rushdi. *Rihlat 'Umr: Tharawat Misr bayna 'Abd al-Nasir wa'l-Sadat*. Cairo: Dar al-Hilal, 2000.

Sajdi, Dana. "Decline, Its Discontents and Ottoman Cultural History: By Way of Introduction." In *Ottoman Tulips, Ottoman Coffee: Leisure and Lifestyle in the Eighteenth Century*, edited by Dana Sajdi, 1–40. New York: Tauris Academic Studies, 2007.

Salibi, Kamal S. *The Modern History of Lebanon.* London: Weidenfeld and Nicolson, 1965.

Salzmann, Ariel. "Measures of Empire: Tax Farmers and the Ottoman *Ancien Régime*, 1695–1807." Ph.D. diss., Columbia University, 1995.

Sampsell, Bonnie M. *A Traveler's Guide to the Geology of Egypt.* New York: American University in Cairo Press, 2003.

Sanders, Paula. *Ritual, Politics, and the City in Fatimid Cairo.* Albany: State University of New York Press, 1994.

Sardar, Ziauddin, ed. *An Early Crescent: The Future of Knowledge and the Environment in Islam.* London: Mansell, 1989.

Säve-Söderbergh, Torgny. *Victoire en Nubie: La Campagne internationale de sauvegarde d'Abou Simbel, de Philae and d'autres trésors culturels.* Paris: UNESCO, 1992.

al-Ṣawāliḥī al-ʿUfī al-Ḥanbalī, Ibrāhīm Ibn Abī Bakr. *Tarājim al-Ṣawāʾiq fī Wāqiʿat al-Ṣanājiq,* edited by ʿAbd al-Raḥīm ʿAbd al-Raḥman ʿAbd al-Raḥīm. Cairo: Institut français d'archéologie orientale, 1986.

Sbeinati, Mohamed Reda, Ryad Darawcheh, and Mikhail Mouty. "The Historical Earthquakes of Syria: An Analysis of Large and Moderate Earthquakes from 1365 B.C. to 1900 A.D." *Annals of Geophysics* 48 (2005): 347–435.

Schade, Abigail E. "Hidden Waters: Groundwater Histories of Iran and the Mediterranean." Ph.D. diss., Columbia University, 2011.

Schamiloglu, Uli. "The Rise of the Ottoman Empire: The Black Death in Medieval Anatolia and Its Impact on Turkish Civilization." In *Views from the Edge: Essays in Honor of Richard W. Bulliet*, edited by Neguin Yavari, Lawrence G. Potter, and Jean-Marc Oppenheim, 255–279. New York: Columbia University Press, 2004.

Schefer, Charles. *Relation de l'Ambassade au Kharezm.* Paris: Ernest Leroux, 1879.

Schimmel, Annemarie. *Islam and the Wonders of Creation: The Animal Kingdom.* London: al-Furqān Islamic Heritage Foundation, 2003.

Schoenfeld, Stuart, ed. *Palestinian and Israeli Environmental Narratives: Proceedings of a Conference Held in Association with the Middle East Environmental Futures Project.* Toronto: York University, 2005.

Scott, James C. *Seeing Like a State: How Certain Schemes to Improve the Human Condition Have Failed.* New Haven, CT: Yale University Press, 1998.

Selaniki, Mustafa. *Tarih-i Selânikî,* edited by Mehmet İpşirli. Ankara: Türk Tarih Kurumu Basımevi, 1999.

Selby, Jan. *Water, Power and Politics in the Middle East: The Other Israeli-Palestinian Conflict.* London: I. B. Tauris, 2003.

Self, Randolph. "Community-based Conservation and Sustainable Development in Tazekka National Park, Morocco." M.A. thesis, Western Washington University, 1997.

Service du Tourisme Gouvernement Général de l'Algérie. *Rapports et Études de la Commission du Tourisme.* Alger: Imprimerie Algérienne, 1920.

Sestier, Jules. *La Piraterie dans l'antiquité.* Paris: Marecq, 1880.

Shafik, Viola. *Popular Egyptian Cinema: Gender, Class, and Nation.* New York: American University in Cairo Press, 2006.

Shaw, Stanford J., ed. and trans. *Ottoman Egypt in the Eighteenth Century: The Niẓâmnâme-i Mıṣır of Cezzâr Aḥmed Pasha.* Cambridge, MA: Center for Middle Eastern Studies of Harvard University, 1964.

Shibl, Yusuf A. *The Aswan High Dam.* Beirut: The Arab Institute for Research and Publishing, 1971.

Shindell, Drew T., et al. "Dynamic Winter Climate Response to Large Tropical Volcanic Eruptions since 1600." *Journal of Geophysical Research* 109 (2004): D05104.

Shirazi, Fazl Allah. *Tarikh-i Zu al-Qarnayn,* edited by Nasir Afsharfar. 2 vols. Tehran: Vizarat-i Farhang va Irshad-i Islami, 2001.

Shokr, Ahmad. "Watering a Revolution: The Aswan High Dam and the Politics of Expertise in Mid-Century Egypt." M.A. thesis, New York University, May 2008.

Shoshan, Boaz. "Grain Riots and the 'Moral Economy': Cairo, 1350–1517." *Journal of Interdisciplinary History* 10 (1980): 459–478.

Shukri, N. M. "Geology of the Nubian Sandstone." *Nature* 156 (July 28, 1945): 116.

Shuval, Hillel, and Hassan Dweik, eds. *Water Resources in the Middle East: Israeli-Palestinian Water Issues, from Conflict to Cooperation.* Berlin: Springer, 2007.

Singer, Amy. *Palestinian Peasants and Ottoman Officials: Rural Administration around Sixteenth-Century Jerusalem.* Cambridge: Cambridge University Press, 1994.

Sinor, Denis. "Horse and Pasture in Inner Asian History." *Oriens Extremus* 19 (1972): 171–183.

Sipihr, Lisan al-Mulk. *Nasikh al-Tavarikh,* edited by Jamshid Kiyanfar. 6 vols. Tehran: Asatir, 1998.

Smil, Vaclav. *Energies: An Illustrated Guide to the Biosphere and Civilization.* Cambridge: Massachusetts Institute of Technology Press, 1999.

Smil, Vaclav. *Energy in Nature and Society: General Energetics of Complex Systems.* Cambridge: Massachusetts Institute of Technology Press, 2008.

Smil, Vaclav. *Energy in World History.* Boulder, CO: Westview Press, 1994.

Smith, G. Rex. *Medieval Muslim Horsemanship: A Fourteenth-Century Arabic Cavalry Manual.* London: British Library, 1979.

Smith, J. Masson, Jr. "Nomads and Ponies vs. Slaves on Horses." *Journal of the American Oriental Society* 118 (1998): 54–63.

Soloman, Steven. *Water: The Epic Struggle for Wealth, Power, and Civilization.* New York: HarperCollins, 2010.

Sood, Gagan D. S. "Pluralism, Hegemony and Custom in Cosmopolitan Islamic Eurasia, ca. 1720–90, with Particular Reference to the Mercantile Arena." Ph.D. diss., Yale University, 2008.

Sosland, Jeffrey K. *Cooperating Rivals: The Riparian Politics of the Jordan River Basin.* Albany: State University of New York Press, 2007.

De Souza, Philip. *Pirates in the Graeco-Roman World.* Cambridge: Cambridge University Press, 2002.

Sowers, Jeannie. "Remapping the Nation, Critiquing the State: Environmental Narratives and Desert Land Reclamation in Egypt." In *Environmental Imaginaries of the Middle East and North Africa*, edited by Diana K. Davis and Edmund Burke III, 158–191. Athens: Ohio University Press, 2011.

Springborg, Robert. *Mubarak's Egypt: Fragmentation of the Political Order*. Boulder, CO: Westview Press.

Springborg, Robert. "Patrimonialism and Policy Making in Egypt: Nasser and Sadat and the Tenure Policy for Reclaimed Lands." *Middle Eastern Studies* 15 (1979): 49–69.

Stearns, Justin K. *Infectious Ideas: Contagion in Premodern Islamic and Christian Thought in the Western Mediterranean*. Baltimore: Johns Hopkins University Press, 2011.

Steinberg, Guido. "Ecology, Knowledge, and Trade in Central Arabia (*Najd*) during the Nineteenth and Early Twentieth Centuries." In *Counter-Narratives: History, Contemporary Society, and Politics in Saudi Arabia and Yemen*, edited by Madawi al-Rasheed and Robert Vitalis, 77–102. New York: Palgrave Macmillan, 2004.

Steinberg, Ted. "Down to Earth: Nature, Agency, and Power in History." *American Historical Review* 107 (2002): 798–820.

Steinhart, Edward I. *Black Poachers, White Hunters: A Social History of Hunting in Colonial Kenya*. Athens: Ohio University Press, 2006.

Sticker, Georg. *Abhandlungen aus der Seuchengeschichte und Seuchenlehre*. Giessen: A. Töpelmann, 1908–1912.

Stokes, Martin. "Listening to Abd al-Halim Hafiz." In *Global Soundtracks: Worlds of Film Music*, edited by Mark Slobin, 309–333. Middletown, CT: Wesleyan University Press, 2008.

Storemyr, Per. "Outline of the Geography and Environmental History of the West Bank at Aswan." In *Quarryscapes Report: Characterisation of Complex Quarry Landscapes, an Example from the West Bank Quarries, Aswan*, edited by Elizabeth Bloxam, Tom Heldal, and Per Storemyr, 9–19. No. 4. Trondheim: QuarryScapes Project, 2007.

Sublet, Jacqueline. "La Peste prise aux rêts de la jurisprudence: Le Traité d'Ibn Ḥaǧar al-'Asqalānī sur la peste." *Studia Islamica* 33 (1971): 141–149.

Sufian, Sandra M. *Healing the Land and the Nation: Malaria and the Zionist Project in Palestine, 1920–1947*. Chicago: University of Chicago Press, 2007.

Sunderland, Willard. *Taming the Wild Field: Colonization and Empire on the Russian Steppe*. Ithaca, NY: Cornell University Press, 2004.

Sutter, Paul. "What Can U.S. Environmental Historians Learn from Non-U.S. Environmental Historiography?" *Environmental History* 8 (2003): 109–129.

Syud, Khwaja Ahmud Shah Nukshbundee. "Narrative of the Travels of Khwaja Ahmud Shah Nukshbundee Syud." *Journal of the Asiatic Society of Bengal* 25 (1856): 344–358.

Szuppe, Maria. "En quête de chevaux turkmènes: le journal de voyage de Mîr 'Izzatullâh de Delhi à Boukhara en 1812–1813." *Cahiers d'Asie centrale* 1–2 (1996): 91–111.

Tabak, Faruk. *The Waning of the Mediterranean, 1550–1870: A Geohistorical Approach*. Baltimore: Johns Hopkins University Press, 2008.

Tal, Alon. *Pollution in a Promised Land: An Environmental History of Israel.* Berkeley: University of California Press, 2002.

Tavakoli-Targhi, Mohamad. *Refashioning Iran: Orientalism, Occidentalism, and Historiography.* New York: Palgrave Macmillan, 2001.

Tezcan, Baki. "Searching for Osman: A Reassessment of the Deposition of the Ottoman Sultan Osman II (1618–1622)." Ph.D. diss., Princeton University, 2001.

Tezcan, Baki. *The Second Ottoman Empire: Political and Social Transformation in the Early Modern World.* New York: Cambridge University Press, 2010.

Thapar, Romila. "The Image of the Barbarian in Early India." *Comparative Studies in Society and History* 13 (1971): 408–436.

Thirgood, J. V. *Man and the Mediterranean Forest: A History of Resource Depletion.* London: Academic Press, 1981.

Tinguely, Frédéric. *L'écriture du Levant à la Renaissance: Enquête sur les voyageurs français dans l'Empire de Soliman le Magnifique.* Geneva: Droz, 2000.

Tomasi, Mari, and Roaldus Richmond. *Men Against Granite.* Shelburne, VT: The New England Press, 2004.

Touchan, R., and M. K. Hughes. "Dendrochronology in Jordan." *Journal of Arid Environments* 42 (1999): 291–303.

Touchan, Ramzi, et al. "Standardized Precipitation Index Reconstructed from Turkish Tree-Ring Widths." *Climatic Change* 72 (2005): 339–353.

Touchan, Ramzi, et al. "A 396-Year Reconstruction of Precipitation in Southern Jordan." *Journal of the American Water Resources Association* 35 (1999): 49–59.

Traboulsi, Fawwaz. *A History of Modern Lebanon.* London: Pluto Press, 2007.

Trotter, H. "Account of the Pundit's Journey in Great Tibet from Leh in Ladakh to Lhasa, and of His Return to India via Assam." *Proceedings of the Royal Geographical Society of London* 21 (1876–77): 325–350.

Tuchscherer, Michel, ed. *Le commerce du café avant l'ère des plantations coloniales: espaces, réseaux, sociétés (XVe-XIXe siècle).* Cairo: Institut français d'archéologie orientale, 2001.

Tucker, William F. "Natural Disasters and the Peasantry in Mamlūk Egypt." *Journal of the Economic and Social History of the Orient* 24 (1981): 215–224.

Turchin, Peter. "A Theory for Formation of Large Empires." *Journal of Global History* 4 (2009): 191–217.

Twigg, Graham. *The Black Death: A Biological Reappraisal.* London: Batsford, 1984.

Twitchell, Karl S. *Saudi Arabia: With an Account of the Development of Its Natural Resources.* 3rd edn. New York: Greenwood Press, 1969.

Uchupi, Elazar, S. A. Swift, and D. A. Ross. "Gas Venting and Late Quaternary Sedimentation in the Persian (Arabian) Gulf." *Marine Geology* 129 (1996): 237–269.

United Arab Republic. *The Year Book, 1965.* Cairo: Information Department Press, 1965.

Utley, Robert. *The Last Days of the Sioux Nation.* New Haven, CT: Yale University Press, 2004.

Vambery, Arminius. *Travels in Central Asia: Being the Account of a Journey from Teheran across the Turkoman Desert on the Eastern Shore of the Caspian to Khiva, Bokhara, and Samarcand Performed in the Year 1863*. London: J. Murray, 1864.

Vassiliev, Alexei. *The History of Saudi Arabia*. London: Saqi Books, 1998.

Venzke, Margaret. "The Question of Declining Cereals' Production in the Sixteenth Century: A Sounding on the Problem-Solving Capacity of the Ottoman Cadastres." *Journal of Turkish Studies* 8 (1984): 251–264.

Vitruvius, Pollio. *The Ten Books on Architecture*, translated by Morris Hicky Morgan. Cambridge, MA: Harvard University Press, 1914.

Voll, Sarah. "Egyptian Land Reclamation Since the Revolution." *Middle East Journal* 34 (1980): 127–148.

Vural, Y., N. Akçar, and C. Schlüchter. "The Frozen Bosphorus and Its Paleoclimatic Implications Based on a Summary of the Historical Data." In *The Black Sea Flood Question: Changes in Coastline, Climate, and Human Settlement*, edited by V. Yanko-Hombach et al., 633–649. Dordrecht: Springer Netherlands, 2007.

Wagstaff, J. M. *The Evolution of Middle Eastern Landscapes: An Outline to A.D. 1840*. London: Croon Helm, 1985.

Wakeem, Najah. *Al-Ayadi al-Soud*. Beirut: Sharikat al-Matbu'at lil-Tawzi' wa al-Nashr, 1998.

Walz, Terence. *Trade between Egypt and Bilād as-Sūdān, 1700–1820*. Cairo: Institut français d'archéologie orientale, 1978.

Wassef, Magda, ed. *Egypte, 100 ans de cinéma*. Paris: Institut du Monde Arabe, 1995.

Waterbury, John. *Hydropolitics of the Nile Valley*. Syracuse, NY: Syracuse University Press, 1979.

Watson, Andrew. *Agricultural Innovation in the Early Islamic World: The Diffusion of Crops and Farming Techniques, 700–1100*. Cambridge: Cambridge University Press, 1983.

Weiss, Barry. "The Decline of Late Bronze Age Civilization as a Possible Response to Climatic Change." *Climatic Change* 4 (1982): 173–198.

Weiss, Harvey. "Beyond the Younger Dryas: Collapse as Adaptation to Abrupt Climate Change in Ancient West Asia and the Ancient Eastern Mediterranean." In *Environmental Disasters and the Archaeology of Human Response*, edited by Garth Bawden and Richard Martin Reycraft, 75–98. Albuquerque, NM: Maxwell Museum of Anthropology, 2000.

Wendorf, Fred. *Desert Days: My Life as a Field Archaeologist*. Dallas, TX: Southern Methodist University Press, 2008.

Whetton, Peter, and Ian Rutherford. "Historical ENSO Teleconnections in the Eastern Hemisphere." *Climatic Change* 28 (1994): 221–253.

White, Gilbert F. "The Environmental Effects of the High Dam at Aswan." *Environment* 30 (1988): 5–11 and 34–40.

White, Lynn Townsend. *Medieval Technology and Social Change*. Oxford: Clarendon Press, 1962.

White, Richard. *Land Use, Environment, and Social Change: The Shaping of Island County, Washington*. Seattle: University of Washington Press, 1980.

White, Richard. *The Middle Ground: Indians, Empires, and Republics in the Great Lakes Region, 1650–1815*. Cambridge: Cambridge University Press, 1991.

White, Sam. *The Climate of Rebellion in the Early Modern Ottoman Empire*. New York: Cambridge University Press, 2011.

White, Sam. "Rethinking Disease in Ottoman History." *International Journal of Middle East Studies* 42 (2010): 549–567.

Whiteman, A. J. "Nubian Group: Origin and Status." *American Association of Petroleum Geologists Bulletin* 54 (March 1970): 522–526.

Willcocks, William. *Egyptian Irrigation*. 2nd edn. London: E. & F. N. Spon, 1899.

Willcocks, W., and J. I. Craig. *Egyptian Irrigation*. 2 vols. New York: Spon and Chamberlain, 1913.

Winegar, Jessica. *Creative Reckonings: The Politics of Art and Culture in Contemporary Egypt*. Stanford, CA: Stanford University Press, 2006.

Wittfogel, Karl A. "The Hydraulic Civilizations." In *Man's Role in Changing the Face of the Earth*, edited by William L. Thomas, Jr., 152–164. Chicago: University of Chicago Press, 1956.

Wittfogel, Karl A. *Oriental Despotism: A Comparative Study of Total Power*. New Haven, CT: Yale University Press, 1957.

Wolf, Eric. *Europe and the People Without History*. Berkeley: University of California Press, 1982.

Wolff, Joseph. *Travels and Adventures of the Reverend Joseph Wolff*. London: Saunders, Otley, and Co., 1861.

World Bank. *New Land Development Project: Project Completion Report*. Report No. 10631. Washington, DC: World Bank, 1992.

World Bank. *Selected Issues in Agriculture, Irrigation, and Land Reclamation*. Report No. 4013. Washington, DC: World Bank, 1983.

Worster, Donald. "Doing Environmental History." In *The Ends of the Earth: Perspectives on Modern Environmental History*, edited by Donald Worster, 289–307. New York: Cambridge University Press, 1988.

Worster, Donald. *Rivers of Empire: Water, Aridity, and the Growth of the American West*. Oxford: Oxford University Press, 1992.

Wulff, Hans E. *The Traditional Crafts of Persia: Their Development, Technology, and Influence on Eastern and Western Civilizations*. Cambridge: Massachusetts Institute of Technology Press, 1966.

Xoplaki, E., et al. "Variability of Climate in Merdional Balkans during the Periods 1675–1715 and 1780–1830 and Its Impact on Human Life." *Climatic Change* 48 (2001): 581–615.

Yaffe, Martin D., ed. *Judaism and Environmental Ethics: A Reader*. Lanham, MD: Lexington Books, 2001.

Yergin, Daniel. *The Prize: The Epic Quest for Oil, Money, and Power*. New York: Simon & Schuster, 1991.

Yule, H., Munphool Pundit, and Faiz Bukhsh. "Papers Connected with the Upper Oxus Regions." *Journal of the Royal Geographical Society of London* 42 (1872): 438–481.

Zachariadou, Elizabeth, ed. *Natural Disasters in the Ottoman Empire*. Rethymnon: Crete University Press, 1999.

Zalla, T., et al. *Availability and Quality of Agricultural Data for the New Lands in Egypt*. Impact Assessment Report No. 12 of the Agricultural Policy Reform Project. Cairo: Ministry of Agriculture and Land Reclamation, 2000.

Zamani, Husayn, ed. *Safarnamah-i Bukhara*. Tehran: Pizhūhishgāh-i ʿUlūm-i Insani va Muʾtalaʾat-i Farhangi vāʾbastah bih Vizarat-i Farhang va Āmūzish-i ʿĀlī, 1994.

Zeʾevi, Dror. *An Ottoman Century: The District of Jerusalem in the 1600s*. Albany: State University of New York Press, 1996.

Zeitoun, Mark. *Power and Water in the Middle East: The Hidden Politics of the Palestinian-Israeli Water Conflict*. London: I. B. Tauris, 2008.

Zerner, Charles, ed. *People, Plants and Justice: The Politics of Nature Conservation*. New York: Columbia University Press, 2000.

Zhang, Jiafeng. "Disease and Its Impact on Politics, Diplomacy and the Military: The Case of Smallpox and the Manchus, 1613–1795." *Journal of the History of Medicine and Allied Sciences* 57 (2002): 177–197.

Zinsser, Hans. *Rats, Lice and History, Being a Study in Biography, which, after Twelve Preliminary Chapters Indispensable for the Preparation of the Lay Reader, Deals with the Life History of Typhus Fever*. London: George Routledge and Sons, 1935.

Index

CPSIA information can be obtained at www.ICGtesting.com
Printed in the USA
BVOW05s2139270714

360599BV00002B/5/P